The Physics of
Microfabrication

The Physics of Microfabrication

Ivor Brodie and Julius J. Muray

SRI International (formerly Stanford Research Institute)
Menlo Park, California

PLENUM PRESS • NEW YORK AND LONDON

Library of Congress Cataloging in Publication Data

Brodie, Ivor.
　The physics of microfabrication.

　Bibliography: p.
　Includes index.
　1. Microelectronics. 2. Thin-film circuits. I. Muray, Julius J. II. Title.
TK7874.B73   621.381'7   82-3835
ISBN 0-306-40863-5   AACR2

©1982 Plenum Press, New York
A Division of Plenum Publishing Corporation
233 Spring Street, New York, N.Y. 10013

Printed in the United States of America

Preface

The Physical Electronics Department of SRI International (formerly Stanford Research Institute) has been pioneering the development of devices fabricated to submicron tolerances for well over 20 years. In 1961, a landmark paper on electron-beam lithography and its associated technologies was published by K. R. Shoulders† (then at SRI), which set the stage for our subsequent efforts in this field. He had the foresight to believe that the building of such small devices was actually within the range of human capabilities.

As a result of this initial momentum, our experience in the technologies associated with microfabrication has become remarkably comprehensive, despite the relatively small size of our research activity. We have frequently been asked to deliver seminars or provide reviews on various aspects of microfabrication. These activities made us aware of the need for a comprehensive overview of the physics of microfabrication. We hope that this book will fill that need.

While there is a special emphasis on silicon microcircuit technology, this book is intended as an introduction for all engineers, scientists, and graduate students who are considering making something very small. The book may also serve as a source of review material for experienced workers in this extensive field, although no specialty is treated exhaustively. Though we have no desire to overlook anyone's contribution, we recognize that we may have failed to reference works or papers of significance. Any oversight is unintentional, and we would be pleased to receive suggestions for the improvement of any of the topics discussed.

†Microelectronics using electron beam activated machining techniques, *Advances in Computers,* **2**, pp. 137–289, edited by F. L. Alt, Academic Press, London and New York (1961).

We wish to acknowledge the assistance provided by past and present members of the Physical Electronics Department scientific staff who have directly or indirectly helped us in the preparation of this book, especially Donald R. Cone, Eugene R. Westerberg, Charles A. Spindt, John B. Mooney, Robert A. Gutcheck, Arden Sher, T. E. Thompson, and John Kelly. We appreciate the support of SRI International through our Laboratory Director, Dr. Fred J. Kamphoefner, and our Division Executive Director, Earle D. Jones, in allowing us to use SRI facilities and materials.

We also wish to express our appreciation to our secretary, Joyce Garbutt, for carrying out the arduous task of typing the manuscript and to our publications group under the direction of Jack Byrne for editing, revisions, and illustrations. We would also like to thank Plenum Press for commissioning this book, and, finally, we would like to thank our wives and families for their patience during the tribulations encountered in the preparation of this book.

I. Brodie
J. J. Muray
Menlo Park, California
April 1981

Introduction

Machines are devised for two basic purposes: to liberate humans from drudgery, and to help perform useful and interesting tasks that we would otherwise be unable to do. As far as size is concerned, machines may be classified into two general types—those with dimensions determined by their function and those whose size is limited only by economic factors. For instance, the dimensions of a plough are determined by the depth and width of the furrow it is to make, but the physical size of the motor that drives the plough or the size of the guidance mechanism that allows a desired ploughing pattern to be maintained are arbitrary and unrelated to the size of the furrow. However, the weight, power consumption, and cost of these ancillary devices are relevant to determining the economics of ploughing.

Both function and economics have supplied the driving forces toward making ever-smaller devices. In particular, the functional need to increase speed and the economic need to reduce the cost of binary storage elements for digital computers have led to the reduction in the size of such elements. What formerly required a flip-flop circuit using two tubes with a largest dimension of perhaps 5 cm, and a cost measured in dollars, has now been replaced with an MOS memory element with a largest dimension of perhaps 15 μm, and a cost measured in millicents.

The continuing demand for inexpensive electronic circuits to feed the second industrial revolution has created the rapidly growing microelectronics industry. The economic pressure for lower cost and lower power consumption and the functional need for higher operational speeds have led to ever larger scales of integration and ever smaller sizes of circuit elements. New fabrication technologies have been and continue to be developed to satisfy the demands of this industry. These technologies include the following: deposition of thin films by sputtering and vacuum evaporation; pattern generation by optical, X-ray, and electron-beam lithographic techniques; and development of appropriate resist materials and methods of applying resists. Devices are routinely made with films as thin as 0.1 μm, with lateral dimensions as small as 5 μm and with an edge sharpness on the order of 0.5 μm. With some of the technologies currently being developed, lateral dimensions of 0.5 μm and edge sharpnesses of 0.1 μm or less are being contemplated.

In addition to providing much of the technology required for fabricating microstructures, the incredible achievements of the microelectronic industry have also removed many of the psychological barriers to thinking small. It is awesome to consider that today's $500 microcomputer has more memory than the first large electronic computer, ENIAC, yet it occupies 1/30,000 of the volume, costs 1/10,000 as much, performs its operations 20 times faster, and consumes the power of a light bulb rather than that of a locomotive. Moreover, we have not yet reached the practical limits to microminiaturization using currently available technologies.

The development of microcircuitry and its related technologies has encouraged people working in widely diverse disciplines to seriously consider making microstructures that would be useful in their fields. For example, in the biological area we may consider fabricating membranes and filters for artificial lungs and kidneys, artificial retinas and cochleas, and eventually even artificial cells. In the optical area, the ability to fabricate minute lenses, prisms, and other optical elements has given rise to integrated optical devices that are currently revolutionizing communications science. In the area of instrumentation we may visualize ultrasmall devices to probe, and manipulate with ever-increasing sensitivity and precision.

Furthermore, we must remember that most of the microstructures we build today are "micro" only in the sense that their physical descriptions differ from those of bulk devices. They are still "macro" in the sense that they are large compared with atoms, and their dimensions are usually limited by some characteristic length associated with the physical principles by which they operate. Characteristic lengths, such as wavelength, Debye screening distances, or depletion depths, are intrinsically large compared with atomic distances, so that the technologies developed for fabricating these devices need not aspire to atomic resolution.

The technologies currently in use are largely based on removing material from a larger block, much as a sculptor carves a statue. In the more distant future we may look forward to molecular engineering technologies[1]† that will

---

†References for the Introduction are given with those for Chapter 1.

enable us to construct our devices atom by atom, much as a builder constructs a house brick by brick. It may be argued that molecular engineering has already been employed for simple applications, such as monomolecular film devices[2,3] and molecular beam epitaxy.[4]

In living cells, nature has given us extraordinarily diverse and intricate devices operating on the molecular level that we may attempt to emulate. The mystery of how these devices function must be solved before molecular engineering can become a reality. Gaining this understanding will require the use of new microinstrumentation techniques[5-8] and a more sophisticated physical interpretation[9] of the measurements so obtained.

The driving forces toward microminiaturization will be dissipated only when the potential of molecular engineering has been fully exploited. However, long before that time arrives we may anticipate that the variety and complexity of microdevices will increase substantially, contributing significantly both to the well-being of mankind and our understanding of nature.

This book is directed to those well educated in an applied science or engineering but who are unfamiliar with all or some of the technologies used in microfabrication. Its purpose is to acquaint the reader with a basic physical understanding of the tools, technologies, and physical models needed to build, analyze, and understand the mechanism of microdevices, to see what physical limitations may be expected, to put numbers to the magnitudes involved, and to provide a guide to the extensive literature available. No attempt has been made to review all the microdevices that are being used or developed today; instead, we have emphasized a particular class of microelectronic devices based on planar silicon technology to illustrate how the techniques of microfabrication may be applied. We have chosen this class of devices because they have had such a major impact on our everyday life and economy and have provided impetus for the evolution of microfabrication technologies.

Our intention, however, is to be more general than required simply to meet the needs of those interested in microcircuit fabrication. We also wish to convince workers in other fields that building microstructures applicable to their needs may not be as difficult as they have imagined.

We have attempted to provide a bibliography that will be useful to the reader who wishes to follow up on a particular subject or who is interested in the historical aspects.

As a general rule, we have used mks units in the equations, unless some other units are specified. We have found that workers in a given specialty tend to develop their own mix of units that are used in common parlance. We have used their units, at least for educational purposes, as appropriate.

In Chapter 1 we present a preliminary survey of the subject matter. Our aim here is to introduce the reader to the types of devices that are microfabricated, particularly planar silicon devices, and provide a first look at the ways in which they are made. Those unfamiliar with microelectronics will also be able to pick up some of the jargon and the "alphabet soup" used to describe the subject matter.

We realize that Chapter 2 on particle beams may be heavy going for the reader. However, we believe that the effort to understand the physics involved in producing, focusing, and deflecting particle beams and their interactions with solid surfaces will be well worthwhile, since so much of the subsequently described technologies are based on these principles. Furthermore, this relieves the subsequent chapters from entering involved descriptions of details that could detract from the main topic under discussion.

Microdevice technology is heavily dependent on our ability to deposit and etch thin films of controlled physical and chemical properties. Chapter 3 reviews this important but arduous field.

While thin films have enabled one dimension of a device to be attained with astounding precision, the other two dimensions have proved to be more difficult. In Chapter 4 we discuss the technologies used to generate two-dimensional patterns in thin films, noting how they have improved to the point where patterns with micron-sized linewidths and a submicron edge sharpness can be routinely generated.

In Chapter 5 some important topics associated with silicon microcircuitry, namely, epitaxy, oxidation, doping, and annealing, are reviewed in more detail.

Having made a microdevice, we may need to "see" it and perform various measurements on it; such topics are discussed in Chapter 6.

Finally, in Chapter 7 we discuss the practical and theoretical limitations on microdevice fabrication and try to anticipate what the future might reasonably have in store in the way of microdevices.

Contents

Chapter 1
PRELIMINARY SURVEY

Chapter 2
PARTICLE BEAMS: SOURCES, OPTICS, AND INTERACTIONS

Chapter 4
PATTERN GENERATION

Chapter 5

SPECIAL PROCESSES DEVELOPED FOR MICROCIRCUIT TECHNOLOGY

Chapter 6
SUBMICRON MICROSCOPY AND MICROPROBES

Chapter 7
FUTURE DIRECTIONS

APPENDICES

1

Preliminary Survey

1.1. Introduction

This chapter is intended to enable those readers with no previous experience in the subject matter to gain a brief overview of the evolution of microfabrication technologies from the needs of the microelectronic industry and of the ways in which such technologies can be applied to nonelectronic devices.

While engineers have always been driven by the need to make small structures, the major impetus in recent times has come from the electronic microcircuit industry, which was heralded by the invention of the transistor in late 1947.[10-12]†

Historically,[13,14] germanium was the first semiconducting material used in the early evolutionary stages of device development. Developments in germanium technology [single-crystal technology,[15] controlled doping,[16] purification of semiconductor crystals,[17] and alloying[18]] made possible a number of devices, including the point-contact germanium transistor and the grown-junction and alloy–germanium transistors.

However, germanium was soon replaced by silicon, which has the attractive property that in an oxidizing atmosphere a thin, strongly bonded, impervious layer of amorphous silica (SiO_2) is formed. The silica film is important because it can be used both as a protective coating over silicon surfaces and as a dielectric insulation between circuit functions.

Gallium arsenide (GaAs) is emerging as an important semiconducting material largely because of its high electron mobility. Progress toward its more

†References 1–9 are cited in the Introduction.

1

general use in microelectronics has been inhibited by technological difficulties in growing large single crystals and in forming insulating layers. Thus, at present microelectronics remains dominated by silicon, except for those classes of devices where the special properties of GaAs are necessary. Conquest of the technological problems with GaAs could change this domination in the future.

The key concept that enabled the next evolutionary steps to take place was that of planar processing, or planar technology. The idea here was to fabricate patterned layers, one on top of the other, made of materials with different electrical properties. Together the multidecked sandwich of patterned layers are made to form various circuit elements such as transistors, capacitors, and rectifiers, and these are finally connected together by a patterned conducting overlayer to form an integrated circuit (IC).

The layers with different electrical properties can be formed by modifying the substrate, for example, by doping or oxidizing it, or by depositing a layer from an external source, for example, by evaporation or sputtering. Patterning is usually accomplished by the process of photolithography. In this process the pattern in the form of a photographic transparency is projected onto a surface that has been previously coated with a photoresist layer. Photoresist materials have two properties: First, when exposed to light, their solubility in one class of solvents is changed, so that after immersion in such a solvent the projected pattern is replicated in the surface; second, the undissolved regions of the resist are completely unaffected by ("resist") a second class of solvents, which are able to etch or modify the underlying material.

If these two properties cannot be found in the same material, it may be necessary to pattern an intermediate layer that does have the required resist properties to a required substrate etching or modifying process. For example, dopants that interact with an organic resist can be diffused into a silicon substrate through open areas in a previously patterned SiO_2 layer, which is impervious to the dopant. This oxide mask enables patterns of doped silicon to be formed in the substrate. Some basic steps of the planar process are illustrated in Figure 1.1. Complete microelectronic devices are made by repetitive application of such processes to build up the layered structures required.

Planar processing is also valuable for economic volume production of integrated circuits. This is because a substantial number of circuits can be fabricated in parallel on a large-area substrate (or "wafer"). After fabrication the circuits (dies) are separated by cutting (or "dicing") the wafer. The external leads are then connected to each circuit and "packaged" for insertion into circuit boards. Wafers 100 cm (4 in) in diameter are conventionally used, with die sides in the range of 3–10 mm.

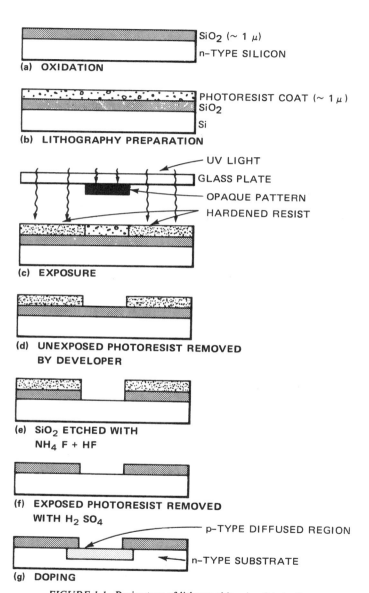

FIGURE 1.1. Basic steps of lithographic microfabrication.

1.2. Microelectronic Devices

The physics of semiconductor device operation and the design of integrated circuits have been well treated in the literature.[19,20] The purpose of this section is simply to describe the more common devices at a level necessary for an understanding of the problems encountered in their fabrication.

Semiconducting solids have important physical properties that are used in the operation of electronic devices. A perfect impurity-free semiconductor at low temperatures has a filled set of energy states (the valence band) separated from the nearest unfilled states (the conduction band) by an energy gap. Given this circumstance, the material is actually an insulator, since all the states in the valence band are filled, so those electrons cannot move in response to an applied electric field, and there are no electrons in the conduction band. However, if an electron is excited from the valence to the conduction band, both the electron in the conduction band (a negative carrier) and the resultant hole in the valence band (which behaves like a positive carrier) are free to respond to an electric field and conduct current. Semiconductors differ from other insulators in three ways. They have moderate to high electron and hole mobilities, their carrier concentrations can be modified in a controlled fashion by the addition of selected impurities (donors or acceptors), and it is possible to make nonrectifying (ohmic) contact to them with metals.

Semiconductors are classified as *n* type or *p* type, depending on whether their conductivity is due to a majority of negative-charge carriers (electrons) or positive-charge carriers (holes). The interface (or junction) between *n* and *p*-type materials is rectifying, since in one direction an applied electric field will drive both carriers toward the junction region where they can recombine, thus allowing current to flow; whereas in the reverse direction the field will drive the majority carriers away from the junction until the internal field balances the applied field, thus limiting the current flow to that due to recombination of the minority carriers. By controlling the purity of a semiconductor material the number density of majority charge carriers within it can be varied from very small (intrinsic) to very large. In silicon number densities greater than $10^{18}/cm^3$ are designated n^+ or p^+. Even in lightly doped silicon the ratio of majority to minority charge carriers is greater than 10^8.

A key device in microelectronic applications is the transistor. This is essentially a device for controlling the current through a circuit by supplying a relatively small current or voltage to an auxiliary electrode.

The bipolar transistor consists of a sandwich of two back-to-back *pn* junctions that share a thin common region called the base (*B*). The outer regions, which are of the same conductivity type, are called the emitter (E) and the collector (C). A voltage is applied between E and C so as to drive the emitter charge carriers toward the collector. If no current is allowed to flow in the E–B circuit, a retarding field builds up at the B–E junction that prevents the majority charge carriers from E penetrating into the B region. However, if a voltage is applied to B such as to allow current to flow in the E–B circuit, the retarding field is diminished and a large number of majority charge carriers can flow from E into the B region. Most of these charges are injected into the B region, where they are minority carriers, are able to diffuse to C before they recombine with the majority carriers flowing into B. Thus a small flow of charge carriers from B to E gives rise to a large flow of (oppositely charged) carriers from E to B. The name "bipolar" derives from the fact that the operation of the device truly depends on the flow of both signs of charge carrier.

The cross section of an *npn* bipolar transistor is shown in Figure 1.2(a). The *p*-type base is formed by diffusing boron into *n*-type silicon. Since boron is trivalent, only three electrons are available to bond to its four adjacent silicon atoms. This local site tends to bind an electron from the valence band, leaving it with a hole or vacancy, thus making the material *p*-type.

The emitter is formed by diffusing atoms with five valence electrons (phosphorus) into the *p*-type base. Since only four of the phosphorus electrons are needed to bond to the four surrounding silicon atoms, one electron is easily freed to conduct electric current (*n*-type semiconductor).

For simplicity in manufacturing, a *pnp* transistor [Figure 1.2(b)] is generally constructed according to a plan in which the E–C current is lateral rather than vertical. The principles of operation are the same as for an *npn* transistor, except that all of the polarities are reversed. Connections to the circuit elements of both *npn* and *pnp* transistors are made through aluminum conductors deposited over an insulating layer of SiO_2.

A significant development in bipolar technology was the ability to form a thin layer of high-resistivity silicon over low-resistivity silicon by epitaxial deposition.[21] The resulting transistor structure has substantially lower collector series resistance and much less stored charge than have conventional diffused devices. Using epitaxial deposition significantly increases the switching speed and high-frequency gain. Figure 1.2(c) shows the cross section of an epitaxial diffused bipolar transistor.

In 1960 the first metal–oxide–semiconductor (MOS) transistor using silicon was reported.[22] The MOS transistor was actually conceived in 1930,[23] but because of processing difficulties, they could not be made economically until the early 1960s.[24]

The MOS field-effect transistor (MOSFET) consists of two small highly conductive regions of the same charge carrier type, called the source (S) and the

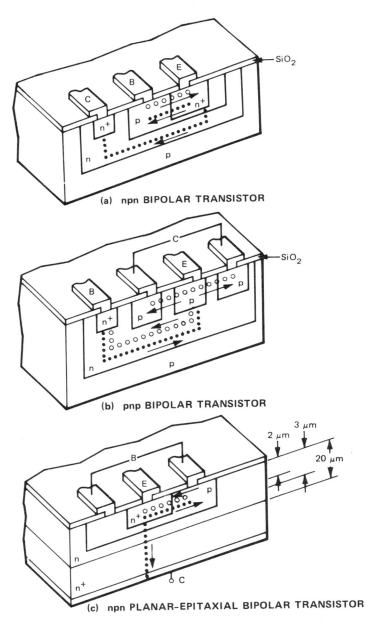

(a) npn BIPOLAR TRANSISTOR

(b) pnp BIPOLAR TRANSISTOR

(c) npn PLANAR–EPITAXIAL BIPOLAR TRANSISTOR

FIGURE 1.2. Bipolar transistors. Note: •, electrons; ○, holes.

drain (D), which are embedded in a substrate of opposite charge carrier type. The region close to the substrate surface between S and D is called the channel. It is covered with a thin layer of insulating material (oxide) that supports a metal electrode called the gate (G). If a voltage is applied between S and D in a direction to drive the charge carriers from S to D, the substrate takes the potential of S, but current cannot flow, as the substrate to D junction is rectifying. In the enhancement mode of operation a voltage is applied between G and S, thus generating an electric field in the channel region. Depending on the direction of the field, charge carriers of either sign may be forced into the thin channel region near the surface. If these charge carriers are of the same sign as those of S and D, current can "channel" between them. Since the oxide is thin, large fields can be generated in the channel region, with relatively small G voltages for switching current between S and D.

The detailed mechanism for the channel formation is complex and may be better understood by considering the case where S and D are *p*-type and the channel region is *n*-type. If the G voltage is negative, the electrons just below G are forced away from the channel region, leaving behind a net positive space charge (a depletion region). The electron energy states are bent upward in this positive space charge region. When the valence band at the semiconductor–insulator interface is raised near the energy of the bulk conduction band edge, electrons can be thermally excited from the valence band at the interface to the bulk conduction band. This leaves a *p*-type channel (called an inversion layer) at the interface that allows a hole current to flow between the *p*-type source and drain.

In the depletion mode of operation, the thin channel region connecting S and D is made of material of the same charge carrier type as that of S and D. Application of the appropriate electric field depletes the channel of charge carriers, thus disconnecting S from D.

A cross section of an *n*-channel or enhancement-mode MOSFET is shown in Figure 1.3(a) and (b). Its starting material is *p*-type silicon, into which two n^+ "pockets" have been diffused to form the source and drain regions of the transistor. The aluminum gate is separated from the silicon by a thin layer of SiO_2, usually between 90 and 120 nm in thickness. It is from this multilayer Al–SiO_2–Si that the term "MOS" (*m*etal–*o*xide–*s*emiconductor) originated.

In MOSFET devices, the thickness and dielectric properties of the SiO_2 film greatly influence the basic characteristics of the transistor. The lack of consistent quality control over these films and the wide resultant distribution of the threshold voltages are the primary reasons why MOS technology was not employed until the 1960s.

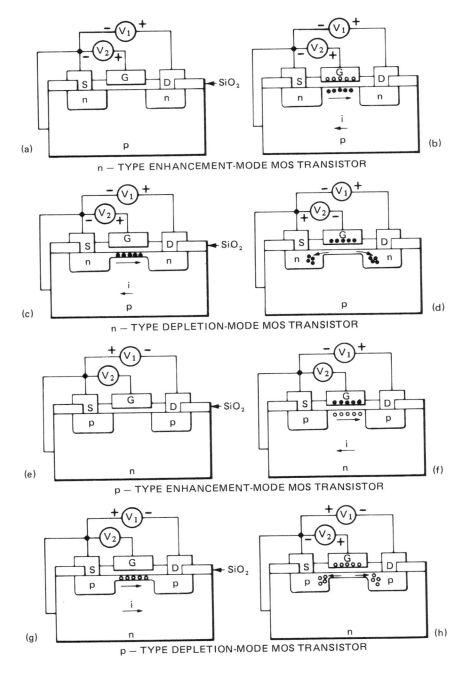

(a)

(b)

n — TYPE ENHANCEMENT-MODE MOS TRANSISTOR

(c)

(d)

n — TYPE DEPLETION-MODE MOS TRANSISTOR

(e)

(f)

p — TYPE ENHANCEMENT-MODE MOS TRANSISTOR

(g)

(h)

p — TYPE DEPLETION-MODE MOS TRANSISTOR

The depletion-mode *n*-MOSFET is shown in Figures 1.3(c) and (d). By forming pockets of *p*-type material in an *n*-type substrate, the corresponding *p*-MOS devices can be constructed, as in Figures 1.3(e) and (h).

In Figure 1.4 the process sequence for the fabrication of a *p*-channel MOS transistor is shown. As a first step, the *n*-type silicon wafer is cleaned to remove contamination and then oxidized to a thickness of approximately 1 μm. A pattern is then cut through the SiO$_2$ by means of lithography, and *p*-type dopant is deposited only through the "windows" in the oxide film. This dopant is then diffused into the silicon and the wafer is again oxidized. After another oxide patterning, a thin (100-nm) gate oxide is grown on the uncovered silicon surface. The oxide mask is then removed in the areas where contact is to be made between the p^+ silicon and aluminum metal conductors. Finally, aluminum is evaporated onto the wafer and patterned to form contacts to the source, gate, and drain.

Because MOS devices do not have to be isolated from each other, MOS transistors can be packed more densely than bipolar transistors, which require isolation. Furthermore, the manufacture of MOS integrated circuits requires fewer steps than do the bipolar ICs (see Figure 1.5). For these reasons, MOS technology has come to dominate the manufacture of large-scale integrated circuits (10^4–10^5 transistors per chip).

A complementary MOS device (CMOS) (Figure 1.6) contains both *n* and *p*-MOS transistors on a single chip of silicon, forming a balanced circuit. Complementary ICs are more difficult to manufacture, but due to their low-power dissipation, there are some applications where their complexity is worth the added cost.[25]

New configurations are always being explored for large-scale integration (LSI) applications. Among recent designs, the most promising structures are the double-diffused (D-MOS) and V-groove or vertical (V-MOS) transistors shown in Figure 1.7. The standard MOS transistor channel length is 4–5 μm,

FIGURE 1.3. Field effect transistors. When no voltage is applied to an *n*-type enhancement-mode MOS transistor (a), the drain to substrate *pn* junction is reverse biased and prevents electrons from flowing from the source to the drain. Applying a negative voltage to the gate electrode (b) induces a "channel" of *p*-type substrate below it. Current can flow from the source to the drain via this channel. In a *n*-type depletion mode MOS transistor (c), there is a continuous channel of *n*-type silicon so that electrons can flow from the source to the drain. The transistor normally conducts. When a negative voltage is applied to the gate electrode (d), electrons are expelled from the channel and the transistor no longer conducts. By doping islands of *p*-type material in a *n*-type substrate, the corresponding *p*-type MOS transistors can be made (e,f,g,h).

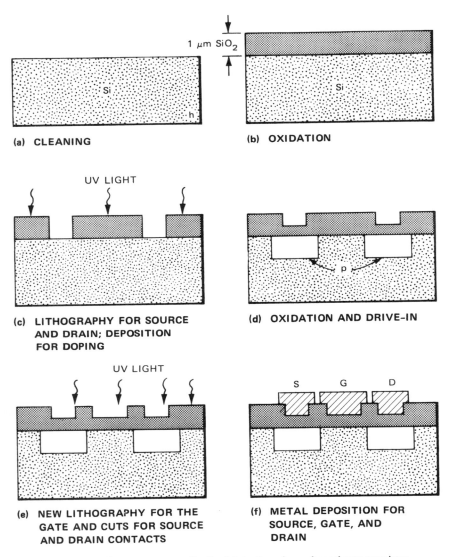

FIGURE 1.4. Process sequence for the fabrication of a *p*-channel MOS transistor.

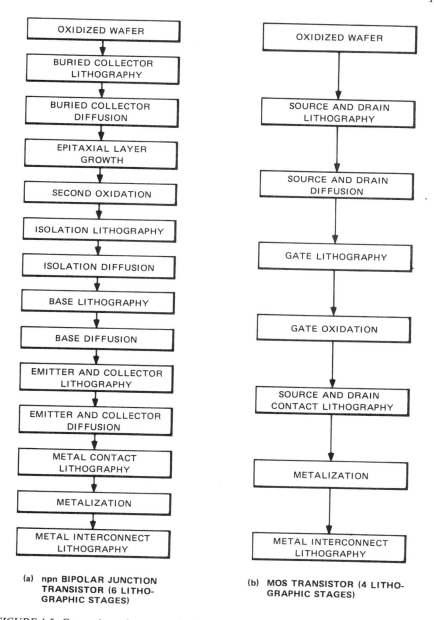

OXIDIZED WAFER	OXIDIZED WAFER
BURIED COLLECTOR LITHOGRAPHY	
BURIED COLLECTOR DIFFUSION	SOURCE AND DRAIN LITHOGRAPHY
EPITAXIAL LAYER GROWTH	
SECOND OXIDATION	SOURCE AND DRAIN DIFFUSION
ISOLATION LITHOGRAPHY	
ISOLATION DIFFUSION	GATE LITHOGRAPHY
BASE LITHOGRAPHY	
BASE DIFFUSION	GATE OXIDATION
EMITTER AND COLLECTOR LITHOGRAPHY	
EMITTER AND COLLECTOR DIFFUSION	SOURCE AND DRAIN CONTACT LITHOGRAPHY
METAL CONTACT LITHOGRAPHY	
METALIZATION	METALIZATION
METAL INTERCONNECT LITHOGRAPHY	METAL INTERCONNECT LITHOGRAPHY

(a) **npn BIPOLAR JUNCTION TRANSISTOR (6 LITHOGRAPHIC STAGES)**

(b) **MOS TRANSISTOR (4 LITHOGRAPHIC STAGES)**

FIGURE 1.5. Comparison of steps required for fabrication of a *npn* bipolar junction transistor and a MOS transistor.

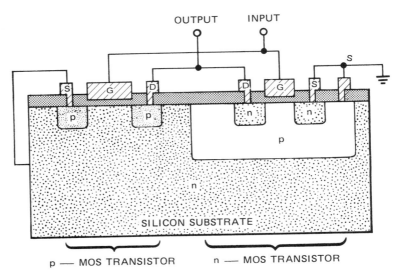

FIGURE 1.6. Complementary MOS integrated circuits.

but in the D-MOS or V-MOS process the channel is only 1 μm. In a D-MOS transistor the substrate is a π-type (lightly doped p-type) material on which p-diffusion is made in selected regions. The lateral dimension of the p-region is the channel width, which can be controlled very accurately.

The fabrication of a V-MOS structure begins with an n substrate on which thin layers (1 μm) of p- and π-type (lightly doped p-type) materials are epitaxially grown sequentially. The dimension of the p-layer corresponds to the length of the channel and can be controlled more easily than in the C-MOS process. An n^+ diffusion is made to form the drain region, then a V-shaped groove is etched through the channel and source regions, using an anisotropic etchant. The process is completed by growing silicon dioxide over the V-groove gate region and then applying the final metal layer.

MOS technology can be used to make high-quality, precisely controlled capacitors. Charge-coupled devices (CCDs) consist simply of closely spaced arrays of MOS capacitors, as shown in Figure 1.8.[26] However, the capacitors are built close enough to one another so that the free charge stored in the inversion layer can be transferred to the channel region of the adjacent device. The exchange of charge is controlled by the voltages applied to the metal gates of the MOS capacitors. CCDs have made significant improvements to memory circuits, signal-processing circuits, and imaging devices.

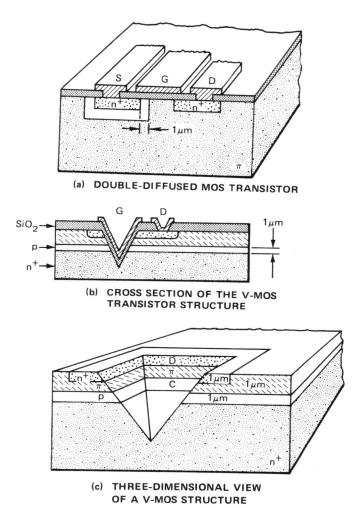

(a) DOUBLE-DIFFUSED MOS TRANSISTOR

(b) CROSS SECTION OF THE V-MOS
TRANSISTOR STRUCTURE

(c) THREE-DIMENSIONAL VIEW
OF A V-MOS STRUCTURE

FIGURE 1.7. Double-diffused (D-MOS) and vertical or V-groove (V-MOS) transistors.

1.3. Planar Technologies Used in Fabricating Microelectronics

Planar technology requires being able to create patterned thin layers of materials with various electrical properties to form a connected set of devices. In this section we introduce the film deposition techniques of epitaxy, oxidation, vacuum evaporation and sputtering, the layer modification techniques of dop-

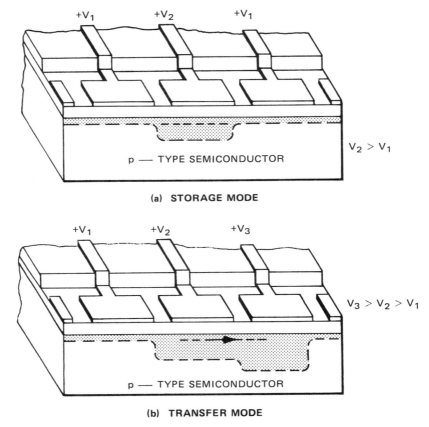

FIGURE 1.8. Transfer mechanism in a charge-coupled device (CCD). In the storage mode (a), charge is held under the center gate. Application of $V_3 > V_2$ on the right-hand gate causes transfer of charge to the right, as shown in (b), the transfer mode.

ing, the pattern generation techniques of lithography, and the selective layer removal techniques of etching.

The fabrication steps of a planar p–n diode are shown in Figure 1.9. Using an n^+ silicon wafer as the starting substrate, a thin layer of n silicon is grown by the epitaxial process. The dopant atoms are added by depositing them onto the silicon surface and then diffusing them into those regions of the silicon that are not protected by a layer of silicon dioxide. The precise deline-ation or patterning of the SiO_2 film is accomplished lithographically. Since the dopant atoms are introduced from the surface and typical diffusions are on the order of a few microns or less, the active region of the diode is within a few microns of the surface of the silicon wafer. The remainder of the wafer thickness (typically 560 μm) serves simply to support the surface structure.

An important aspect of planar technology is that each step of the fabrication process is applied to the entire wafer at the same time. This simultaneous fabrication of several semiconductor devices in the same piece of semiconductor results in an IC in one silicon chip and many ICs (several hundred) per wafer.

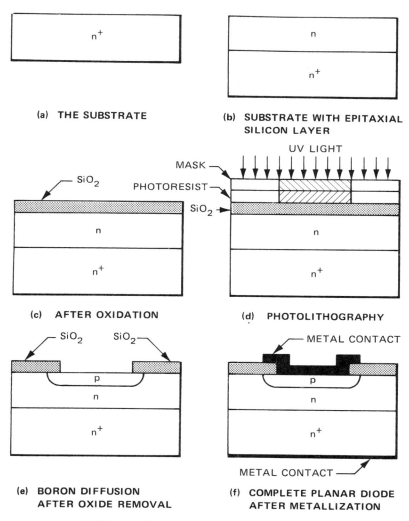

FIGURE 1.9. Fabrication steps of a planar $p-n$ diode.

1.3.1. Epitaxy

The word "epitaxy" is from the Greek word meaning "arranged upon."
It describes the growth technique of arranging atoms in single-crystal fashion
upon a crystalline substrate so that the lattice structure of the newly grown
film duplicates that of the substrate. The key reason for using this growth technique is the ability to obtain an extremely pure film while retaining control of
the dopant level. The dopant in the film can be *n*- or *p*-type and is independent
of the substrate doping. The three types of epitaxial growth processes currently
employed are vapor-phase (VPE), liquid-phase (LPE), and molecular beam
epitaxy (MBE).

A vapor-phase epitaxial growth system is shown in Figure 1.10(a). Hydrogen gas containing a controlled concentration of $SiCl_4$ is fed into a reactor con-

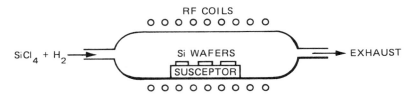

(a) SCHEMATIC OF A VAPOR-PHASE EPITAXIAL GROWTH SYSTEM

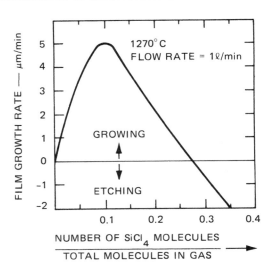

**(b) GROWTH RATE OF EPITAXIAL FILM AS A FUNCTION
OF PERCENTAGE CONCENTRATION OF $SiCl_4$ IN GAS**

FIGURE 1.10. Vapor-phase epitaxy (VPE).

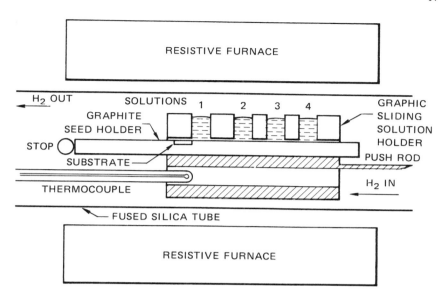

FIGURE 1.11. Schematic of liquid-phase epitaxial reactor.

taining silicon wafers in a graphite susceptor. The graphite is inductively heated by a radio-frequency (rf) coil to a high temperature ($< 1000°C$). The high temperature is necessary to permit the deposited atoms to find their proper position in the lattice in order to maintain a single-crystal film. The basic reaction is given by $SiCl_4 + 2H_2 \rightleftharpoons Si(solid) + 4HCl$. This reaction is reversible. The forward reaction produces an epitaxial film on the silicon. The reverse reaction removes or etches the substrate.

The growth rate of the film as a function of the concentration of $SiCl_4$ in the gas is shown in Figure 1.10(b). Notice that the growth rate reaches a maximum and decreases as $SiCl_4$ concentration is increased. This effect is caused by the competing chemical reaction $SiCl_4 + Si(solid) \rightarrow 2SiCl_2$. Therefore, etching of silicon occurs at high concentrations of $SiCl_4$.

Impurity atoms are introduced in the gas stream to grow a doped epitaxial layer. Phosphine (PH_3) is used for *n*-type doping and diborane (B_2H_3) is used for *p*-type doping.

For depositing multilayers of different materials on the same substrate, LPE is used. Figure 1.11 shows an LPE apparatus for the epitaxial growth of four different layers. In operation, the sliding solution holder is moved to bring the substrate in contact with the solute. With this method, junctions of different materials (Ge–Si, GaAs–GaP), that is, heterojunctions, can be fabricated where the layer thicknesses are less than 1 μm.

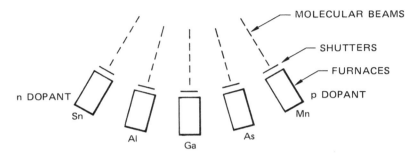

FIGURE 1.12. Schematic of molecular beam epitaxial furnaces.

MBE achieves crystal growth in an ultrahigh vacuum (UHV) environment through the reaction of multiple molecular beams with a heated single-crystal substrate. This process is illustrated in Figure 1.12, which shows the essential elements for MBE of doped ($Al_xGa_{1-x}As$). Each furnace contains a crucible, which in turn contains one of the constituent elements of the desired film. The temperature of each furnace is chosen so that the vapor pressures of the materials are sufficiently high for free evaporation generation of thermal-energy molecular "beams." The furnaces are arranged so that the central portion of the beam flux distribution from each furnace intersects the substrate. By choosing appropriate furnace and substrate temperatures, epitaxial films of the desired chemical composition can be obtained. Additional control over the growth process is achieved by individual shutters interposed between each furnace and the substrate. Operation of these shutters permits abrupt cessation or initiation of any given beam flux to the substrate.

One of the distinguishing characteristics of MBE is the low growth rate: approximately 1 μm/h or, equivalently, 1 monolayer/s. The molecular beam flux at the substrate can therefore be readily modulated in monolayer quantities, with shutter operation times below 1 s. MBE brings to microfabrication almost two orders of magnitude improvement in structural resolution in the direction of growth over techniques of LPE and VPE.

MBE has been used to prepare films and layer structures for a variety of GaAs and $Al_xGa_{1-x}As$ devices. These include varactor diodes having highly controlled hyperabrupt capacitance–voltage characteristics, impatt diodes, microwave mixer diodes, Schottky-barrier field-effect transistors (FETs), injection lasers, optical waveguides, and integrated optical structures. The potential of MBE for future solid-state electronics is greatest for microwave and optical solid-state devices and circuits in which submicrometer layer structures are essential. The inherent adaptability of the process to planar technology and integration also offers significant design opportunities.

Possible longer-term implications of MBE for solid-state electronics are related to its capability of growing extended layer sequences with alternating composition, such as GaAs and AlAs. Such superlattice structures with periodicities of 50 to 100 Å show negative resistance characteristics attributed to resonant tunneling.

1.3.2. Oxidation

A silicon dioxide layer is usually formed on the wafer by the chemical combination of silicon atoms in the semiconductor with oxygen that is allowed to flow over the silicon wafer surfaces while the wafer is heated to a high temperature (900–1200°C) in a resistance-heated furnace, as illustrated in Figure 1.13. The oxidizing ambient can be either dry or wet oxygen. The chemical reaction takes place according to one of the following equations:

Dry oxidation: $Si(solid) + O_2 \rightarrow SiO_2(solid)$

Steam oxidation: $Si(solid) + 2H_2O \rightarrow SiO_2(solid) + 2H_2$

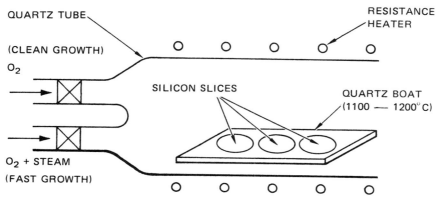

FIGURE 1.13. Thermal oxidation of silicon dioxide layer.

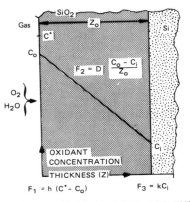

(a) FLUX CONDITIONS FOR THE SiO₂-Si SYSTEM

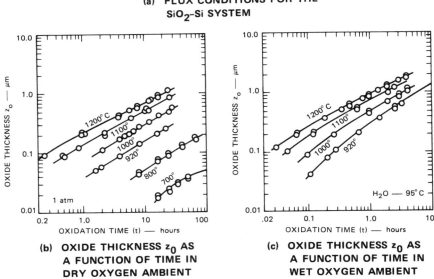

(b) OXIDE THICKNESS z_0 AS A FUNCTION OF TIME IN DRY OXYGEN AMBIENT

(c) OXIDE THICKNESS z_0 AS A FUNCTION OF TIME IN WET OXYGEN AMBIENT

FIGURE 1.14. Thermal oxidation in wet- and dry-oxygen ambients. Source: ref. 27.

Oxidation proceeds much more rapidly in a wet-oxygen ambiance, which is used for the formation of thicker protective layers. The physical theory of thermal oxidation[27,28] can be explained with the help of the simple model shown in Figure 1.14(a).

Oxidation takes place at the Si–SiO₂ interface, so the oxidizing species must diffuse through any previously formed oxide and then react with the silicon at this interface. According to Henry's law, the equilibrium solid concentration is proportional to the bulk gas partial pressure P,

$$C^* = HP$$

where C^* is the maximum oxidant concentration for a given P_1 and H is Henry's law coefficient. The oxidant concentration in this nonequilibrium case is less than C^* on the solid surface. The flux F_1 is determined by the difference in oxidant concentrations,

$$F_1 = h(C^* - C_0)$$

where C_0 is the oxidant surface concentration and h is the main transfer coefficient.

The oxidant concentration C_0 at the oxide surface is determined by the temperature, the gas flow rate, and the solid solubility in the oxide. To determine the oxide growth rate we consider the fluxes of the oxidant in the oxide (F_2) and at the oxide–silicon interface (F_3). From Fick's law, the flux across the oxide is given by the gradient of the oxidant concentration:

$$F_2 = -D \frac{dC}{dz} = \frac{D(C_0 - C_i)}{z_0} \tag{1.1}$$

where C_i is the oxidant concentration, in molecules/cm^3, at $z = z_0$; D is the diffusion constant at a given temperature; and z_0 is the oxide thickness.

The flux (F_3) at the oxide–silicon interface is determined by the surface reaction rate constant K and is given by

$$F_3 = KC_i \tag{1.2}$$

Under steady-state conditions, these fluxes must be equal so that $F_3 = F_1 = F_2 = F$. Therefore, we can equate Eqs. (1.1) and (1.2) to yield C_i and C_0 in terms of C^*:

$$C_i = \frac{C^*}{1 + K/h + Kz_0/D} \tag{1.3a}$$

and

$$C_0 = \left(1 + \frac{Kz_0}{D}\right) C_i \tag{1.3b}$$

To describe the rate of oxide growth, the flux equations at the SiO$_2$–Si interface is written in the form

$$N_i \frac{dz_0}{dt} = F_3 = \frac{KC^*}{1 + K/h + Kz_0/D} \tag{1.4}$$

The oxide growth rate is determined by the flux (F_3) and the number of oxidant molecules (N_i) needed to form a unit volume of oxide. Since there are 2.2×10^{22} SiO_2 molecules/cm³ in the oxide, we need 2.2×10^{22} molecules/cm³ of O_2, or 4.4×10^{22} molecules/cm³ of H_2O.

The following integral defines the relationship of z_0 and t:

$$N_i \int_{z_i}^{z_0} \left(1 + \frac{K}{h} + \frac{Kz_0}{D} \right) dz = KC^* \int_0^t d\tau \tag{1.5}$$

Integrating gives

$$z_0^2 + Az_0 = B(t + \tau) \tag{1.6}$$

where

$$A \equiv 2D \left(\frac{1}{K} + \frac{1}{h} \right) \approx \frac{2D}{K} \qquad \text{(since generally } h \gg K\text{)}$$

$$B \equiv \frac{2DC^*}{N_i}$$

$$\tau \equiv \frac{z_i^2 + Az_i}{B}$$

where

$$z_i \approx \begin{cases} 200\text{Å for dry } O_2 \\ 0 \text{ for wet } O_2 \end{cases}$$

and z_i is the initial value of the oxide thickness at $t = 0$. Solving for the oxide thickness z_0,

$$z_0 = \frac{A}{2} \left[\left(1 + \frac{t + \tau}{A^2/4B} \right)^{1/2} - 1 \right] \tag{1.7}$$

Equation (1.7) reduces to

$$z_0 \approx \frac{B}{A}(t + \tau) \approx \frac{KC^*}{N_i}(t + \tau) \qquad \text{for } (t + \tau) \ll \frac{A_2}{4B} \tag{1.8}$$

and

$$z_0 \approx (Bt)^{1/2} \approx \left(\frac{2DC^*}{N_i} t \right)^{1/2} \quad \text{for } t \gg \frac{A^2}{4B} \quad (1.9)$$

Therefore, for short oxidation times [Eq. (1.8)], the oxide thickness is determined by the surface reaction rate constant (K) and is linearly proportional to the oxidation time. If the oxidation time is long [Eq. (1.9)], the oxide growth is determined by the diffusion constant D. The oxide thickness in this case is proportional to the square root of the oxidation time.

Using experimentally determined values of A, B, and τ, we may plot the oxide thickness as a function of time for several commonly used oxidation temperatures for both dry and wet oxidation. [See Figures 1.14(b) and (c).]

Oxide thicknesses of a few tenths of a micron are often used, with 1–2 μm being the upper practical limit for conventional thermal oxidation. A significant recent advance was the addition of chlorine-containing chemicals during oxidation. This resulted in improved device threshold voltage (V_T) stability, increased dielectric breakdown strength,[29,30] and an increased rate of silicon oxidation. The primary function of chlorine in SiO_2 films (10^{16}–10^{20}/cm^3) is to passivate or electrically inactivate impurity ions (such as sodium and potassium) that have inadvertently been incorporated into the SiO_2.

1.3.3. Lithographies

Figure 1.15 shows the types of lithographies used for microstructure fabrication.[31] Photolithography [Figure 1.15(a)] is the most important technology in the microelectronics industry; it is used routinely down to 2–3 μm linewidths. Electron-beam lithography [Figure 1.15(b)] is currently used for mask making. Electron backscattering limits its practical minimum linewidth to 0.5 μm for high-density microstructures. X-ray lithography [Figure 1.15(c)] is usable down to near a few hundred angstroms, but requires a complex absorber mask or a thin-film support structure. Ion-beam lithography [Figure 1.15(d)] offers patterned doping capability and very high resolution ($\sim$ 100 Å).

The different technologies for lithography encompass two major areas: mask making and image transfer from mask to wafer. As will be discussed later, electron-beam lithography is capable of writing directly on the wafer without use of a mask. Figure 1.16 shows the different combinations currently applied for mask making and image transfer.[31]

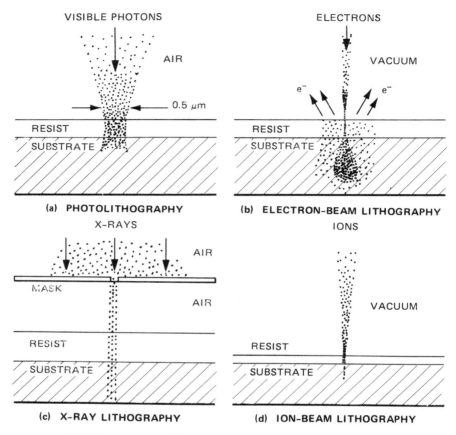

FIGURE 1.15. Types of lithographies used in microstructure fabrication.

1.3.3.1. Photolithography

In photolithography it is first necessary to produce a mask or transparency of the pattern required.

Mask making begins with a large-scale layout called the artwork. Once the designer has completed a circuit design, the locations of all the circuit components on the surface of the chip must be determined. Although the chip dimensions are from 0.5–5 mm on a side, the artwork must be made many times the actual chip size to avoid large tolerance errors and to be of a reasonable size for human operators to handle. The layout is usually made in the form of a drawing showing the position of the windows that are required for a particular step of the fabrication sequence. Six or more layout drawings are required for a typical circuit. For complex circuits, the layout process can be performed by use of computer-aided graphics, and the computer can generate the drawing.

Next, the artwork is photographed by a large camera. Typically the original artwork will be as much as 500 times the size of the final circuit chip. Consequently, for a chip 2.5 mm on a side, the artwork may be 125 cm on a side, and a large camera is indeed required. Successive photographs are taken to reduce the artwork first to 100 times, then to 10 times, and finally to exact size on a master plate. The master plate is used in a precision step-and-repeat printer to produce multiple sequential images of the layout on a high-resolution photographic plate. This plate is used as the mask in the photoresist operation to transfer the layout pattern to the wafer surface.

To illustrate the photoresist procedure we consider the case where small openings are to be made in the silicon dioxide layer covering a silicon wafer.

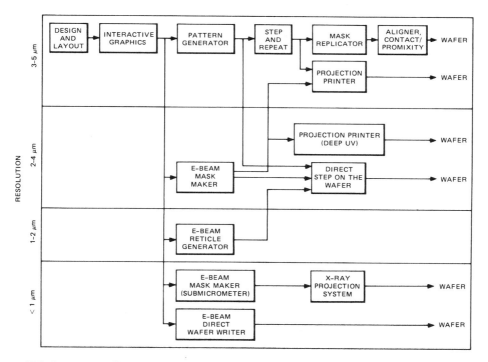

FIGURE 1.16. Different combinations of electron-beam lithography for mask making and image transfer.

Photoresist is coated on the oxide layer, as shown in Figure 1.17(a), placed in contact with the glass mask containing the pattern of the oxide to be removed, and exposed to light, as shown in Figure 1.17(b). During the development process, the unexposed coating is dissolved, leaving an opening in the coating, as shown in Figure 1.17(c). The photoresist coating that remains is chemically resistant to the buffered acid solution used to etch through the oxide layer, which produces an opening in the oxide as shown in Figure 1.17(d). The remaining photoresist coating is then removed from the wafer, and the wafer is ready for the next step.

This method of generating patterns on semiconductor wafers is called contact printing. Contact with the photoresist wears the mask and eventually causes defects. Spacing the mask away from the substrate prevents contact and the defects that result from contact, but causes other problems. The larger spacing increases diffraction of the transmitted light, thus reducing resolution and blurring the individual photoresist features. The degree to which this occurs depends on the actual mask-to-wafer spacing, which may vary across the wafer. Wafer flatness variations (which are especially troublesome at small spacings) and diffraction effects (which are troublesome at large spacings) limit proximity printing with visible light to features no smaller than 7 μm.

1.3.3.2. Projection Printing

In projection printing an image of the photomask is projected directly onto the photoresist-covered wafer by means of a high-resolution lens between the mask and the wafer. In this system the mask life is potentially limited only by handling damage. In one class of projection printers the entire wafer is illuminated in a single exposure, with a pattern originating in a mask that is usually the same size as the wafer (5–10 cm in diameter). One-to-one projection printers are commercially available that have an image resolution in the range of 2–3 μm and an overall registration accuracy of about 0.3–0.6 μm.[32,33]

In the other class of projection printers only part of the wafer is exposed to a pattern originating in a mask (or reticle) that is 5 or 10 times larger than the projected image. The wafer is then stepped to a new position, where another part of the wafer can be exposed. By stepping and repeating, the entire wafer is covered with the reticule pattern. Step-and-repeat systems using 9 to 81 steps are also commercially available, and they typically have a resolution in the range of 1–2 μm and an overall registration accuracy of about 0.25–0.5 μm.[34–38]

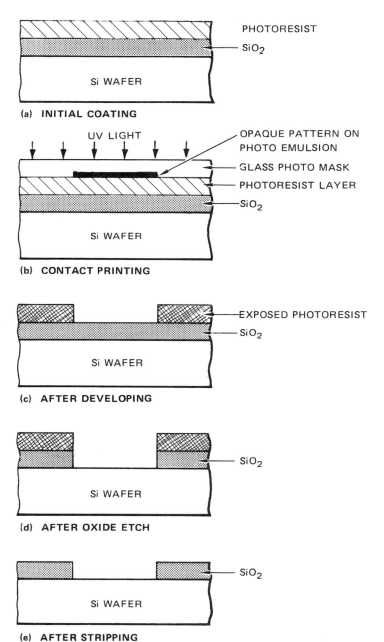

FIGURE 1.17. Steps in the photoresist process.

1.3.3.3. Electron-Beam Lithography

The smallest features that can be formed by the conventional photolithographic process are ultimately limited by the wavelength of light. Current technology can routinely reproduce elements a few micrometers across, and it appears possible to reduce the smallest features to about 1 μm. Electron beams and X rays, however, have wavelengths measured in nanometers and even smaller units, and they are thus capable of producing extremely fine features.

Electron beams are attractive for lithography for additional reasons other than their short wavelengths:

- Electrons can be imaged to form either a pattern or a small point $\leq$ 100 Å, as opposed to 5000 Å for light.
- Electron beams can be deflected and modulated with speed and precision by electrostatic or magnetic fields.
- The energy and dose delivered to the resist-coated wafer can be controlled precisely.

Electrons are used either by scanning the beam to generate patterns directly from computer programs or by electron imaging through special masks. Electrons from a source can be formed into a pencil-like beam that can be deflected over an electron resist coated substrate and modulated to draw a desired pattern. The beam can be imaged to a submicron spot, with sufficient current to expose the resist in less than 10^{-7} s. Since up to 10^{10} spots are typically required for a wafer having 0.5-$\times$-0.5-cm^2 chips and 0.25-μm^2 picture elements (pixels), this extremely high speed is important.

Accurately positioning 10^{10} spots requires a precision of about 1 part in 10^6 along each coordinate axis. This places severe demands on the electron deflection system. Difficulties arise from inaccuracies in digital-to-analog conversion, nonlinearities in the electron deflection system, uncertainties in the surface position of the substrate, wafer bow, and electrostatic and magnetostatic disturbances. All these factors limit the practical range of deflection to about 1 mm. Since this is less than the width of the chip, a mechanical table is used to move the substrate to allow the full field to be exposed. Insuring that the relative position of the beam and table remain within the required accuracy has led to two divergent approaches. In each, the electron column is used as a scanning electron microscope to locate registration features on the substrate for alignment purposes. In one approach, alignment on separate registration marks is accomplished for each table position. In the other, following electron-beam registration, laser interferometers are used to precisely measure the table movement. Further electron-beam registration is used occasionally on a single mark to guard against beam drift.

Two scanning systems are used: raster scan and vector control. The raster-scan systems methodically cover the entire area of the pattern to be generated, turning the electron beam on or off as demanded by the requirements of the pattern being recorded. The vector-control systems deflect the beam to follow paths specified by the needs of the pattern. A block diagram of an electron-beam generator developed for microcircuit fabrication is shown in Figure 1.18. The electron optics associated with electron-beam pattern-generation equipment are similar to those used in electron microscopy. An electron source, normally a heated cathode, supplies free electrons by thermionic emission. These free electrons are subsequently accelerated by electrostatic fields and focused by electromagnetic fields. They are controlled and deflected by combinations

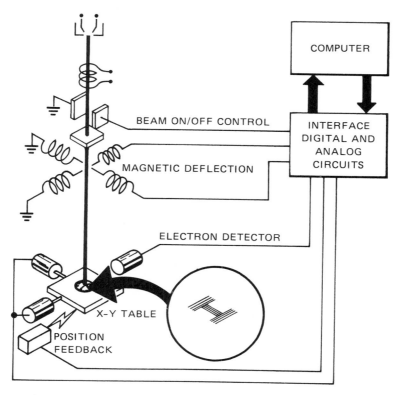

FIGURE 1.18. Schematic illustration of an electron-beam lithography system.

of electromagnetic and electrostatic fields to achieve the final result of tracing out a well-defined pattern.

Electrons can also be imaged to form a complete pattern at one time. An electron image-projection system (ELIPS) has been described[39,40] in which a photocathode is deposited on the patterned surface of an optical mask. Ultraviolet (uv) light illuminates the photocathode layer through the substrate in the transparent regions of the mask, causing a patterned emission of electrons from the photocathode. These electrons are imaged by uniform coaxial electrostatic and magnetic fields onto the facing resist-coated substrate. The system is capable of submicron resolution over the full wafer area.

In the past few years considerable work has been done to apply electron lithography to the fabrication of microelectronic devices. Most of this work has used the scanning electron beam because of its ability to create high-resolution patterns (linewidth $\leq$ 5000 Å), its programmability, its large depth of focus ($\sim$ 10 μm), and its capability of providing focus and registration via scanning electron microscopy.

There are two distinct uses of scanning-electron-beam lithography in microelectronic device fabrication: using the scanning beam to expose directly a resist on a device substrate or using the scanning electron beam to create a mask whose pattern can then be transferred onto device substrates.

1.3.3.4. X-Ray Lithography

The X-ray lithography technique is illustrated in Figure 1.19. The mask consists of an X-ray transparent membrane which supports a thin patterned film made of a material that strongly absorbs X rays. It is placed over a substrate coated with a radiation-sensitive resist. A distant "point" source of X rays produced by a focused electron beam illuminates the mask, thus projecting the shadow of the X-ray absorber onto the polymer film. This is the only feasible exposure scheme, since efficient X-ray lenses and mirrors for collimation cannot be made. The inset of Figure 1.18 illustrates the penumbral shadowing δ, which results from the finite size d of any practical X-ray source. In any given exposure situation, the magnitude of δ can be made as small as required by a proper choice of the parameters s, d, and D.

Figures 1.20 and 1.21 show the processes used to fabricate micrometer surface relief and doping structures for simple and layered substrates after lithography.[41] Generally, following exposure, a development step removes either the exposed regions (positive resist) or the unexposed regions (negative

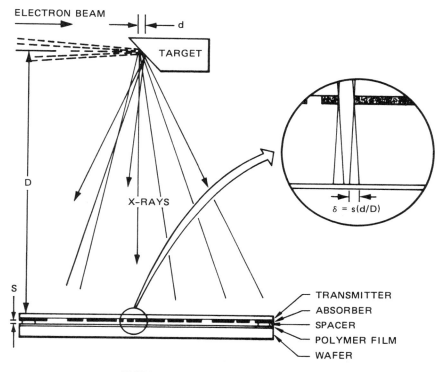

FIGURE 1.19. X-ray lithography technique.

resist), thereby leaving a resist pattern in relief on the substrate surface. Following the creation of the relief structure in the resist, the substrate is processed in one of the following ways:

- By etching a relief structure in it
- By growing material
- By doping
- By depositing material through the open spaces in the resist pattern (lift-off process)

Since the X-ray wavelength used is on the order of 10 Å, diffraction effects are generally negligible, and proximity masking may be used. A partial list of the lithographic tools and associated performance criteria is presented in Table 1.

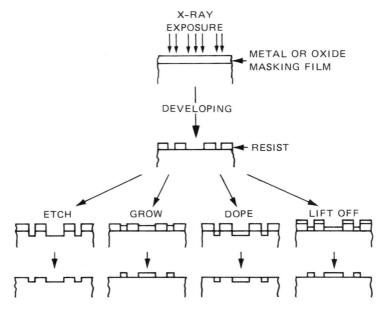

FIGURE 1.20. Process for fabricating micrometer surface relief and doping structures for simple substrates.

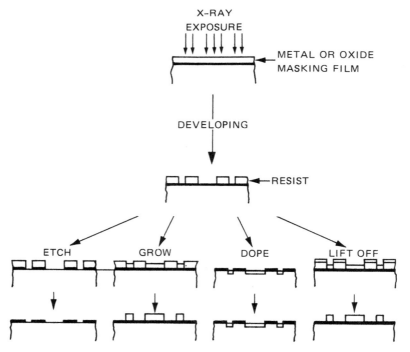

FIGURE 1.21. Process for fabricating micrometer surface relief and doping structures for layered substrates.

TABLE 1. *A Partial List of Lithography Tools*

System	Wavelength or Acceleration Potential	Minimum Linewidth (μm)	Registration Method and Accuracy (μm)	Throughput (Wafers/hr) [a]	References
Optical systems					
Near-contact printers	−4000 Å	2	Automatic or manual ±0.5	50–100	32
Deep uv contact printers	−2500 Å	0.5		5–10	33,34
1:1 projection printers	−4000 Å	2–3	Manual ±0.8	40–60	35
Reduction projection systems	−4000 Å	1.25	Manual	10–50	36,37,38
X-ray systems	4–10 Å	0.2	0.1 for lab models	1–10	39,40
Scanning-electron-beam systems					
EBES (Bell Labs)	15 kV	0.2	Automatic ±0.2	2	41
VS (IBM)	25 kV	0.1	Automatic ±0.1	2–5	42,43,44
ELI (IBM)	25 kV	2.5	Automatic ±0.25	20	45
EBM (TI) printers	15 kV	0.25	Automatic ±0.25	2–5	46

[a]Note: Throughput numbers are for 52-mm wafers and include wafer handling and pumpdown time for electron-beam exposure tools.

1.3.4. Etching

After a resist pattern has been generated by the lithographic process (photo, electron, or X-ray), the film beneath the resist is usually delineated by an etching process. In this section we briefly review the wet-chemical and dry etching (ion etching, plasma etching, and reactive plasma etching) processes.

1.3.4.1. Chemical Etching

Table 2 lists the typical etchants used for thin-film materials in the microelectronic industry. Wet-chemical or solution etching presents several problems. Photoresists often lose their adhesion to underlying films when exposed to hot acids. Also, as etching proceeds downward, it also proceeds laterally, thereby undercutting the photoresist film and broadening lines. A final consideration with wet-chemical etching is the trend toward submicron geometries.

TABLE 2. Thin-Film Materials and Liquid Etchants Used during Integrated
Circuit Manufacture

Film	Etchant	Temperature of Etch Bath ($^\circ$C)
SiO$_2$	HF	20–25
Si$_3$N$_4$	H$_3$PO$_4$	160–180
Al	H$_3$PO$_4$/HNO$_3$/HC$_2$H$_3$O$_2$	40–50
Crystalline/polycrystalline Si	HNO$_3$/HF	20–25

Solvent etching becomes more difficult because the surface tension of the solutions tends to cause the liquid to bridge the space between two strips of photoresist, thereby precluding etching of the underlying film. Because of these difficulties in delineating high-resolution patterns by wet etching, dry etching processes using ionized gases have been developed,[42] as discussed in Sections 1.3.4.3. and 1.3.4.4.

1.3.4.2. Anisotropic Etching

Anistropic etching of a single-crystal material results from the different etch rates to chemical solvents of its crystallographic planes. The diamond cubic structure of silicon and the Miller indices of two of its planes are illustrated in Figure 1.22. The $\langle 111 \rangle$ plane of silicon [Figure 1.22(b)] is more closely packed then is the $\langle 100 \rangle$ plane [Figure 1.22(c)] and etches at a much lower rate. This concept has been used to fabricate a variety of active and passive three-dimensional structures[43] and surface devices.

Figure 1.22(d) shows the cross section of an anisotropically etched square pyramidal hole in $\langle 100 \rangle$ silicon. The dimensions of the hole are given by the expression

$$W_0 = W_{Si} - \sqrt{2}\,t_{Si}$$

where W_0 is the side of the square apex, W_{Si} is the side of the square base hole in the wafer surface, and t_{Si} is the etched depth. This technique is based on the fact that the etch rate of the $\langle 100 \rangle$ plane is much higher than that for the $\langle 111 \rangle$ plane.

A membrane-type circular orifice (Figure 1.23) can be fabricated by the

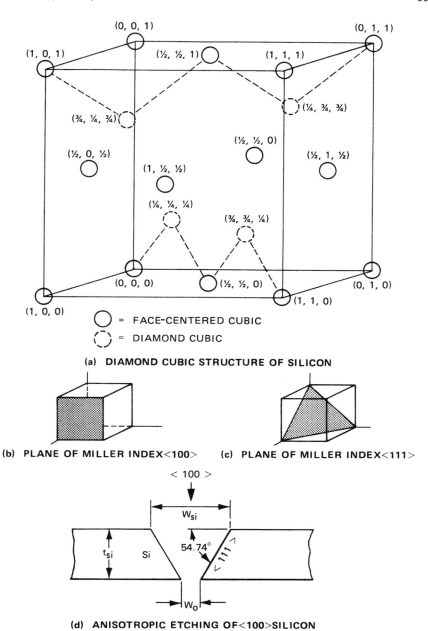

(a) DIAMOND CUBIC STRUCTURE OF SILICON

(b) PLANE OF MILLER INDEX<100>   (c) PLANE OF MILLER INDEX<111>

(d) ANISOTROPIC ETCHING OF<100>SILICON

FIGURE 1.22. Structural characteristics relevant to anisotropic etching of silicon.

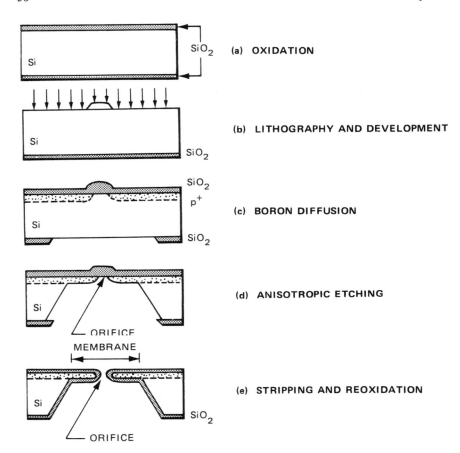

FIGURE 1.23. Fabrication of a membrane nozzle.

anisotropic etching process in combination with a process that takes advantage of the etch resistance of heavily doped p^+ silicon in P–ED, an anisotropic etching solution.[44] At an impurity concentration $N_A \sim 10^{19}/cm^3$, the etch rate of Si in P–ED drops sharply and practically reaches zero at $N_A \geq 7 \times 10^{19}/cm^3$. When a silicon wafer with a heavily doped p^+ surface layer is etched in P–ED, the undoped Si is removed and a p^+ membrane is left, with a thickness equal to the depth of the surface layer and a concentration $N_A \geq 10^{19}/cm^3$. This property has been used to fabricate various device structures incorporating membranes ranging in thickness between 1 and 10 μm.

1.3.4.3. Ion Etching

In ion etching the surface is eroded by bombardment with energetic ions. In one form of ion etching the substrates are laid directly on the cathode of a discharge tube. It is called ion-beam milling if the ion beam is generated in a plasma source separated from the substrates being milled.

The key parameter in ion etching is the sputtering yield, which is defined as the number of sputtered target atoms per incident ion. The sputtering yield depends on the material being etched, the bombarding atom species, its energy, the angle of incidence, and in some instances the composition of the background gas. The sputtering yield bears a relation to the binding energy of the target atoms. It is low for elements such as carbon, silicon, and titanium, and high for gold and platinum. The sputtering yield increases with increasing energy and usually saturates above a few keV, since the penetration of an ion into a target increases without a corresponding increase in the energy lost close to the surface. For this reason, ion etching is usually done with energies below a few kV. The etch rate

$$\dot\delta(\theta) = 9.6 \times 10^{25} \frac{S(\theta)}{n} \cos\theta \left| \frac{\text{Å/min}}{\text{mA/cm}^2} \right| \tag{1.10}$$

where $S(\theta)$ is the sputtering yield, n is the atomic density of the target material in atoms/cm³, and the $\cos\theta$ term accounts for the reduced current density at an angle θ off the normal.

Since the sputtering yield is related to the binding energy of the atoms in the material being etched, it is possible to vary its value by introducing reactive gases. These gases can react with the etched surface and vary the binding energy, and consequently the etch rate. For example, oxygen will adsorb on fresh surfaces of materials like titanium and silicon during ion etching, form oxides, and reduce the etch rate.[45] On the other hand, activated chlorine or fluorine-containing species react with the above materials to form loosely bound or even volatile compounds, and thus increase the etch rate.[46]

A system for ion-beam milling is shown in Figure 1.24. It contains a gun, a neutralizer, and a substrate stage. The gun generates ions in a confined plasma discharge and accelerates them in the form of a beam towards the sample. The neutralizer, usually a hot filament, emits a flux of electrons to keep the sample neutral. Noble gases are usually used in the ion-milling machines because they exhibit higher sputtering yields relative to other atoms and also

HEATED CATHODE

ARGON GAS INLET

ANODE

MAGNETIC FIELD PROVIDES
HELICAL ELECTRON PATH
TO INCREASE SOURCE
IONIZATION

PLASMA COMPLETELY
CONFINED TO DIS-
CHARGE SOURCE
CHAMBER

ELECTRON SUPPRESSOR
GRID

EXTRACTION GRID AT
GROUND POTENTIAL

NEUTRALIZATION
FILAMENT

ION BEAM

SHUTTER

SUBSTRATE
TABLE

WORK CHAMBER

VACUUM
SYSTEM

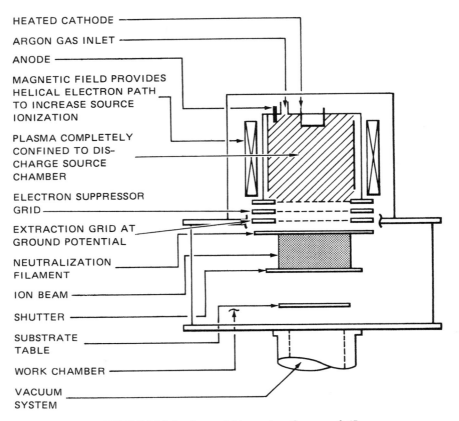

FIGURE 1.24. Ion-beam etching system. Source: ref. 47.

avoid chemical reactions. Table 3 gives the etch rates for argon-ion beams at normal beam incidence.

1.3.4.4. Plasma Etching

The etching of a sample exposed to a reactive gas plasma in which no high-voltage ions impinge on the sample is generally referred to as plasma etching. In plasma etching undercutting can be controlled to some degree by optimizing the mean free path of the etchant molecules by pressure variation, and the adhesion remains good.

An important consideration with respect to using plasma etching techniques involves the selectivity of an etchant for one film material compared to another. Whereas liquid etchants can be completely selective, such good fortune does not extend to plasma etching. Typical plasma reaction chambers are illustrated in Figures 1.25(a) and (b). They usually consist of a fused-silica tube in which the samples can be stacked vertically or horizontally. A rf potential is applied between the electrodes in the vacuum station. The rf field (≈ 15-mHz frequency) causes the free electrons present in the plasma to oscillate, resulting in collisions with gas atoms, a certain fraction of which result in ionization of the gas atoms and sustain the plasma. The pressure required to sustain the plasma is from 2 to 20 Torr. The resultant plasma contains chemically active species that etch the sample material. The etching is not a sputtering phenomenon, but a chemical reaction that converts the etched material to a volatile compound. For example, oxygen plasma is used to remove photoresist material by oxidizing the hydrocarbon-based photoresists to volatile products.

The addition of a chemically active gas to the ion plasma can have the

TABLE 3. *Etch Rates with Argon Ions,*
Normal Beam Incidence[a]

Target Material	Å/min
Silicon	360
Gallium arsenide	2,600
SiO_2 (thermal)	420
SiO_2 (evaporated)	380
KTFR photo resist	390
AZ 1350 photo resist	600
PMM electron resist	840
Silver	2,000
Gold	1,600
Platinum	1,200
Copper	1,000
Palladium	900
Nickel	540
Aluminum	440
Zirconium	320
Niobium	300
Iron	320
Ferrous oxide	660
Molybdenum	400
Titanium	100
Chromium	200
Alumina	130

[a] 1 keV energy, $j = 1.0$ mA/cm^2.

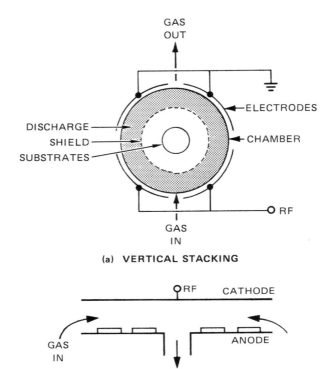

GAS
OUT

ELECTRODES

DISCHARGE
SHIELD
SUBSTRATES

CHAMBER

RF

GAS
IN

(a) VERTICAL STACKING

RF   CATHODE

GAS
IN

ANODE

GAS
OUT

(b) HORIZONTAL STACKING

FIGURE 1.25. Plasma reaction chambers.

effect of changing the etching rates through the chemical interaction between the substrate and the added gas. The addition of oxygen[48] has the effect of reducing the sputtering yield of metals that oxidize readily (Cr and Al), but it has relatively little effect on inert metals (Au and Pt). The use of a halocarbon (CCl_2F_2 and CCl_2FCClF_2) plasma[49] increases the sputter etch rate of various materials (Si, SiO_2, Al, and photoresist).

We see, therefore, that plasma etching and reactive plasma etching are processes in which the material removal is mainly caused by a chemical reaction that results in a loosely bound compound. Because of the chemical nature of reactive etching, it can be more efficient than ion etching and also allow a higher degree of control over the relative etch rate of different materials. However, undercutting and edge profiles are more difficult to control than in ion etching.

Ion etching, the process in which material is removed by means of physical bombardment, has a very high resolution capability, but is inefficient and offers little flexibility in preferential etching of different materials.

1.3.5. Doping

Fabrication of circuit elements requires a method for selective *n*- or *p*-type doping of the silicon substrate. Until the early 1970s, this task was accomplished by diffusion techniques[50-53] in a furnace of 950–1280°C, as shown in Figure 1.26(a). The diffusion process involves two steps. First, the dopant atoms are placed on or near the surface of the wafer by deposition from the

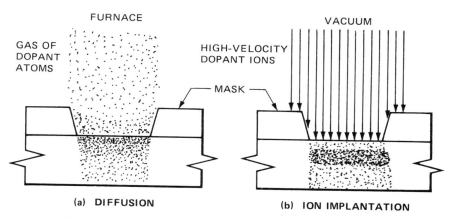

Process Parameter	Diffusion	Ion Implantation
Process Control Dopant Uniformity and Reproducibility	±5% on Wafer ±15% overall	±1% overall
Contamination Danger	High	Inherently Low
Delineation Techniques	Refractory Ins. Refractory Metals Poly Silicon	Refractory and Non- Refractory Materials: Insulators, Metals, Polymers (Beam Writing)
Environment	Furnace	Vacuum

FIGURE 1.26. Comparison of diffusion and ion-implantation process for selective introduction of dopants into the silicon lattice.

gas phase or by coating the wafer with a layer containing the desired dopant impurity. This is followed by drive-in diffusion, which moves the dopant atoms further into the wafer. The shape of the resulting dopant distribution is determined primarily by the manner in which the dopant is placed near the surface, while the diffusion depth depends primarily on the temperature and time of the drive-in diffusion.

Since the early 1970s, selective introduction of dopants into a silicon lattice has been performed at room temperature by ion implantation,[54,55] as shown in Figure 1.26(b), and more recently by neutron transformation. These methods lead to better control than that obtained by diffusion.

1.3.5.1. Diffusion

Diffusion is used to introduce controlled amounts of dopants into the semiconductor. Most impurity atoms in silicon are situated substitutionally in the lattice sites. These impurities can be relocated whenever empty lattice sites or vacancies exist next to them. At a high temperature (e.g., ~ 1000°C) many silicon atoms have moved out of their lattice sites and a high density of vancancies exists. If a concentration gradient exists, the impurity atoms move by way of the vacancies, as shown in Figure 1.27(a), and solid-state diffusion takes place. When the crystal is cooled after the diffusion, the vacancies disappear and the impurity atoms that occupy lattice sites are fixed.

Impurity atoms that occupy the voids between atoms are known as interstitial impurities. These impurities move through the crystal lattice by jumping from one interstitial void (or site) to another. At a high temperature, the spac-

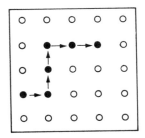
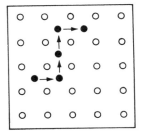

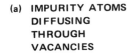

(a) IMPURITY ATOMS DIFFUSING THROUGH VACANCIES

(b) IMPURITY ATOMS DIFFUSING THROUGH INTERSTITIAL SITES

FIGURE 1.27. Diffusion of impurities through a crystal lattice.

STEP	BOUNDARY CONDITIONS	SOLUTIONS	MODEL
PREDEPOSITION	$N = N_0$ at $z = 0$ $N = 0$ at $z = \infty$	$N(z,t) =$ $N_0 \, \text{erfc} \dfrac{z}{2\sqrt{Dt}}$ $Q(t) = \displaystyle\int_0^\infty N(z,t)\, dz$ $= \dfrac{2}{\sqrt{\pi}} \cdot \sqrt{Dt} \cdot N_0$	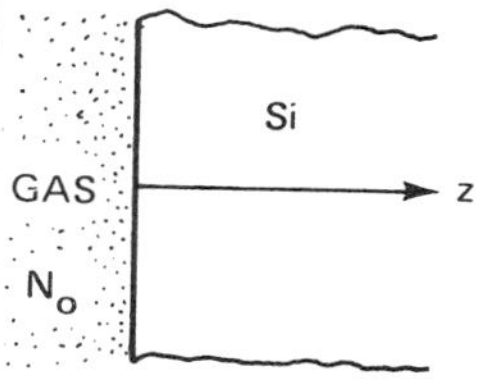
DRIVE-IN	$\left.\dfrac{\partial N}{\partial z}\right|_{(0,t)} = 0$ $N(\infty,t) = 0$	$N(z,t) =$ $\dfrac{Q}{\sqrt{\pi Dt}}\, e^{-\frac{z^2}{4Dt}}$	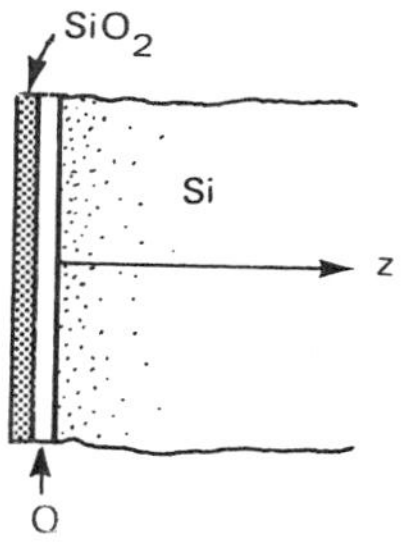

FIGURE 1.28. Diffusion process for impurity atoms.

ing between atoms is wider, so that impurities can diffuse via interstitial sites, as seen in Figure 1.27(b). When the crystal is cooled, interstitial atoms may return to substitutional sites and become electronically active. Substitution is the diffusion mechanism of boron, phosphorus, and most impurities used in silicon. An important exception is gold, which diffuses primarily by the interstitial mechanism.

The two steps in the diffusion process, the predeposition and the drive in, are described by the diffusion equation

$$\frac{\partial N(z,\,t)}{\partial t} = D\,\frac{\partial^2 N(z,\,t)}{\partial z^2} \tag{1.11}$$

and the corresponding boundary conditions as shown in Figure 1.28.

The complementary error function, erfc, is tabulated in Appendix A, and the solid solubilities of several common dopants in silicon as functions of temperature are shown in Figure 1.29. Figure 1.30 shows the diffusivities of the dopants as a function of temperature.

A $p-n$ junction can be formed when a p-type impurity is diffused into a n-type material. In this case the impurity distribution becomes

$$N(z,\ t)\ =\ N_0\ \text{erfc}\ \frac{z}{2(Dt)^{1/2}}\ -\ N_c \tag{1.12}$$

where N_c is the background doping density of the n-type silicon. The junction depth z_j from the surface can be obtained by setting $N = 0$:

$$z_j\ =\ 2(Dt)^{1/2}\ \text{erfc}^{-1}\ \frac{N_c}{N_0} \tag{1.13}$$

This junction depth is an important device parameter.

Figure 1.31 shows the impurity profiles for a $p-n$ junction and *npn* transistor fabricated by the diffusion process. The dopant impurity profile changes every time the wafer is reheated for another process step such as oxidation or diffusion. This means that during the planning of a fabrication process, each

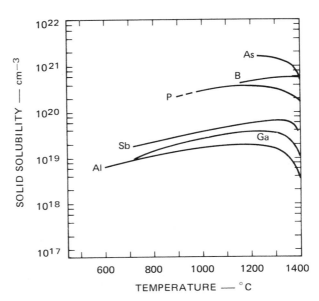

FIGURE 1.29. Solid solubilities of several common dopants in silicon, expressed as functions of temperature.

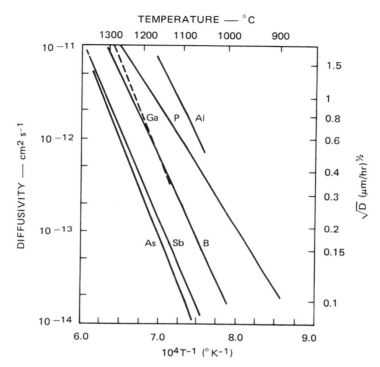

FIGURE 1.30. Diffusivities of several common dopants as functions of temperature.

proposed heat treatment must be taken into account to achieve the desired overall junction depths and dopant profiles.

1.3.5.2. Ion Implantation

In this technique, the dopant atoms are ionized and then accelerated to high energies (30–350 keV) by an electric field, as shown in Figure 1.32. A mass-separating magnet eliminates unwanted ion species. After passing through the deflection and focusing control, the ion beam is aimed at the semiconductor target so that the high-energy ions can penetrate the semiconductor surface. The energetic ions lose their energy through collisions with the target nuclei and electrons and finally come to rest. The total distance that an ion travels before coming to rest is called its range (R); the projection of this distance onto the direction of incidence is called the projected range (R_p).

The doping profile is usually characterized by the projected range and its

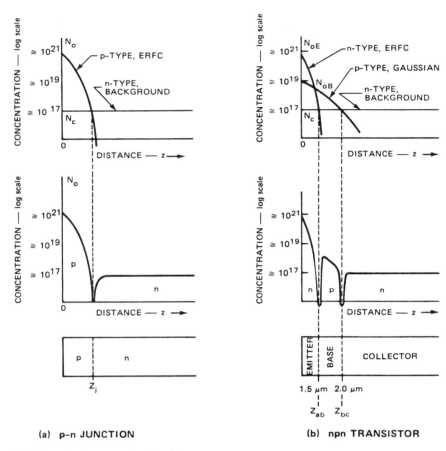

FIGURE 1.31. Impurity of profiles for a *p–n* junction and *npn* transistor fabricated by the diffusion process.

standard deviation ΔR_p. Figure 1.33 shows R_p and $R_p + \Delta R_p$ for boron, phosphorus, and arsenic ions implanted into silicon. Provided that the semiconductor is not aligned in a major crystallographic direction with the ion beam, profiles resulting from implants are nearly Gaussian in shape. The dopant concentration as a function of the distance from the silicon surface is given as

$$N(z) = N_{max} e^{\dfrac{-(z - R_p)^2}{\Delta R_p^2}} \qquad (1.14)$$

where

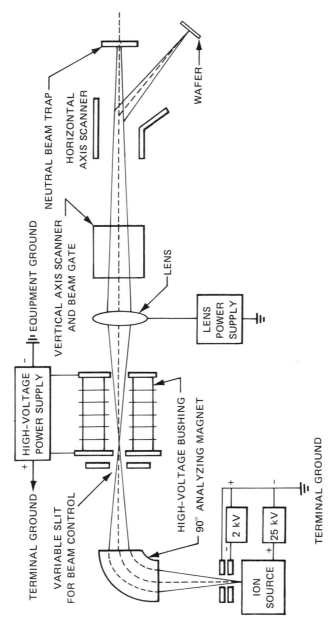

FIGURE 1.32. Ion-implantation technique for selective introduction of dopants into a silicon lattice.

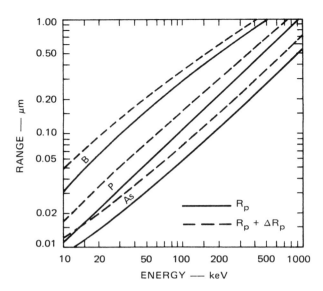

FIGURE 1.33. Projected range R_p and the standard deviation in projected range ΔR_p as a function of the implanted-ion energy.

$$N_{max} = \frac{N_D}{2.5 \, \Delta R_p}$$

and

$$N_D = \frac{\text{number of implanted atoms}}{\text{cm}^2} \qquad (1.16)$$

However, if the ion enters the Si crystal parallel to a major crystal axis or plane, the penetration depth is increased by channeling, giving a profile as illustrated in Figure 1.34. Spatially selective doping is achieved by the use of a patterned silicon dioxide or photoresist film that serves as a mask.

The primary advantage of ion implantation is the process control it offers. By measuring the ion current and implant time, the exact number of dopant ions incorporated in the silicon can be determined. Also, wide variations in impurity concentrations can be achieved, since the amount of dopant introduced is independent of solubility considerations. Furthermore, some control of the distribution of impurity species prior to high-temperature diffusion is possible by regulation of the accelerating voltage.

As the accelerated ions collide with silicon in the lattice, a significant amount of damage is imparted to the crystal lattice. Fortunately, most of this damage can be eliminated by annealing the wafers at 700–1000°C, which allows the displaced atoms to move back to their equilibrium lattice positions, thereby restoring the single-crystal structure. Although some diffusion of the dopant takes place during annealing, the temperatures are usually lower and the times shorter than for typical diffusion cycles.

Recently, postdoping high-intensity laser irradiation has been used to eliminate crystal damage of ion implantation.[56] Lasers are used in two modes for annealing: pulsed and continuous wave (cw). In the pulsed mode, the Q-switched laser creates very rapid (30–50 ns) heating of a narrow surface

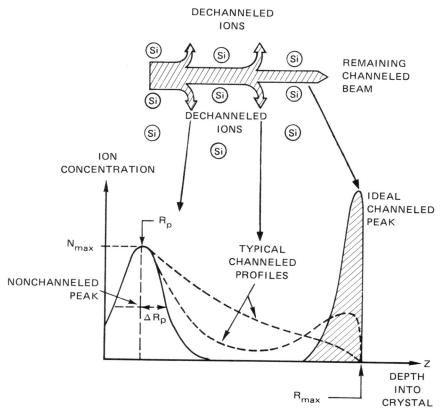

FIGURE 1.34. Channeling effect of an ion entering the Si crystal parallel to a major crystal axis or plane.

region. This region melts and recrystallizes from the substrate outwards, forming a high-quality crystal layer and fully annealing the implanted-impurity profile. In the case of cw annealing, the laser scans across the wafer and heats the Si to a temperature below the melting point. Here the damaged implanted regions regrow via the mechanism of solid-phase epitaxy toward the surface, without observable impurity diffusion.

1.3.5.3. Neutron Transmutation Doping (NTD)

An interesting new process utilized in doping crystals is the neutron transmutation of some of the silicon atoms to phosphorus by placing the silicon in a nuclear reactor and passing a stream of thermal neutrons ($E_n \sim 0.025$ eV) through it.[57] The consecutive thermal neutron reactions during NTD of Si can be summarized as

$$^{30}\text{Si} + n(\text{thermal}) \longrightarrow {}^{31}\text{Si}$$
$$^{31}\text{Si} \xrightarrow{\;t_{1/2} = 2.6 \text{ hr}\;} {}^{31}\text{P} + \beta^-$$

$$^{31}\text{P} + n(\text{thermal}) \longrightarrow {}^{32}\text{P}$$
$$^{32}\text{P} \xrightarrow{\;t_{1/2} = 14.3 d\;} {}^{32}\text{S} + \beta^-$$

From these reactions it is clear that the NTD process is complicated by the fact that ^{31}P formed by the primary transmutation of ^{30}Si can also capture a neutron to give rise to a beta emitter ^{32}P. While the beta decay of ^{31}Si is very short lived and exhibits no residual radioactivity, that of ^{32}P is moderately long lived and can result in measurable amounts of radioactivity. The amount of ^{32}P present depends primarily, of course, on the amounts of ^{31}P produced and, to some degree, on neutron flux and the amount of phosphorus originally in the silicon.

Doping silicon by NTD processing can, of course, only form *n*-type phosphorus-doped material, but it has the advantage of providing an extremely uniform phosphorus concentration in the silicon. Some radiation damage exists after processing in the nuclear reactor, and therefore an anneal is required to restore the lattice and resistivity.

Neutron transmutation-doped silicon has primarily been used in high-power applications, but recent interest has been shown in using it for integrated circuits and semiconductor optical sensors.

1.3.6. Thin-Film Deposition

For the deposition of the thin films used in microelectronics, low-pressure processes such as sputtering and evaporation are preferred[58] over the older, conventional chemical methods such as electroplating, which produce poorer-quality films.

1.3.6.1. Cathode Sputtering

In this thin-film deposition process, the material to be deposited is used as a cathode in a system in which a glow discharge is established in an inert gas (Ar or Xe) at a pressure of 10^{-1}–10^{-2} Torr and a voltage of several kilovolts. The substrate to which the film is to be deposited is placed on the anode. The positive ions of the gas created by the discharge are accelerated toward the cathode and arrive with the energy gained in the cathode fall region (see Figure 1.35). Under the ion bombardment material is sputtered from the cathode, mostly in the form of neutral atoms, but partly in the form of ions. The liberated components condense on surrounding areas, including the substrates on the anode.

The amount of the material sputtered, Q, in a unit of time under constant conditions is inversely proportional to the gas pressure p and anode–cathode distance d:

$$Q = \frac{kVi}{pd} \tag{1.17}$$

where k is the constant of proportionality, V is the working voltage, and i is the discharge current.

1.3.6.2. Vacuum Evaporation

Vacuum evaporation is the most widely used method for the preparation of thin films. The method is comparatively simple, but it can, under proper experimental conditions, provide films of extreme purity and known structure.

The process of film formation by evaporation proceeds by several physical stages:

- Transformation of the material to be deposited by evaporation or sublimation into the gaseous state.

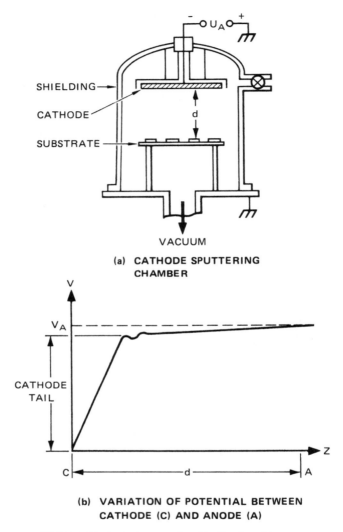

SHIELDING

CATHODE

d

SUBSTRATE

VACUUM

**(a) CATHODE SPUTTERING
CHAMBER**

V

V_A

CATHODE
TAIL

Z

C   d   A

**(b) VARIATION OF POTENTIAL BETWEEN
CATHODE (C) AND ANODE (A)**

FIGURE 1.35. Cathode sputtering thin-film deposition process.

- Deposition of these particles on the substrate.
- Rearrangement or modification of the gaseous particles for bonding on the surface of the substrate.

For most materials, this means that an elevated source temperature is necessary. The rate of film growth depends on the rate of arrival of evaporant particles and the probability of a given particle being retained on the surface. The latter is measured by a parameter called the "sticking" coefficient. The most used substances for thin-film formation are elements or simple compounds whose vapor pressures range from 1 to 10×10^{-3} Torr in the temperature interval 600–1200°C. The pressure and composition of background gases are important if the film properties are sensitive to impurity atoms.

The number of particles with molecular weight M leaving a source at temperature T per unit time per square centimeter can be determined using elementary kinetic theory and is given as

$$N_e = \frac{P_e}{(2\pi MkT)^{1/2}} = 3.513 \times 10^{22} \frac{P_e}{(MT)^{1/2}} \qquad (1.18)$$

where P_e is the vapor pressure in Torr.

Since the substrate is colder than the source, the liberated particles travel in space, with their thermal velocities along a straight line, until collision with another particle. To ensure a straight path for them between the source and substrate, the residual particle concentration in the space must be low, that is, the space must be sufficiently evacuated. The fraction of particles scattered by collisions with atoms of residual gas is proportional to $1 - \exp(-d/\lambda)$, where d is the source–substrate distance and λ is the mean free path of the particles.

The earliest sources were refractory metals such as tungsten, which could be heated by passage of a current. Later, boats of various materials such as molybdenum, graphite, and boron nitride were developed. In addition to direct electrical heating, inductive methods are also in use. By using rf induction and large-diameter inert crucibles, evaporation of a number of different metals in a single evaporator cycle is possible. The major difficulties with this method are contamination of the deposit by the source and source failure.

To circumvent these problems, a focused electron beam can be used to heat a portion of the material while the crucible containing the material is water-cooled.

1.3.7. Interconnect Materials

After the individual circuit elements have been fabricated in the silicon wafer, they must be interconnected to form an IC. This is accomplished by the process of metalization, in which a metal film (~ 0.5–2 μm) is deposited on the wafer and subsequently patterned so that the appropriate transistors, diodes, capacitors, and resistors are interconnected.

Currently, aluminum or aluminum alloys are used for metalization because they satisfy most of the requirements (good adhesion, low resistivity, and low cost) for metalizing materials. Table 4 is a summary of the more common metals used on these films for metalization.

Most of the shortcomings of aluminum metalization arise from the ever-decreasing size of IC elements. Smaller and faster devices require shallower $p-n$ junctions. These junctions cannot withstand heat treatments ($< 500°C$) performed after metalization because of the interdiffusion of Si and Al.[59]

Electromigration [60] caused by momentum transfer from the flowing electrons to the stationary metal atoms presents another problem for aluminum metalization. Because of this phenomenon, voids and hillocks are produced along the length of the conductor, which eventually leads to loss of continuity and circuit failure. High-current-density application ($> 1 \times 10^5$ A/cm^2) requires the use of other metals (e.g., gold) that are not prone to electromigration. With the decrease in feature sizes, high current densities are becoming the rule, rather than the exception.

Although superficially gold appears to be an attractive material for metalization because of its high conductivity, low corrosion, and electromigration

TABLE 4. Properties of Metals Employed as Thin Films

Metal	Limiting Current Density (A/cm^2 $\times 10^5$)	Resistivity ($\times 10^{-6}$ $\Omega \cdot$ cm)	Remarks
Ag	4.0	1.59	Poor adhesion
Cu	4.0	1.67	Poor adhesion; corrosion
Au	7.0	2.35	Silicon eutectic at 370°C; poor adhesion
Al	0.5	2.65	Silicon eutectic at 577°C; electromigration
Al + Cu	2.0		
Mg	—	4.45	Extremely reactive
Rh	—	4.51	Poor adhesion
Ir	—	5.3	Poor adhesion
W	20.0	5.6	Difficult etching
Mo	10.0	5.7	Corrosion susceptibility
Pt	—	9.8	Poor adhesion
Ti	—	0.55	

TABLE 5. *Thin-Film Materials Used in Microelectronics (Classified by Area of Application)*

Interconnections/ Terminals	Insulation/ Passivation	Encapsulation	Resistive	Capacitive	Semiconducting	
Al	$SiO_2 \cdot P_2O_5$	SiO_2	Ta	SiO_2	Si	InAs
Al alloys (Cu, Si)	SiO_2	$SiO_2 \cdot P_2O_5$	TaN_2	Ta_2O_5	Se	InSb
Cu	Si_3N_4	Al_2O_3	$Cr \cdot SiO$	HfO_2	Te	Pbs
Mo–Au	Al_2O_3	Parylene	NiCr	ZrO_2	SiC	PbTe
Ti–Ag	BN	$PbO \cdot B_2O_3 \cdot SiO_2$	SnO_2	$PbTiO_3$	GaAs	Cds
Cr–Ag	SiO		Kanthal		GaP	CdSe
Pt–Au					AIN	ZnSe
Cr–CuAu						
Pb–Sn						

resistance, it cannot be used generally. This is because the adhesion of gold to silicon dioxide is poor, and the use of gold establishes a temperature limit (380°C) for subsequent processing because it forms a eutectic with silicon. It is, however, used in conjunction with other metal films to improve contact adhesion.

At the present time, refractory metal silicides are being considered as interconnect materials because they can be highly conducting and yet withstand the temperatures and reagents encountered during processing.[61] Tantalum, molybdenum, and tungsten silicides are presently favored, although they have resistivities at room temperatures 10 to 30 times that of aluminum.

Corrosion resistance and scratch protection of ICs are achieved by a final "passivation" coating. Materials for passivation and other processes[62] are shown in Table 5.

1.3.8. Summary of Process Steps in Planar Technology

We have now introduced the more important processes that are used in planar technology. For high-volume manufacturing it is desirable to start with a large substrate area so that as many individual circuits (dies) can be made as possible. This places heavy demands on the accuracy of the masks, the uniformity with which the films are deposited, and the precision of pattern over-

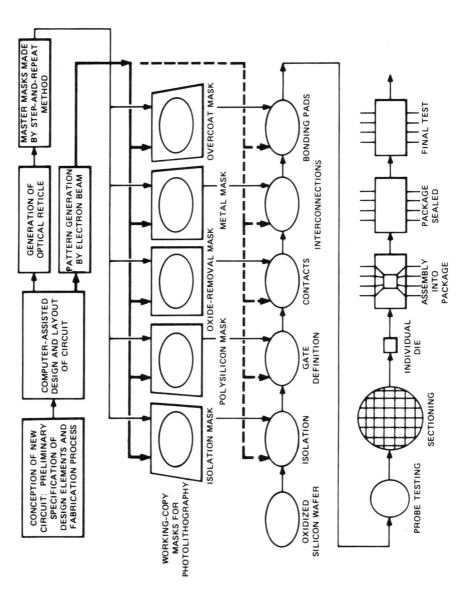

FIGURE 1.36. Process steps in large-scale integration (LSI) applications.

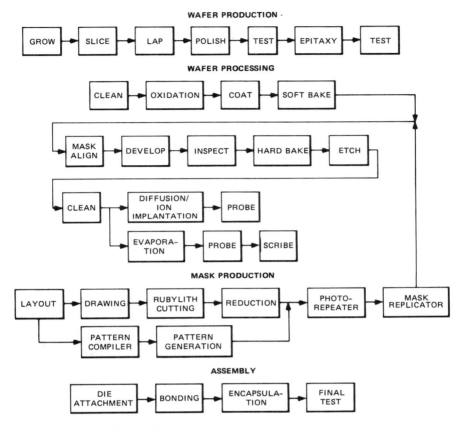

FIGURE 1.37. General semiconductor production process.

laying. Nevertheless, wafer sizes up to 6 in. in diameter are used, although at present a 4 in. diameter appears to be a standard.

In Figures 1.36 and 1.37 we show schematically the sequence of process steps used for large-scale integration (LSI) and, more generally, for semiconductor device manufacture.

1.4. Microstructures for Other Applications

The evolution of microfabrication technology has had widespread consequences in other sciences, including physics, materials science, and biology. In this section we introduce some other types of microstructures and some of the special fabrication techniques associated with them.

1.4.1. Physics

Of interest to physicists are the microstructures used in integrated optics and X-ray optics, as well as other micro- or nanodevices, such as superconducting junctions and electron and ion sources.

1.4.1.1. Integrated Optics

Integrated optics are characterized by devices that combine optical and electrical components using thin-film structures on a single substrate. Such integrated devices (waveguides, couplers, lasers, and modulators) are smaller in size, lower in cost, and more complex than the currently used bulk optical components.

The most promising material in integrated optics is the GaAs–GaAlAs heterostructure,[63] because by varying the alloy composition both the index of refraction and the band gap can be changed. This material also has high electro-optic and acousto-optic coefficients,[64] making it usable for modulation and switching. Heterostructures are also used in injection lasers, in which the stimulated emissions are produced by the recombination of carriers injected across a $p–n$ junction.

Most fabrication techniques, including photon- and electron-beam lithographies, molecular beam epitaxy, ion milling, epitaxy, and ion implantation, are the same as used for Si. This means that fabrication of the optical and electrical circuits can be performed simultaneously on the same chip.

The optical waveguard is the basic integrated optical component. The light wave is usually guided in the middle layer of a three-layer structure of three different material compositions (see Figure 1.38). The necessary condition for waveguiding in these layers is that the middle layer have a higher refractive index than those of the outer layers so that the wave is totally reflected at the boundary. These structures can be manufactured easily using molecular beam epitaxy.

In integrated optics it is necessary for light to be coupled from one waveguide to another. For this purpose, three types of couplers are utilized: prism, grating, and taper. Figure 1.39 shows the structure of an output coupler based on a bulk prism and a prism–output coupler integrated into a heterostructure laser.[65]

In the integrated prism coupler [Figure 1.39(a)], a portion of the light wave propagating in the waveguide extends into the region of the prism if the air gap is small enough (typically 1000 Å). If the refractive index of the prism n_p is higher than the equivalent refractive index of the guided mode $n_{eq} < n_2$,

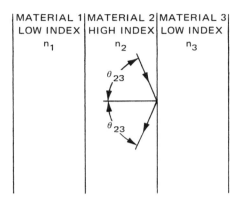

FIGURE 1.38. Heterostructure optical
waveguide.

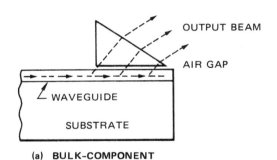

(a) BULK-COMPONENT

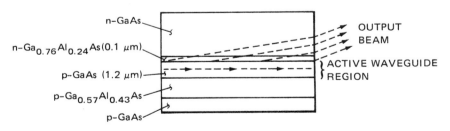

(b) PRISM COMPONENTS INTEGRATED
INTO A DOUBLE HETEROSTRUCTURE
DIODE LASER

FIGURE 1.39. Prism output couplers.

then light will "tunnel" through the air into the prism. Furthermore, if $n_p >$ n_{eq}, that wave radiates in the prism at an angle to the original propagation direction and is therefore effectively coupled out. The light may be coupled into the waveguide by launching a beam in the opposite direction into a prism.

In the prism-output coupled diode laser [Figure 1.39(b)] the thin (0.1-μm) n-Ga$_{0.76}$Al$_{0.24}$As heterostructure layer replaces the air gap. This region has a lower refractive index ($\sim$ 3.4) than that of either the n-GaAs substrate (n $\sim$ 3.6), which corresponds to the prism, or the p-GaAs active waveguide region ($n \sim$ 3.6). Thus, again because of the thin n-Ga$_{0.76}$Al$_{0.24}$As layer, the evanescent tail of the guided mode overlaps the high-index substrate region, and the light is coupled out at an angle given by

$$\sin \theta = (n_s^2 - n_{eq}^2)^{1/2}/n_s \qquad (1.19)$$

where, in terms of the wavelength in the guide, $n_{eq} = \lambda_0/\lambda_g$. The prism-coupled output beam illuminates the facet over a large area.

Schematic diagrams of a bulk grating coupler and an integrated-output grating coupler are shown in Figure 1.40. Here the light propagating in the waveguide interacts with the grating and is diffracted out. For output coupling from GaAs/GaAlAs lasers, which emit light with a free-space wavelength of $\sim$ 8500 Å and have an equivalent refractive index n_{eq} of $\sim$3.6, a grating period of $\sim$ 2400 Å is required if the light is to be diffracted out at right angles to its direction of propagation. Gratings with these short periods are fabricated by electron-beam lithography or holography.[66-68]

Taper couplers are used to couple a guided mode from one optical waveguide to another. In a taper coupler, one optical waveguide becomes progressively thinner, forcing the wave into a parallel guide. Figure 1.41 shows the structure of a taper coupler integrated into a double heterostructure laser. The output waveguide is coupled into a light modulator.

Optical modulators are electrically controlled devices that vary the amplitude, phase, or frequency of light. p–n junctions can be used as modulators because applying a reverse voltage to the junction sweeps out the free carriers and increases the refractive index, thus causing a phase shift. This phase modulation can be converted to amplitude modulation by the use of an external polarizer.

Power propagating in a planar waveguide can be diffracted by a periodic variation in the refractive index (Figure 1.42). Here the periodic index change is induced by a voltage applied to the electrodes.

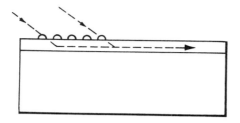

(a) **BULK COMPONENT GRATING COUPLER**

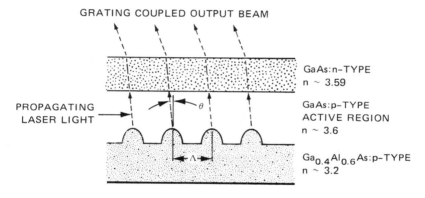

GRATING COUPLED OUTPUT BEAM

PROPAGATING LASER LIGHT

GaAs:n-TYPE
n ~ 3.59

GaAs:p-TYPE
ACTIVE REGION
n ~ 3.6

$Ga_{0.4}Al_{0.6}As$:p-TYPE
n ~ 3.2

(b) **INTEGRATED OUTPUT**

FIGURE 1.40. Grating output couplers.

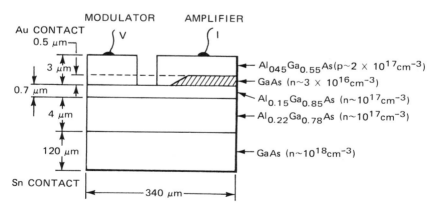

MODULATOR   AMPLIFIER

Au CONTACT
0.5 μm
3 μm
0.7 μm
4 μm
120 μm

Sn CONTACT

$Al_{045}Ga_{0.55}As$ (p~2 × $10^{17}cm^{-3}$)
GaAs (n~3 × $10^{16}cm^{-3}$)
$Al_{0.15}Ga_{0.85}As$ (n~$10^{17}cm^{-3}$)
$Al_{0.22}Ga_{0.78}As$ (n~$10^{17}cm^{-3}$)
GaAs (n~$10^{18}cm^{-3}$)

340 μm

FIGURE 1.41. Taper coupler integrated into a double heterostructure laser.

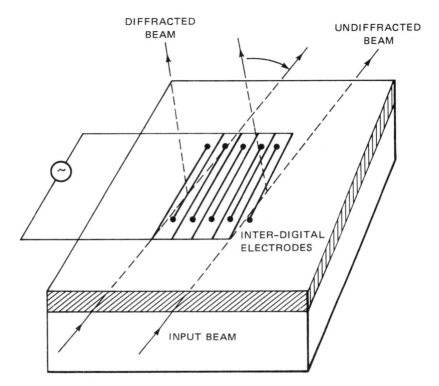

FIGURE 1.42. Diffraction-type light modulator.

1.4.1.2. X-Ray Optics

Considerable effort is currently being expended developing X-ray optical components such as waveguides, mirrors, lenses, and zone plates. These components are needed for X-ray instruments for use in microscopy, materials analysis, and microstructure fabrication. The techniques currently used for the fabrication of these X-ray optical components include electron-optical demagnification,[69] sputter deposition of multilayers,[70] and molecular beam epitaxy. Table 6 summarizes the present and future fabrication for these components.

1.4.1.3. Superconducting Junctions

A pair of superconductors separated by a thin insulating barrier forms a "Josephson junction." A Josephson junction can be viewed as a weak spot in

TABLE 6. *Present and Estimated Future Fabrication Capability for X-Ray Optical Components*

Fabrication Parameter	Present	Future
Microroughness	50 Å	10 Å
Zone-plate resolution	10^5 lines/cm	2×10^6 lines/cm
Thin-film thickness	1 Å	1 Å
Waveguide smoothness	Atomically smooth	Atomically smooth

a superconducting circuit at which a localized difference in quantum phase can be established between two superconducting regions.[71] This phase difference can be monitored by observing the "supercurrent" through the junction. This supercurrent responds to externally applied electric and magnetic fields, and it is this response that leads to a variety of interesting and useful Josephson devices.[72]

Josephson devices operate at liquid-helium temperatures (1–4°K) and are used to measure low-frequency voltages and magnetic fields. They are also used as detectors, mixers, thermometers, and computer elements. Josephson junctions can be constructed in several ways, using microfabrication techniques.[73] The most important planar structures in which microfabrication techniques can be used are shown in Figure 1.43.

Figure 1.43(a) shows the tunnel junction of the type originally proposed by Josephson. Here two superconducting films are coupled via coherent tunneling of bound electron pairs through an insulating oxide film between the layers.

In a Dayem bridge [Figure 1.43(b)], the two superconductors are coupled by a narrow constriction (weak link). The bridge must be extremely narrow ($\sim 1 \mu$m) to be coupled weakly enough to exhibit Josephson behavior. Figure 1.43(c) shows the structure of a Dayem bridge fabricated by electron-beam lithography and ion implantation.[74] The implanted nitrogen ions lower the transition temperature (T_c) from ~ 9 to 2.7–4.1°K and increase the resistivity by a factor of 3 to 26, depending on the dose.

Since Josephson junctions can be switched between different voltage states in less than 10^{-9} s and can also be packed densely, they could conceivably form the basis for logic elements for a small, fast computer. As the speed of a computer increases, its dimensions must decrease, since the transmission velocity of signals is limited by the speed of light. For example, if a computer has a

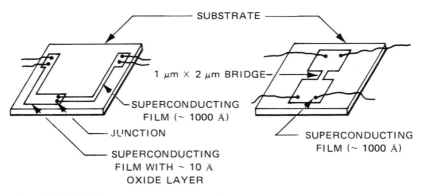

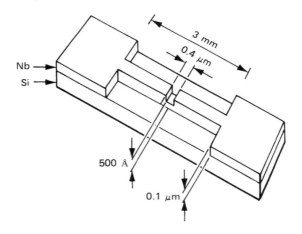

(c) **DAYEM BRIDGE FABRICATED
BY ELECTRON BEAM LITHOGRAPHY
AND ION IMPLANTATION**

FIGURE 1.43. Microfabricated Josephson junction structures.

cycle time of 1 ns and the dielectric constant and its transmission line is 4, then
it must have a linear dimension less than or equal to 5 cm. Because of this
small size, the heat dissipation must be low so that the component (active or
passive) density can be high. Josephson junctions satisfy these conditions and
are thus suitable for the purpose.

1.4.2. Electron and Ion Sources

Microstructures have also been used for making electron and ion sources based on electron emission and ionization phenomena that take place in intense electron fields.

One problem with point field-emission sources currently in use is that they need a high voltage (2–10 kV) to produce the high fields (10^7–10^8 V/cm) at the tip necessary for field emission. The high voltages give ions formed by electron impact with the residual gas sufficiently high energies so that they can sputter or erode the tip. By reducing the radius of the tip to 500 Å, one can reduce the voltage to 50–100 V, that is, as close to or below the sputtering threshold for the tip material, and thus produce long-life devices.

Figure 1.44 shows scanning electron micrograph (SEM) photographs of microfabricated field-emission cathode sources.[75,76] Figure 1.44(a) shows part of a a regular array of cathode tips arranged on 6.35-μm centers. Typical cathode arrays have over 20,000 tips in an area of about 1 mm². A magnified view of one of the field-emission tips is shown in Figure 1.44(b). The radius of the tip is about 500 Å, and the diameter of the hole in the anode film is slightly larger than 1 μm. The top anode film and cone are made of electron-beam-evaporated molybdenum, and the cone rests on a silicon substrate. Electrical insulation is provided by a thin film of thermally grown SiO_2.

The cutaway view in Figure 1.44(c) shows the cone in the silicon (dark area) and the hole in the SiO_2. In Figure 1.44(d), the entire top molybdenum film has been removed to clearly show the perfection of the underlying structure. Emitter arrays have been operated at current densities as high as 20 A/cm², averaged over the area occupied by the array, and are being life-tested at lower currents (3 A/cm²) for operating times exceeding 50,000 hr. These field-emission sources are among the smallest three-dimensional structures ever fabricated.

The sharp tips shown in Figure 1.45(a) are grown at the intersections of a 1000-mesh screen to form part of a field ionization array.[77] As can be seen in Figure 1.45(b), the tip diameters are substantially less than 5000 Å. If a second 1000-mesh screen is registered so that its openings are directly above the tips and voltage applied, this array becomes a field ionization source. Molecular gases can be passed through the bottom support screen for ionization in the vicinity of the sharp tips. Field ionization generally does not fragment the molecules and is thus valuable for mass spectroscopy. The ionization efficiency can be increased by growing needlelike whiskers on the tips of the cones to augment the ionization area.

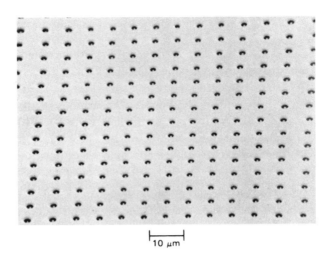

(a) **REGULAR ARRAY OF CATHODE TIPS**

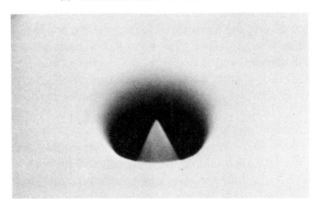

0.00004 inch
1 μm

(b) **MAGNIFIED VIEW OF FIELD EMISSION TIP**

FIGURE 1.44. Scanning electron micrographs of microfabricated field-emission cathode sources.

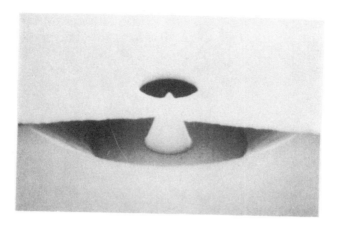

├─────┤
1 μm

(c) CUT-AWAY VIEW OF FIELD EMISSION TIP

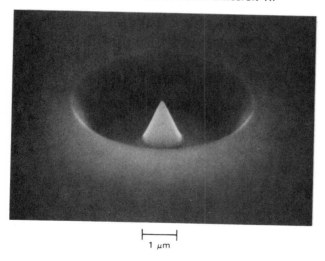

├───┤
1 μm

(d) TIP WITH MOLYBDENUM FILM REMOVED

50 µm

(a) SHARP TIPS GROWN AT INTERSECTIONS OF 1000 MESH SCREEN

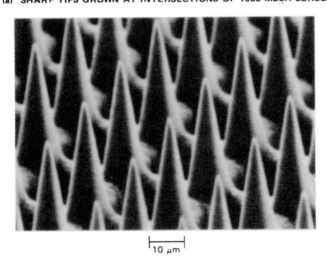

10 µm

(b) MAGNIFICATION OF ABOVE TIPS

FIGURE 1.45. Scanning electron micrographs of field ionization array.

1.4.3. Materials Science

There is a symbiotic relationship between materials research and micro-fabrication technology:

- Microfabrication techniques open the possibility of structuring new materials on a molecular scale.
- New materials may possess new and previously unavailable properties for fabricating new microdevices.

One example of the first kind is the technique of "graphoepitaxy," developed at MIT Lincoln Laboratory.[78] This technique produces a layer of crystalline silicon, about 0.5 μm thick, with a predetermined crystallographic orientation. To produce the crystalline layer, a layer of amorphous silicon is deposited onto a slab of fused silica that has a shallow grating etched into it by photolithography. The grating has a square-wave cross section with a 3.8-μm spatial period, a 1000-Å depth, and corner radii of about 50 Å (see Figure 1.46). The amorphous silicon is then fused by heating it with a laser beam. It then recrystallizes as a single crystal, with the $\langle 100 \rangle$ direction perpendicular to the plane of the substrate and the $\langle 001 \rangle$ direction along the grating.

This technique for growing an oriented semiconductor film on an amorphous substrate in turn opens new opportunity in the fabrication of microelectronic devices. One can have different crystal orientations on the same substrate or fabricate three-dimensional devices by growing a second semiconductor layer on the top of an amorphous overlayer, such as SiO_2.

Molecular beam epitaxy (MBE) is another very promising technology for producing new semiconductor compounds. The potential of MBE for future solid-state electronics is greatest for microwave and optical solid-state devices and circuits where submicrometer layer structures are essential.

Possible longer-term implications of MBE for microdevices are related to the capability of growing layer sequences with alternative composition, for example, layers of GaAs and AlAs. Such superlattice structures with periodicities of 10 to 100 Å show negative resistance characteristics attributed to resonant tunneling into the quantized energy states, which are associated with the narrow potential wells formed by the layers.[79]

Very thin films in the 10-Å region can be fabricated by the state-of-the-art deposition techniques. However, some solids crystallize in structures of stacked monolayers spontaneously. Examples of layered compounds are prevalent, such as graphite,[80] nearly all clay minerals, and many metal oxides.

An important property of layered solids is their ability to intercalate (i.e., to include within their bulk in a two-dimensional fashion) various guest species.

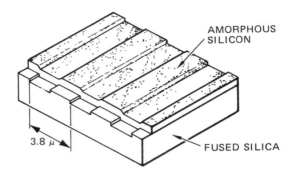

**(a) AMORPHOUS SILICON BEFORE
LASER IRRADIATION**

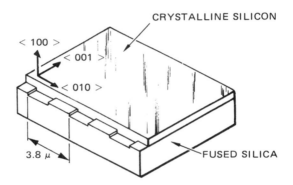

**(b) CRYSTALLINE SILICON SURFACE
AFTER LASER IRRADIATION**

FIGURE 1.46. Method for depositing an oriented silicon crystal on an amorphous silicon.
Source: ref. 78.

As a result of this intercalation process, the properties (conductivity, superconducting transition temperature, and mechanical properties) of both guest and
host substrates are changed to some degree.

Intercalation commonly proceeds via a process of staging, in which inter-
layer regions periodically fill with guest molecules (see Figure 1.47). The number of host layers between the guest molecular layers is defined as the "stage"
of the compound. As the overall concentration increases from zero to saturation, the stage number falls from infinity to unity. Such materials can be considered as molecular analogs of stacked thin films and suggest the possibility
of tailored architecture on the nanometer scale.

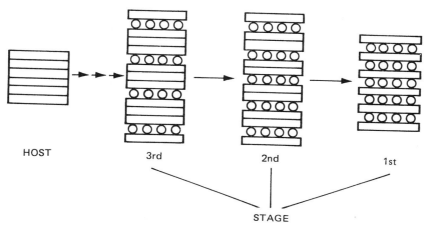

FIGURE 1.47. Intercalation of guest species.

1.4.4. Unconventional Electronic Devices

In this section we briefly discuss just a few of the novel microdevices used in electronics science. These include new concepts for particle-beam memories,[81] semiconductor memory devices, and topologically structured thin-film devices.

1.4.4.1. Particle-Beam Memories

Particle-beam memories permit very large permanent stores, ranging from 10^{12} to 10^{15} bits, with millisecond access times. With such recording schemes, bit spacings on the order of 0.1 μm become feasible at the data rate of 10,000 bit/s.

The particle-beam memory elements are written with focused ion or electron beams and read with an electron beam. The writing method is to produce small (0.1-μm) alloyed or doped regions near the surface of a semiconductor p–n junction (see Figure 1.48). These amorphous regions have a higher resistance than does the rest of the surface. The stored information can be read by focusing on an electron beam (Figure 1.49) on a storage element (pixel) and reading the current across the p–n junction.

This type of memory system with an active storage area of 10 cm^2 has a storage capacity equivalent to 400 rolls of magnetic tape. The disadvantage of this system is the complex beam system required for writing and reading.

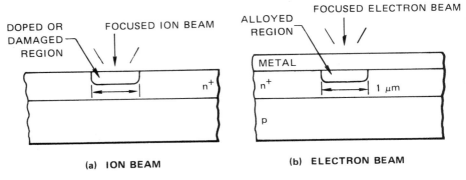

FIGURE 1.48. Writing in beam memory systems. Source: ref. 81.

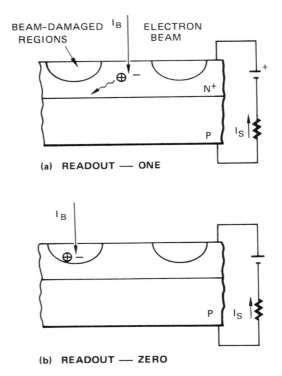

FIGURE 1.49. Reading a particle-beam memory. Source: ref. 81.

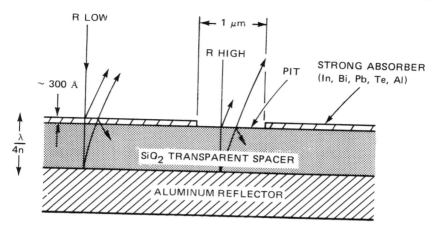

FIGURE 1.50. Laser recording structure and recording parameters.

λ(nm) RECORDING LASER	THICKNESS (nm)		ABSORPTION (%)
	SiO$_2$	Ti	
488	80	5	95
800	128	5	93

Another interesting archival digital memory system that requires laser microfabrication is the optical memory.[82] Here the recording medium (metal-coated dielectric film and multilayers of dielectric and metal films) has a reflectivity that can be changed by the absorption of laser radiation and read by optical detection techniques. The bit storage area (spot) is usually in the micron range for high packing density.

The simplest thermal recording process is the ablation of a thin film on a layered structure (see Figure 1.50) with an intensity-modulated laser. For readout, one can detect the reflected light from the spots by using a less intense laser. In the case of a recorded track on a disk, the reflected light is intensity modulated by the presence of the recorded spots.

1.4.4.2. Nonvolatile Semiconductor Memory Devices

Nonvolatile memories[83] are designed to be able to withstand power interruption without losing the stored data. Nonvolatility of semiconductor memory devices is usually achieved by storing a charge in the gate structure and thereby changing the threshold voltage for the device so that two stable states can be detected electrically.

Charge-storage devices may have the charge stored in the following:

- Traps at the interfaces of the multilayer insulator gate structure (charge-trapping devices)
- A conducting or semiconducting layer between insulators (floating-gate devices)

1.4.4.3 Charge-Trapping Devices

Among the charge-trapping devices, the metal–nitride–oxide semiconductor (MNOS) device has received the most attention. A typical MNOS memory FET (p channel) is shown in Figure 1.51(a); it is a conventional MOSFET in which the oxide is replaced by a double layer of nitride and oxide. The oxide layer is very thin, typically 50 Å thick, and the nitride layer is about 500 Å thick. For low values of negative voltage applied to the gate, the MNOS transistor behaves like a conventional p-channel MOSFET. Upon application of a sufficiently high positive-charging voltage V_c to the gate, electrons will tunnel from the silicon conduction band to reach traps in the nitride–oxide interface and nitride layer, resulting in negative charge accumulation there and a change in the threshold voltage.

To switch to the low-threshold voltage, a large negative voltage pulse is applied to the gate, and electrons are driven out of the interface traps and returned to the silicon substrate. The two different threshold voltage levels represent the two memory states "1" and "0," which may be designated arbitrarily.

The stacked-gate MNOS tetrode [Figure 1.51(b)] utilizes hot electron injection to improve the switching speed and reduce the high switching field of the conventional MNOS structure. The control gate controls the channel current. In a n-channel device, electrons are "eaten up" as they drift in the depletion region under the offset gate toward the drain. The offset gate is biased more positive than is the drain to favor electron injection into the gate structure, and the electrons are trapped in the nitride layer as stored information. To erase the stored information, a positive voltage much higher than that used

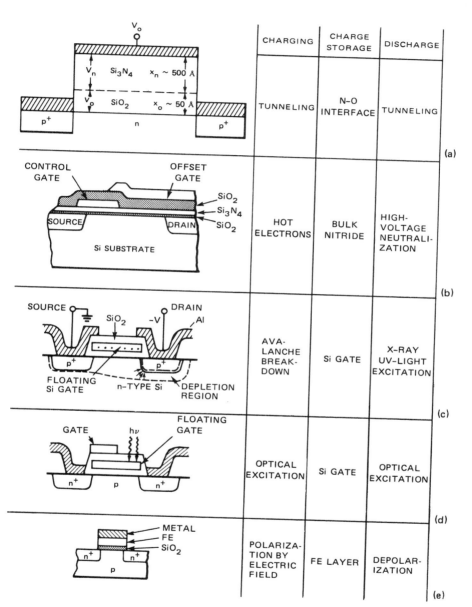

FIGURE 1.51. Comparison of charge-trapping devices.

for writing is applied to the offset gate to remove or neutralize the electrons stored in the nitride.

1.4.4.4. Floating-Gate Devices

To avoid the leakage problem existing in the thin (<50 Å) SiO_2 layer of the MNOS device, the floating-gate avalanche-injection MOS (FAMOS) was developed. This device [see Figure 1.51(c)] uses thick oxide (~ 1000 Å) to insulate a polycrystalline silicon floating gate. Since direct tunneling through 1000 Å of oxide is impossible, these devices use avalanche injection for charging. To charge the floating gate of the FAMOS device, the electrons are accelerated in the depletion region and some electrons are injected from the silicon substrate into the silicon dioxide. These electrons are attracted to the floating gate because of the field set up by the voltage divider action. The negative charge accumulated at the floating gate creates a p channel and gives rise to a low threshold voltage. Under steady-state conditions, the electrons are trapped in a deep potential well. FAMOS devices have a very long retention time. Discharge of the floating gate requires a uv light source with enough energy to excite the trapped electrons into the conduction band of the SiO_2–gate dielectric; therefore, erasing is not convenient, so the FAMOS is used solely in read-only memory systems.

A floating gate FET structure that can be optically switched is shown in Figure 1.51(d). By properly biasing the source, drain, and top gate, the floating gate can be charged and discharged by optical excitation. Figure 1.51(e) shows the structure of a ferroelectric FET (FEFET). By polarizing the ferroelectric material, the device is switched to high- and low-threshold voltages. The FEFET device is a low-voltage, fast (1 μm) memory element free from the usual fatigue effects observed in charge-storage devices.

1.4.4.5. Topologically Structured Thin Films

Topologically structured thin film devices[84,85] are thin-film (1–25 μm) membranes attached to a substrate at one end and extending over a shallow pit, as shown in Figure 1.52. This pit is realized by etching silicon out from under the deposited film with a carefully but easily controlled etching procedure. Using an electric field, it is possible to move the film down toward the substrate or up away from it. An array of these membranes can function as light modulators (Figure 1.53). The light is directed onto the thin film, and by

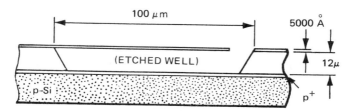

FIGURE 1.52. Metal-coated oxide membrane. Source: ref. 84.

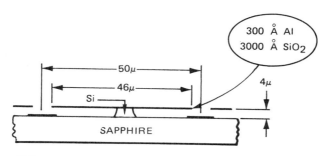

FIGURE 1.53. Light modulator (field controlled). Source: ref. 84.

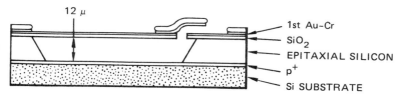

FIGURE 1.54. Micromechanical switch. Source: ref. 84.

changing its position the angle by which the light is deflected changes. Figure 1.54 shows a micromechanical switch that can be open or closed, depending on the electric field applied. Moving thin-film structures can also be used as accelerometers.

In the fabrication processes of these devices, EDP silicon[86] etchant is used because of its unique capability as an anisotropic etchant of silicon. It has a very low etch rate for SiO_2 ($\sim$ 100 Å/hr) and other insulators, compared to silicon ($\sim$ 50 μm/hr). It also does not appreciably attach Cr or Au, so metalization can be patterned prior to the Si etching. The etching rate of Si highly doped with boron is reduced by at least a factor of 50.

1.4.5. Biology

In this section we offer a few speculative examples for microstructure usage in biology and medicine.

With the decrease in the sizes of circuits and sensors, it might be possible to make instruments and transmitters so small that they could be swallowed or even inserted in particular portions of the circulatory system. These instruments might measure blood pressure at a given location, map the arteries, or even perform local analytical measurements (infrared spectroscopy, optical spectroscopy, chromatography, cell counting, blood gas analysis, radiation measurement, etc.). They might also be used as "assist devices" (e.g., pacemakers, electrical stimulators) or substitutes for natural bilogical microstructures, such as membranes, lungs, or kidneys. For example, microfabrication techniques could be used to build a microchannel membrane oxygenerator that could be used for long-term cardiopulmonary bypass or lung failure emergencies.

An erythrocyte (a blood cell that contains the hemoglobin and transports it through the bloodstream) is about 8 μm in diameter and 2 μm thick. Thus, a membrane with 3-μm-diameter holes should easily enable the corpuscles to contact the oxygenating gas without escaping through the holes.[87] Using electron-beam lithography and thin-film technology, it is practical to produce an array of holes covering an area of 10×10 cm and have hole sizes 1 μm in diameter.

When we also consider that the volume of a normal human nerve cell is many cubic micrometers, and that an individual memory cell can be packed on a scale that is small compared to the nerve cells, we begin to feel that microstructures can perhaps improve nerve behavior by use of amplifiers, "memory cells," and switches, or even replace defective cells.

These are speculations, but they do not seem improbable considering the ever-expanding capabilities of microfabrication technologies.

Particle Beams: Sources, Optics, and Interactions

2.1. Introduction

A key element in our ability to view, fabricate, and in some cases operate microdevices has been the availability of tightly focused particle beams, particularly of photons, electrons, and ions. Consideration of diffraction effects leads to the general rule that if one wishes to focus a beam of particles into a spot of a given size, the wavelength associated with the particles must be less than the required spot diameter. In Table 1 are listed the wavelengths (in μm) of three particles (photons, electrons, and protons) at various energies.

From Table 1 we see that photons in the visible range (1.6–3.5 eV) are useful for resolutions down to perhaps a micron and for feature sizes of a few microns. This is improved by the use of photons in the uv to soft X-ray range (5–1000 eV). Photon energies greater than 1000 eV rapidly lose their usefulness as a result of the increasing range of scattered photons. We are obviously unconcerned about wavelength limitation to electron beams over the energy range in which they are normally used (10^2–10^5 eV), even if we are seeking to resolve atoms that are a few angstroms (10^{-10} m) in diameter. Again, the limitation lies in the range of the scattered electrons. With ions, even very-low-energy beams are not wavelength limited: Because they are of a size comparable to the atomic arrays on which they are impinging, their range is always quite small.

In this chapter we treat electron, ion, and photon beams from the viewpoints of particle sources, particle-beam formation, optics, focusing, deflection,

TABLE 1. Particle wavelengths (μm)

Particle	Particle Energy E_0 (eV)						
	1	10	10^2	10^3	10^4	10^5	10^6
Photons $\lambda = \dfrac{1.2399}{E_0}\ \mu m$	1.24	1.24×10^{-1}	1.24×10^{-2}	1.24×10^{-3}	1.24×10^{-4}	1.24×10^{-5}	1.24×10^{-6}
Electrons $\lambda_e = \dfrac{1226}{E_0^{1/2}}\ 10^{-6}\ \mu m$	1.23×10^{-3}	3.88×10^{-3}	1.23×10^{-4}	3.88×10^{-5}	1.23×10^{-5}	3.88×10^{-6}	1.23×10^{-6}
Protons $\lambda_p = \dfrac{28 \times 10^{-6}}{E_0^{1/2}}\ \mu m$	2.87×10^{-5}	9.07×10^{-6}	2.87×10^{-6}	9.07×10^{-7}	2.87×10^{-7}	9.07×10^{-8}	2.87×10^{-8}

and energy analysis. The interactions of particle beams with surfaces are discussed, since these form the basis of the microfabrication functions desired. Electrons are treated in greater detail because of their versatility and emerging importance in submicron lithography, surface analysis, structural analysis, and microscopy. The concepts of electron optics were derived by analogy with classical photon optics and carry over directly to ion optics simply by interchanging the mass and charge of the ionic particle for that of the electron in the relevant equations.

This chapter is fairly comprehensive, since the manipulation of particle beams has formed the basis of much of the microdevice technologies used today. Descriptions in the subsequent chapters rely heavily on the specifics discussed here.

2.2. Electron Sources

2.2.1. Elementary Theory of Thermionic and Field Emission

The conductivity of a metal is due to electrons that move freely within the crystal lattice. At the surface of the crystal there are intense electric fields which, because of the asymmetric spatial distribution of the surface ions, form a potential barrier that prevents the conduction electrons from escaping. The energy distribution of the conduction electrons is given by Fermi–Dirac statistics. The energy required to remove an electron with the Fermi energy inside the conductor to a field-free region outside the conductor is termed the work function ϕ and is usually measured in electron volts. If an electric field E is applied at the surface in a direction so as to help an electron escape, the work function is lowered by an amount

$$\left(\frac{eE}{4\pi\epsilon_0} \right)^{1/2} \text{eV}$$

where e is the electronic charge and ϵ_0 is the permittivity of free space. This is due to the reduction in the amount of work required to overcome the image force between the electron and the conductor surface and is termed the Schottky effect. The value of the work function is in part determined by the lattice structure and, as such, also changes with temperature. It is generally assumed, with good experimental evidence, that the work function is linearly dependent on the temperature with coefficient α; so in general we may write

$$\phi = \phi_0 + \alpha T - \left(\frac{eE}{4\pi\epsilon_0} \right)^{1/2} \qquad (2.1)$$

where ϕ_0 is termed the zero-field work function.

The work function may also be substantially modified by a single atomic layer of foreign atoms adsorbed on the metal surface. Chemisorbed atoms tend to give up or accept an electron from the substrate lattice to form a strong electrostatic bond and a dipole layer at the surface. Electropositive atoms such as thorium and barium are adsorbed as positive ions and make it easier for electrons to escape from the crystal; that is, they lower the work function. Electronegative atoms such as oxygen and fluorine are adsorbed as negative ions and make it more difficult for electrons to escape; that is, they increase the work function.

The number of conduction electrons packed per unit volume within a metal is extremely large, being on the order of $10^{22}/cm^3$, making it an attractive possible source of electrons. To extract electrons from the potential well of the crystal lattice in which they are trapped, one must seek means to give sufficient energy to the conduction electrons so that they may overcome the potential barrier. Alternatively, by applying an intense external electric field to the surface, one can provide conditions for quantum mechanical tunneling of electrons through the potential barrier.

In the first category we may simply heat the metal, since the number of conduction electrons with energies sufficient to overcome the work-function barrier increases exponentially with temperature according to Fermi–Dirac statistics. This is the thermionic emission on which most practical vacuum electronic devices depend as their source of electrons. Bombardment of the metal with energetic particles (photons, electrons, and ions) gives rise to an energy exchange with both conduction and core electrons in the lattice, whereby some may acquire sufficient energy to escape from the surface. Such effects are now rarely used as sources for electron guns, since the energy distribution of the emitted electrons is too broad for most practical applications. The second category, termed field emission, is very attractive in theory because very large current densities can be obtained with a small energy spread. However, it has proved to be most difficult to implement such sources because of practical difficulties that will be discussed in Section 2.2.3.2.

By heating a cathode to near thermionic temperatures, a substantial number of energy levels above the Fermi level are occupied by the conduction electrons. The tunneling probability of these electrons under the action of a strong surface electric field is substantially greater than that of electrons at the Fermi level or below, and the cathode is said to be operating in the thermal-field or TF mode.

The classical theory of thermionic emission has been well summarized by Nottingham,[1] and that of field-emission by Good and Müller.[2] A brief outline is given below.

If the energy associated with the velocity component of an electron moving in a direction perpendicular to a surface is designated ϵ, then the current escaping from a unit area of the surface is given in general by

$$j = e \int_{-\infty}^{+\infty} N(T, \epsilon) D(E, \epsilon, \varphi) \, d\epsilon \qquad (2.2)$$

where $N(T, \epsilon) \, d\epsilon$ is the number of electrons incident upon the surface per unit area per second with velocity component perpendicular to the surface in the associated energy range ϵ to $\epsilon + d\epsilon$, when the crystal is at temperature T. $N(T, \epsilon)$ is called the supply function. $D(E, \epsilon, \varphi)$ is the probability of an electron with energy ϵ penetrating a potential barrier of work function ϕ with external field E applied; $D(E, \epsilon, \varphi)$ is called the transmission function.

The supply function for conduction electrons is calculated from Fermi–Dirac statistics and is given by

$$N(T, \epsilon) = \frac{4\pi m k T}{h^3} \ln \left[1 + \exp \left(\frac{-\epsilon}{kT} \right) \right] \qquad (2.3)$$

where m is the mass of an electron, h is Planck's constant, and k is Boltzmann's constant. In the case of thermionic emission with small external electric fields, the transmission function $D = 0$ for

$$\epsilon \leq \phi - \left(\frac{eE}{4\pi\epsilon_0} \right)^{1/2}$$

and $D = 1 - r(\epsilon)$ for

$$\epsilon > \phi - \left(\frac{eE}{4\pi\epsilon_0} \right)^{1/2}$$

where $r(\epsilon)$ is the reflection coefficient for electrons of energy ϵ arriving at the surface. Since $r(\epsilon)$ is essentially unknown, it is usually assumed to be zero, or at least an average value $\bar{r}$ is assumed so that the integration may be carried out. This results in Richardson's equation for thermionic emission:

$$
\begin{aligned}
j &= \left(\frac{4\pi m k^2 e}{h^3} \right) T^2 (1 - \bar{r}) \exp \left[-\left(\frac{\phi_0 + \alpha T - |eE/4\pi\epsilon_0|^{1/2}}{kT} \right) \right] \\
&= \left[120 \cdot 4(1 - \bar{r}) \exp \left(-\frac{\alpha}{k} \right) \right] T^2 \\
&\quad \times \exp \left[-\left(\frac{\phi_0 - |eE/4\pi\epsilon_0|^{1/2}}{kT} \right) \right] A/cm^2
\end{aligned}
\tag{2.4}
$$

Figure 2.1 shows plots of j as a function of T for various ϕ, assuming that r, α, and $E = 0$.

For field emission at $T = 0$, it has been shown[2] that

$$
D \approx \exp \left(\frac{4(2m)^{1/2} |\epsilon_0^3|}{3\hbar eE} v(y) \right)
\tag{2.5}
$$

where $y = (eE/4\pi\epsilon_0)^{1/2}/\phi_0$ and $\hbar = h/2\pi$. Putting this value of D in Eq. (2.2) and integrating leads to the Fowler–Nordheim equation for field emission:

$$
j = \frac{e^3}{8\pi h} \frac{E^2}{\phi t^2(y)} \exp \left[-\left(\frac{4(2m)^{1/2}}{3\hbar e} \frac{\phi^{3/2}}{E} v(y) \right) \right]
\tag{2.6}
$$

where $v(y)$ and $t(y)$ are the Nordheim elliptic functions, for which computed values are available. To a good approximation we may write $t^2(y) = 1.1$, and $v(y) = 0.95 - y^2$, leading to

$$
\begin{aligned}
j = 1.5 \times 10^{-6} \frac{E^2}{\phi} \exp \left(\frac{10.4}{\phi^{1/2}} \right) \\
\times \exp |-6.44 \times 10^{-7} \phi^{3/2}/E| \ A/cm^2
\end{aligned}
\tag{2.7}
$$

Figure 2.2 shows plots of j as a function of E for various ϕ.

For field emission at moderate temperatures, Müller[2] has shown that

$$
j(T) = j(0) \frac{\pi kT/d}{\sin (\pi kT/d)}
\tag{2.8}
$$

where $j(0)$ is given by Eq. (2.6), and

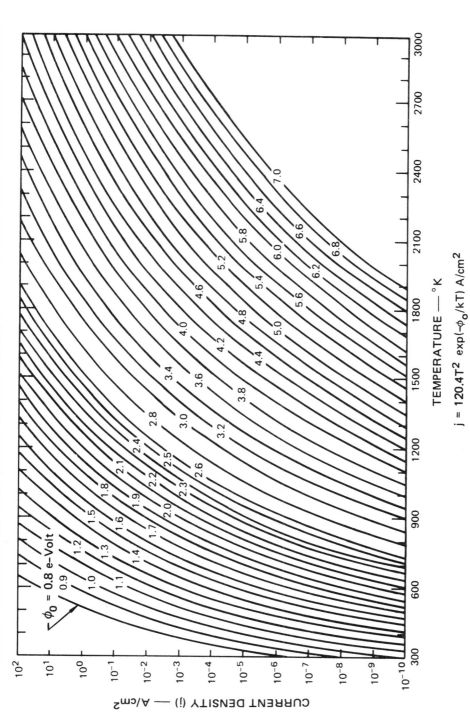

FIGURE 2.1. Plots of Richardson's equation for current density (j) as a function of temperature (T) with ϕ_0 as a parameter.

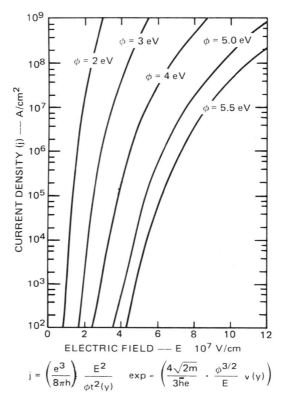

$$j = \left(\frac{e^3}{8\pi h}\right) \frac{E^2}{\phi t^2(y)} \quad \exp - \left(\frac{4\sqrt{2m}}{3\hbar e} \cdot \frac{\phi^{3/2}}{E} v(y)\right)$$

FIGURE 2.2. Plots of the Fowler–Nordheim equation for current density (j) as a function of energy (E), with ϕ as a parameter.

$$d = \frac{\hbar e E}{2(2m\phi)^{1/2} t(y)} \tag{2.9}$$

For field emission at higher temperatures (TF emission), it has been shown[3] that

$$j = \frac{E}{2\pi}\left(\frac{kT t(y)}{2\pi}\right)^{1/2} \exp\left(-\frac{\phi}{kT} + \frac{E\theta(y)}{24(kT)^3}\right) \tag{2.10}$$

where

$$\theta(y) = \left(\frac{3}{t^2(y)} - \frac{2v(y)}{t^3(y)}\right) \simeq 1.1 + y^2 \tag{2.11}$$

Equation (2.10) gives thermionic and field emissions as limiting cases.

2.2.2. Cathode Requirements

Electron sources for devices requiring a beam of electrons for their operation are called cathodes and are subject to a number of requirements depending on the device specifications. These relate to the following:

- The uniformity of emission over the cathode surface
- The current density required from the cathode
- The energy distribution of the emitted electrons
- The current fluctuations as a function of time
- The evaporation of material from the cathode surface
- The ability of the cathode to operate in a given environment
- The lifetime of the cathode under given operating conditions

For applications in microscience, we usually need to focus a beam of electrons of a given energy into the smallest possible spot and with the highest possible current. The Langmuir limit to the focused spot size in electron optics (see Section 2.3.3) requires that the current density drawn from the cathode be as large as possible and that the energy spread of these electrons be as small as possible.

Thermionic cathodes are usually operated space-charge limited for the following reasons:

- For practical polycrystalline cathodes, the work function over the emitting surface varies from point to point on a macroscopic scale, depending on the crystal face presented to the surface. The presence of space charge creates a potential minimum of value V_m a distance d_m in front of the cathode surface. If d_m is greater than the dimensions of the polycrystals forming the surface, the effects of different current densities emerging from the crystallites forming the surface and the small constant potential differences between them are blurred out. The emitting surface, as far as electron optics are concerned, appears as a uniformly emitting virtual cathode a distance d_m from the cathode surface and at a potential $-V_m$ below that applied to the cathode. The value of V_m is given by the increase in the average work function of the cathode that would be necessary to decrease the zero-field emission j_0 from the cathode to that required for space-charge-limited flow j_{SCL}; that is,

$$e^{V_m/kT} = j_0 - j_{SCL} \qquad (2.12)$$

The value of d_m can then be calculated from the full theory of space-charge-limited flow.[1] For a planar diode of interelectrode spacing d with applied voltage V, we may use the approximation

$$j_{\text{SCL}} = \tfrac{4}{9}\epsilon_0 \left(\frac{2e}{m} \right)^{1/2} \frac{(V - V_m)^{3/2}}{(d - d_m)^2} \tag{2.13}$$

By measuring j_0, j_{SCL}, V, T, and d, an estimate of d_m can be made using Eqs. (2.12) and (2.13).

- The space-charge-limited current density is only slightly dependent on the cathode temperature, whereas the saturated thermionic emission is exponentially dependent on temperature. The temperature stability requirements for constant emission are usually too stringent for the cathode to be used in the saturated mode. This effect is further aggravated by the fact that the work-function distribution of the emitting surface may change with time and with any changes in the vacuum environment in which the cathode is operating.
- The space charge serves to reduce—by several orders of magnitude—current fluctuations due to both the discrete nature of the electron (shot noise) and random changes in the emission capability of different points on the cathode surface (flicker noise).[4]

The energy distribution of thermionically emitted electrons is given by Maxwell–Boltzmann statistics. This gives rise to tails of electrons with energies both above and below the energy given to them by the electric fields through which they are accelerated. The mean spread, however, is very narrow, being of the order of kT. In field emission the maximum of the energy distribution is at the Fermi level, but at normal temperatures there are very few electrons in states above the Fermi level. Thus, essentially a single tail exists, containing only electrons below the energy to which they are accelerated. By arranging to cut off the low-energy tail, a very narrow energy spread may be obtained. The sharp front of the low-temperature field-emission distribution is especially useful for measuring effects whose appearance occurs at sharply defined electron energies.

Flicker noise is prominent in cathodes in which the work function is lowered by an adsorbed monolayer of electropositive atoms. Such atoms move over the surface as a two-dimensional gas until they are lost by evaporation. Lost atoms are replenished from the body of the cathode material by a chemical dispensing mechanism. An atomic site at the surface of the cathode spends some portions of its time in an emitting state when an electropositive atom is adsorbed on it, and the remaining time in a nonemitting state when there is no electropositive atom adsorbed on it. The constant on-and-off twinkling of the atomic sites gives rise to the flicker noise. The flicker noise spectrum depends on the mean times during which the sites are emitting and nonemitting.

If a high current density is required from a given thermionic cathode, its temperature of operation has to be increased. Natural limits occur when the cathode material melts or the vapor pressure of the evaporation products exceeds some tolerable limit for the device in which it is operated. Because of such limits, very few practical thermionic cathodes can be operated continuously at current densities exceeding 10 A/cm^2.

Another limit lies in the cooling of the cathode surface when high current densities are drawn from it. As the electrons leave, each carries away an average energy $\phi e + 2kT$. Thus, when drawing j A/cm^2, the number of watts per square centimeter that must flow to the surface to maintain its temperature of operation, assuming no conduction losses, is given by

$$w = j(\phi + 1.725 \times 10^{-5} T) + \xi \sigma T^4 \text{ W/cm}^2 \qquad (2.14)$$

where ξ is the emissivity and σ is Stefan's constant. When the work-function cooling loss becomes substantially greater than the radiation loss ($\xi \sigma T^4$), it becomes increasingly difficult to maintain the surface temperature without melting or obtaining substantial evaporation from the cathode support structure. While continuous current densities from cathodes up to 20 A/cm^2 have been demonstrated, it is likely that an absolute limit to continuously operated thermionic cathodes lies between 20 and 100 A/cm^2. The problem can be somewhat alleviated by drawing current for very short times (ns to μs) and allowing relatively long times for the cathode to thermally recover, but this restricts such usage to a narrow range of applications.

Cathodes that depend on the dispensing of an active material to their surface for their operation can be rendered inactive (poisoned) by minute amounts of substances that change the chemical nature of the cathode surface or inhibit the flow of dispensed materials.

In field emission, it has proved possible to obtain current densities exceeding 10^8 A/cm^2. However, such current densities can only be obtained over very small areas (less than a few μm^2). It has been found that with macroscopic surfaces, even those of a single crystal face, cleaned and prepared by the best methods known, electrons are emitted from them when macroscopic fields above 10^5 V/cm are applied rather than the anticipated 10^7–10^8 V/cm required for field emission. These electrons have been shown to originate at relatively few tiny whiskers, or projections, where field-enhancement factors of over 100 are easily obtained. These whiskers range in height from 10 to 1000 atoms tall, with 1 to 100 atoms in their tips.[5]

Wires etched to extremely sharp points are erratic in their emission characteristics unless, as was first demonstrated by Müller in 1936, they are heat-treated to form a single crystal at their tips. These crystals are usually 0.2–1 μm in radius and exhibit various crystal faces. The lowest-work-function faces emit more profusely, and in using such cathodes for an electron gun it is necessary to find means for selecting only electrons from a particular face. The currents are unstable and cathode lifetimes are short, mainly owing to sputtering of the surface by ions formed in the residual gas close to the cathode surface. It has been shown[6] that a residual pressure of 10^{-14} Torr would be necessary to avoid sputtering damage for long, although useful lives of a few hours may be obtained at 10^{-10} Torr to perform short-term experiments. For these reasons, some workers have turned to using TF emission, since heating to these temperatures has the effect of allowing the tip to repair itself after damage by the sputtering action. The increased stability, however, is gained at the price of increasing the energy spread of the emitted electrons.

Field-emission electrons tunneling from below the Fermi level give up energy to the lattice, thereby heating the surface (Nottingham effect). Those that tunnel from above the Fermi level cool it down. Hence, when field-emission current is drawn, the cathode surface will move toward the temperature at which the energy carried away by electrons from above the Fermi level is exactly compensated by the energy given up by electrons from below the Fermi level. This is known as the transition temperature and is given approximately[7] by

$$T_c = 5.32 \times 10^{-5} E \frac{\phi^{1/2}}{t(y)} \tag{2.15}$$

The actual surface temperature T_s will be given by a complex balance between the various forms of heat loss and generation,[8] but if T_s exceeds the melting point, a sudden disruption of the emitting tips may be expected.

2.2.3. Practical Cathodes

2.2.3.1. Thermionic Cathodes

Although there have been attempts to utilize an extraordinary variety of cathodes and cathode structures to overcome the difficulties associated with particular devices, relatively few cathodes have met with general acceptance.[9] Of the pure metals, tungsten is the favorite. Tantalum is less favored because although it operates at a slightly lower temperature than tungsten for a given emission, it also melts at a lower temperature (2996°C compared with 3410°C for tungsten).

Thoriated tungsten (containing about 2% by weight thorium oxide) may be considered the first dispenser-type cathode, since the work function of the tungsten is lowered by an adsorbed monolayer of thorium.[9] In order to dispense thorium to the surface at a rate sufficient to maintain the low work function at the operating temperature in a normal vacuum environment, the cathode is carburized; that is, a layer of tungsten carbide is formed at the surface by heating it to a high temperature in a hydrocarbon atmosphere or in contact with powdered carbon. In operation, the tungsten carbide reacts with the thorium oxide, forming free thorium and carbon monoxide. The presence of carbon also enables the cathode to operate at relatively high partial pressures of oxygen (10^{-5} Torr) without poisoning, since the oxides of carbon do not remain on the emitting surface. These cathodes can give up to 3 A/cm^2 when operated at 1720°C.[10] The main disadvantage of thoriated tungsten emitters lies in the fact that they constantly evolve carbon monoxide gas. This can cause problems in UHV devices and devices sensitive to the presence of ions. However, in cases in which a rugged cathode is necessary, such as a demountable system or electron-beam lithography systems used for mask or wafer mass production, this cathode performs extremely well.

Lanthanum hexaboride also relies on a dispensing mechanism for its operation.[11] The crystal structure is unusual in that the boron atoms are linked by strong valence bonds and form a three-dimensional cagelike structure around the lanthanum atom. The lanthanum, however, is not bonded to the boron. The strong boron bonds make the crystal refractory and the lanthanum valence electron makes it conductive. When heated, lanthanum diffuses to the surface, forming a low-work-function monolayer. Lanthanum lost by evaporation from the surface is replenished from the interior. The material, however, is difficult to support, since it reacts or alloys with most refractory metallic substrates at operating temperatures. Cathodes were originally made with sintered powders, but single crystals are currently favored. The crystal tips, usually in the form of a pointed rod, can be heated by electron bombardment while the support structure and electrical contact are attached at the cooler base.[11] Such cathodes appear to be used in the saturated (Schottky) mode to obtain the highest brightness, since the emission is obtained in "lobes" that must be steered into the electron-optical path.

The oxide-coated cathode is attractive because of its low temperature of operation.[12] It consists of loosely packed crystals of the oxides of barium, strontium, and calcium on a nickel substrate. The oxide crystals, which are hygroscopic in air, have to be formed in the vacuum by heating the carbonates to drive off the excess carbon dioxide. Care must be taken so that the mixture of carbonates and oxides does not melt during this process, thereby losing cathode porosity. The nickel substrate contains an impurity such as aluminum or zirconium, which at the operating temperature (750°C) reacts with the barium oxide, forming free barium. The free barium contributes both to increasing the conductivity of the crystals and in forming a low-work-function monolayer on the surfaces. In addition to electronic conduction through the crystals, charge is transported from the base nickel to the emitting surface by space-charge-limited flow in the interstices and electrolysis of the crystals. Continuous currents up to about 0.5 A/cm² can be obtained with long life, and pulsed currents over 100 A/cm² have been utilized.

Unfortunately, oxide cathodes are highly sensitive to their vacuum environment and their emission properties can easily be permanently destroyed (poisoned). Thus they are unsuitable for the demountable, high-voltage systems usually associated with electron microscopy and lithography, although they are the first choice for cathode-ray display tubes and camera tubes. To overcome these effects, attempts have been made to enclose the cathodes in some form of nickel matrix. While such cathodes are more rugged, they operate at higher temperatures, and their performance is limited by the temperature at which the nickel evaporation rate is too high to tolerate (1000°C).

Tungsten-based barium-dispenser cathodes have been highly successful for high-current-density applications. They consist of a tungsten sponge with interconnected pores made by powder metallurgy. Barium is dispensed through the pores to the emitting surface, where it forms a low-work-function monolayer. In the preferred configuration the pores are filled with barium/calcium aluminate compounds by impregnation in the molten phase.[13] In operating the aluminates react with the tungsten, forming the free barium, which diffuses easily through the matrix. Current densities of 5 A/cm², with the cathode operating at 1100°C, are routinely utilized in sealed-off power tubes. If the emitting surface is precoated with osmium[14] (M cathode) or iridium,[15] the work function is substantially lowered and the cathodes can deliver up to 20 A/cm² before the barium evaporation rate becomes intolerable. While barium-dispenser cathodes are sensitive to their environment, they are not usually permanently poisoned by high pressures of ambient gases.

2.2.3.2. Field-Emission Cathodes

For the purposes of obtaining a small focused spot as is required in electron microscopy and lithography, the necessarily small emission area of a field-emission cathode tip (radius in the range of 0.2–1 μm) is attractive, since the electron optics are required to produce an only slightly demagnified image of the tip. A crossover where the electron density is large enough to cause energy exchange between electrons that increases their effective temperature (see Section 2.3.4) is necessary for thermionic electron optics in which large demagnifications are required, but can be eliminated in field-emission electron optics, which utilizes a virtual crossover.

The major disadvantages of field-emission are the following:

- The emission from the tip is emitted in "lobes" because of the various crystal faces present in the surface
- The instability of the emission

The crystal faces in the tip are usually symmetrically arranged around a central face, which usually has a high work function, that is, is nonemitting, unless special steps are taken to ensure that this is not so. One approach is to use tungsten wire made from a single crystal oriented so that the low-work-function $\langle 110 \rangle$ face is axial. Another utilizes adsorbed material to lower the work function of the central face. However, such cathodes are only useful in devices for which the pressure of the vacuum environment can be maintained sufficiently low ($<10^{-10}$ Torr) to prevent rapid erosion of the tip from sputtering of the surface by ions formed by interaction of the electron beam with the residual gas. One approach to overcome this problem has been to operate the field-emission tips at a temperature sufficiently high to enable it to continuously repair itself by recrystallization as it is damaged. This brings the cathode in the TF mode of operation. This heating usually causes the crystal to reorient itself continuously, presenting different faces to the electron-optical axis. Despite these difficulties, certain workers[16] have found that zirconium adsorbed on $\langle 100 \rangle$-oriented tungsten operated at 1350 $\pm$ 50°K can give 100-μA emission with 2500–3000 V applied for 1000 hr, provided that the pressure is less than 2×10^{-8} Torr.

A radically different approach[17] has been to bring the anode so close to the tip that the applied voltage required to produce a field of 5×10^7 V/cm at the cathode surface for field emission is sufficiently low that ions formed in the interelectrode region cannot gather sufficient energy to damage the tip. Using microfabrication technology, molybdenum tips of 50-nm radius set in the center of an anode hole of 0.6-μm radius have been fabricated. Currents in the range of 1–150 μA can be drawn with voltages in the range of 25–250 V, and cathodes can be operated in pressures of up to 10^{-5} Torr. Since the tips are formed by vacuum evaporation, they are not single crystals and their surfaces are not atomically clean. Emission appears to proceed from a few atomic sites on the surface that are constantly changing, giving rise to lobe and current fluctuations. Operating the cathodes at about 400°C substantially reduces these fluctuations.

Under the action of an intense electric field, liquid metals can be extracted from tiny nozzles. The tips of the liquid metal cones can be of extremely small radii, and when the field is in the correct direction electrons are field-emitted from the tips. It has been demonstrated[18] that with an indium–gallium alloy pulsed currents as high as 250 A can be obtained (see Section 2.4.5.).

It is clear that although substantial progress has been made in operating field-emission cathodes for particular devices, they cannot at the time of this writing be considered a fully satisfactory emission source for microdevice fabrication or usage.

Table 2 summarizes the properties of the more practical electron sources.

2.3. Characteristics and Design of Electron Guns

2.3.1. Gun Requirements and Limitations

An electron gun is a device that generates, accelerates, and focuses a beam of electrons into a small spot. In this section we discuss the requirements and limitations for the type of electron gun structure used in microfabrication equipment (electron-beam lithography) and instruments (electron microscopes).

In most electron-optical instruments the electron gun is required to give as high as possible a current density over a prescribed area. The angular divergence of electrons from the focal spot has to be small to minimize lens aberrations or to obtain electron-optical coherence. Thus, the electrons should travel through the focal spot with the highest possible current density per unit solid angle. We regard a gun as "good" if there is a high current density in the focal area, if the focal area is small, and if the aberrations from its optical elements are small.

The "brightness" of a gun is defined as the current density (j) per unit solid angle (the solid angle of a cone of semiangle θ) is $2\pi(1 - \cos\theta) \approx \pi\theta^2$ for small θ, and is denoted by β or R (from the German term *Richtstrahlwert*):

$$\text{i.e., } \beta = \frac{j}{\pi\theta^2} \tag{2.16}$$

As shown in Section 2.3.3, β has an upper limit (Langmuir limit) given by

$$\beta_{\max} = \frac{jeV}{\pi kT} \tag{2.17}$$

TABLE 2. Cathodes and Their Properties

Type of Emission	Type of Cathode	Emission (A/cm²)	Operating Temperature (T_c) (°K)	Upper Pressure Limit (Torr)	Brightness (β) (A/cm²·sr⁻¹ at 20 kV)
Thermionic	Tungsten	0.6	2470	10^{-4}	1.8×10^4
Thermionic	Tantalum	7.3	2700		1.9×10^5
Thermionic	Thoriated tungsten	0.5	2300	10^{-5}	1.6×10^4
Thermionic	Oxide coated	1–3	2000	5×10^{-6}	$3.75 \times 10^4 \rightarrow 1.1 \times 10^5$
Thermionic	Barium dispenser	0.5	1050	10^{-6}	3.4×10^4
Thermionic	Lanthanum hexaboride	0.5–6	1150–1400	5×10^{-6}	$3.3 \times 10^4 \rightarrow 3.2 \times 10^5$
Thermionic		20.4	2100	10^{-6}	9.5×10^5
Field	Single-crystal tungsten	Up to 10^6	Room	10^{-10}	10^8
Temperature field	Zirconated tungsten		1400–1800 (1.5 eV)	10^{-9}	10^{10}
Photo	Pd	2×10^{-5}	Room (2 eV)	10^{-7}	2×10^{-1}
Photo	CsI	5×10^{-6}	Room (0.5 eV)	10^{-4}	2×10^{-1}

where j is the current density at the cathode, T is the cathode temperature, $-e$ is the electron charge, and k is Boltzmann's constant.

This theoretical limitation derives from the fact that electrons emitted from the cathode have a Maxwellian energy distribution, with a most probable energy of kT and an average energy of $2kT$. Electrons emitted into a free-field space follow Lambert's law:

$$j(\theta) = \frac{j}{\pi} \cos \theta \qquad (2.18)$$

where $j(\theta)$ is the emission current density in any direction making an angle θ with the normal and j is the total current density into the hemisphere (see Figure 2.3). If the electrons are accelerated by a uniform electrostatic field parallel to the optical axis, their velocity component (v_z) that is parallel to the field increases, but the perpendicular component (v_r) remains unchanged. The semiangle of divergence is thus reduced from π to the ratio of perpendicular and parallel velocities θ ($\cong v_r/v_z$). If the electrons have an initial energy $eV_0 = kT$ and they are accelerated to an energy eV, then the new semiangle θ' (Figure 2.3) can be written in terms of energies; that is,

$$\theta' = \left(\frac{kT}{eV} \right)^{1/2}$$

and the brightness, using Eq. (2.16), is

$$\beta = \frac{j}{\pi \theta'^2} = \frac{jeV}{\pi kT}$$

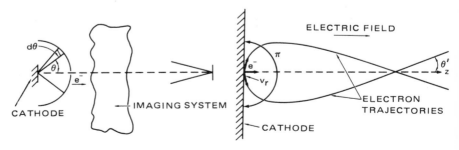

FIGURE 2.3. Electron trajectories from a flat-surface emitter.

This expression can be applied irrespective of any focusing system that may follow the gun and, in fact, no focusing system can increase the brightness beyond that of the electron-emitting surface.

2.3.2. Image Formation

Fermat's law of geometrical optics states that among a number of possible paths between two points A and B, a light ray will take that path that has the shortest transit time. This statement can be expressed mathematically by requiring the variation of the transit time to be zero; that is,

$$\delta \int_A^B \frac{ds}{v} = \delta \int_A^B \frac{n}{c} \, ds = 0$$

where v is the velocity of propagation of the light wave, $v = c/n$ (where n is the index of refraction of the medium through which the light passes), and ds is an element of that path.

A closely corresponding principle governs the motion of material particles in conservative force fields. This is the principle of least action or the principle of Maupertuis:

$$\delta \int_A^B mv \, ds = 0 \qquad (2.19)$$

Since $mv = (2meV)^{1/2}$, from $\frac{1}{2}mv^2 = eV$, this equation is identical with Eq. (2.18) if the refractive index is taken as

$$n = \sqrt{V} \qquad (2.20)$$

assuming that V is zero at a point of zero electron velocity.

The use of Eq. (2.20) may be illustrated by the example of the refraction of an electron beam (see Figure 2.4). Suppose that an electron traveling with uniform velocity v through a space of constant potential V passes a potential step into a space with another homogeneous potential V', so that the path of the electron suddenly changes its direction. Assuming that the potential $V' > V$, the normal velocity component v_y of the electron is increased, but the tangential component and v_x remains unchanged ($v_x = v'_x/x$). Moreover,

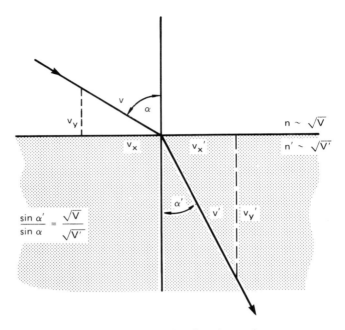

FIGURE 2.4. Refraction of an electron beam.

$$\sin \alpha = v_x/v$$ (2.21)
$$\sin \alpha' = v'_x/v$$

Therefore

$$\frac{\sin \alpha'}{\sin \alpha} = \frac{\sqrt{V}}{\sqrt{V'}}$$

which is Snell's law for electron optics.

In light optics the Abbe sine law states

$$ny \sin \theta = n'y' \sin \theta'$$

where y and y' are the distances from the axis of an object point and its corresponding image point, θ and θ' are the semiaperture of rays at the object and image (see Figure 2.5), and n and n' are the refractive indices of the object and image space. In electron optics, when the object is in a region of potential V and the image is a region of constant potential V', the sine law can be written as

$$\sqrt{V}y \sin \theta = \sqrt{V'}y' \sin \theta' \qquad (2.22)$$

Since $\sqrt{V} \sin \theta$ is proportional to the transverse velocity (v_y) of the electrons, this law is consistent with Liouville's theorem that states that the (yv_y) product is an invariant of the motion. In electron optics, just as in light optics, Abbe's sine law implies that the image formation is paraxial if θ is very small. Now since $\sin \theta = \theta - \theta^3/3! + \theta^5/5!$, Eq. (2.22) can be written for small θ as

$$\sqrt{V}y\theta = \sqrt{V'}y'\theta' \qquad (2.23)$$

When $\sin \theta$ can be approximated by the first term, it is called "Gaussian optics" or first-order theory, as distinguished from third-order theory, which includes the second term in the series.

Relation (2.22) is referred to as the Helmholtz–Lagrange theorem. If there is no loss of electron current in an axially symmetric optical system, and i represents the total emission current of the source and j the current density, then the electron current density in the image is given by the relationship

$$\frac{j'}{j} = \frac{i}{\pi r'^2} \frac{\pi r^2}{i} = \frac{V' \sin^2 \theta'}{V \sin^2 \theta} \qquad (2.24)$$

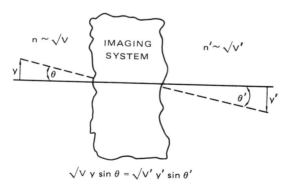

FIGURE 2.5. Image formation in electron optics.

2.3.3. Beam Characteristics

Consider the characteristics of electron beams generated by guns restricted by the following conditions:

- They are of circular symmetry.
- The emitter has a Maxwellian velocity spectrum, both radially and axially.
- The image and object spaces are related by the Helmholtz–Lagrange relation.
- Space-charge forces are neglected.

Figure 2.6 shows a simplified gun configuration used in microfabrication equipment and instruments.[19] Electrons are extracted from the cathode when suitable potentials are applied to the succeeding electrodes (grid and anode), and these electrons pass through a "crossover," which serves as the "object" in the optical system. After passing through the crossover, the rays diverge into a roughly conical bundle. An aligned circular aperture serves to pass only a small central portion of the total ray cone. This accepted portion of the whole beam is then focused by the lens onto the target or fluorescent screen.

The cathode, grid, and anode are collectively called the "triode." If the lens is assumed to be "thin," and if the potential of the screen is equal to the potential on the anode, then the lens images the crossover with a magnification $M = b/a$. By simply changing the b/a ratio, it is apparent that the final spot

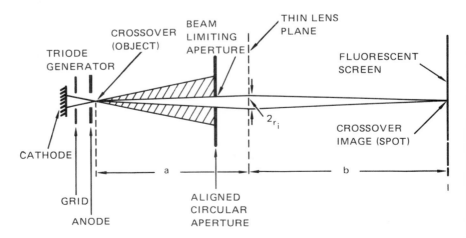

FIGURE 2.6. Gun configuration for beam fabrication.

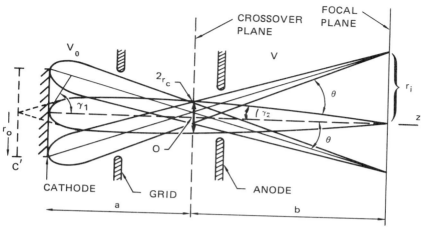

FIGURE 2.7. Electron trajectories in the near cathode region.

size is controllable, and changes in this magnification constitute the basic way of varying focus quality in gun design.

The distinctive feature of such guns is that the cross section of the beam at the screen is determined to a first order not by the area and shape of the emissive surface of the cathode, but by the crossover radius (r_c), which can be much smaller than the radius of the cathode. The first lens of the gun is an immersion objective that forms the crossover and controls the beam current. The crossover is the object for the second lens.

Figure 2.7 shows the configuration and electron trajectories in the near cathode region. If the electrons leave the cathode with zero initial velocity, they would intersect the optical axis at the same point 0, that is, the crossover radius r_c would be zero; electrons with initial velocities result in paths that intersect the axis farther than the point 0, depending on the direction of the initial velocity. [The electrons appear to originate from the surface of the virtual cathode (C').]

To determine the crossover radius (r_c) for electrons emerging from the cathode with a velocity v_0 corresponding to an energy eV_0, we use the Helmholtz–Lagrange theorem in the following form:

$$r_0 \sqrt{V_0} \sin \gamma_1 = r_i \sqrt{V} \sin \gamma_2 \qquad (2.25)$$

where r_0 and r_i are the radii of the cathode and image, respectively, V_0 and V are the potentials at the cathode and image sides, and γ_1 and γ_2 are the aperture angles.

From Figure 2.7 we find that the crossover radius $r_c = b \tan \gamma_2 \cong b \sin \gamma_2$ for small γ_2, $r_i \cong b \tan \theta$, and using Eq. (2.25), we obtain

$$r_c = \frac{r_0}{\tan \theta} \left(\frac{V_0}{V} \right)^{1/2} \sin \gamma_1 \qquad (2.26)$$

If the direction of the initial velocity is assumed to be equiprobable in the whole range of γ_1 from 0–90°, the maximum value of $\sin \gamma_1$ is equal to unity and the maximum crossover radius (r_c) is given by

$$r_c = a \left(\frac{V_0}{V} \right)^{1/2} \qquad (2.27)$$

where $a = r_0/\tan \theta$ is the distance from the cathode surface to the crossover plane.

Equation (2.27) shows that to a first approximation the crossover radius does not depend on the area of the emissive surface and is determined only by the ratio of the initial energy of the electrons (eV_0) to the energy of the electrons in the crossover region (eV).

Equation (2.27) was derived assuming that all the electrons emerging from the cathode have the same initial energy (eV_0), resulting in a crossover having a sharp boundary radius r_c. However, this assumption is rather crude, since the electrons emitted by the cathode actually have a Maxwellian velocity distribution. Electrons having energies lower than eV_0 will intersect the crossover plane inside the crossover circle r_c, but electrons with higher initial energies may cross the plane outside of r_c. Each group of electrons with the same initial energy lies within a specific circle in the crossover plane, and for higher initial electron energies the radius of the crossover circle is greater. In the case of a Maxwellian distribution the electron density in the crossover circle will not have a sharp boundary but will decrease rapidly with the distance from the axis. For this reason the crossover radius is usually defined as the radius of the circle that contains 90% of the electrons crossing the plane.

To estimate this crossover radius one has to know the current distribution in the crossover as a function of the radial distance from the optical axis. According to Maxwell's energy distribution law, the number of electrons (N) emitted per unit area of cathode surface per unit time per unit solid angle and having an initial energy between eV_0 and $e(V_0 + dV_0)$ is given by

$$N(V_0) \, dV_0 = N_0 \frac{eV_0}{kT} e^{-eV_0/kT} \, d\left(\frac{eV_0}{kT} \right) \qquad (2.28)$$

where N_0 is the number of electrons having all possible energies emitted by the cathode per unit area per unit time per unit solid angle. The value N_0 can be obtained from the emission current density of the cathode (j_c), since

$$j_c = e \int_{V_0=0}^{\infty} \int_{\gamma=0}^{\pi} N(V_0) \, dV_0 \, d\gamma = e \pi N_0 \qquad (2.29)$$

The current emitted from the area A with electron energies in the range eV_0 to $e(V_0 + dV_0)$ into the solid angle from γ to $\gamma + d\gamma$ is given by

$$di_\gamma = 2\pi AN(V_0) \, dV_0 \, e \sin \gamma \cos \gamma \, d\gamma$$

This current flows into an annular ring of radius r (see Figure 2.8), where the current density is

$$j_r = \frac{di_\gamma}{d(\pi r^2)} = \frac{AN(V_0) \, dV_0 \, e \sin \gamma \cos \gamma \, d\gamma}{r \, dr}$$

Using Eq. (2.26) to calculate $r \, dr$, one obtains

$$j_r = \frac{AN(V_0) \, dV_0 \, e}{a^2} \frac{V}{V_0} \qquad (2.30)$$

where V is the potential at the crossover plane. Equation (2.30) shows that in the annular ring ($r = r_c$) the current density produced by the electrons having energies between eV_0 and $e(V_0 + dV_0)$ is independent of γ, the initial angle of the emitted electrons.

To calculate the total current density $j(r)$ at a radius r in the crossover

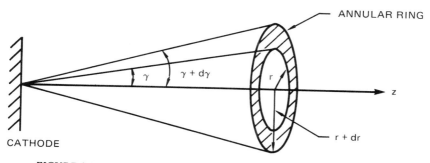

FIGURE 2.8. Correlation between the crossover radius and the energy spread.

plane it is necessary to integrate the current densities for all annular rings formed by all electrons with energies larger than eV_0, for which $r_c > r$. Here we assume that all the crossovers are formed in a single plane independently of the initial energy of the emitted electrons; that is, there is no chromatic aberration in the gun's electron-optical system. Integrating Eq. (2.30), we see that the total current density as a function of the radial distance is given by

$$j(r) = \frac{Ae}{a^2} \int_{eV_0}^{\infty} \frac{V}{V_0} N(V_0) \, dV_0$$

where eV_0 is the energy corresponding to a crossover radius of r. Using the Maxwellian energy distribution [Eqs. (2.28) and (2.27)], we obtain

$$j(r) = \frac{AeN_0}{a^2} \left(\frac{eV}{kT} \right) e^{-(eV/kT)(r^2/a^2)} \tag{2.31}$$

or

$$j(r) = j_0 e^{-r^2/\rho_0^2} \tag{2.32}$$

where

$$j_0 = \frac{AN_0 e}{a^2} \frac{eV}{kT}$$

equals the current density at the crossover center ($r = 0$), and

$$\rho_0^2 = \frac{a^2 kT}{eV}$$

is a constant for a given operating condition of the gun.

The current density at the crossover center (j_0) is associated with the density of the cathode emission current j_c and leads to the Langmuir limit equation

$$j_0 = j_c \left(\frac{eV}{kT} \right) \sin^2 \theta \tag{2.33}$$

This is obtained from Eq. (2.32) by replacing a^2 by $r_0^2/\tan^2 \theta \cong r_0^2/\sin^2 \theta$ and taking into account that $A = \pi r_0^2$ and $j_c = e\pi N_0$ from Eq. (2.29).

From Eq. (2.33) it follows that an increased current density in the crossover center can only be obtained by increasing the cathode emission current density (j_c) or decreasing its operating temperature. It is clear that in the case of thermionic cathodes these requirements are conflicting, since a decrease in the cathode temperature causes a decrease in the emission current density.

For small θ Eq. (2.33) can be written as

$$j_0 = \pi\beta\theta^2 \tag{2.34}$$

which is identical to Eq. (2.16). This relation is quite general; no assumptions concerning the configuration of fields or beam concentration system were made in its derivation.

The Langmuir limit [Eq. (2.33)] gives the maximum current density in the spot that is theoretically possible; it does not indicate that such a value is actually attainable. In practice, spot current density values, roughly 50% of those predicted by the Langmuir limit equation, can be obtained with well-designed optical systems.[19]

2.3.4. Space-Charge Effects

The simplified theory of forming a crossover is based on the assumption that the electrostatic field in the near-cathode region is uniquely determined by the potentials of the gun electrode configuration. The field created by the space charge of the electrons is not taken into account. The effect of space charge can be ignored only for low-perveance (P) guns where P is defined by

$$P \equiv \frac{i}{V^{3/2}} \text{ perv} \tag{2.35}$$

for i in amperes and V in volts.

The influence of the internal electric field of the beam has a perceptible effect on the motion of electrons when $P \geq 10^{-8}$ perv. For the electron beams used in microlithography P is less than this value so that space-charge effects are not important in the image formation. However, for ion beams used for implantation or annealing, space-charge effects have to be taken into account.

In the near cathode region, where the electron velocity is low, and in the crossover region, where the current density is very high, the Coulombic interaction between the electrons is much greater than in the final image spot. Such interactions can have a substantial effect on the beam formation. Qualitatively, the presence of space charge in the cathode region results in a radial drift of electrons across the laminar trajectories and a change in the current density distribution.

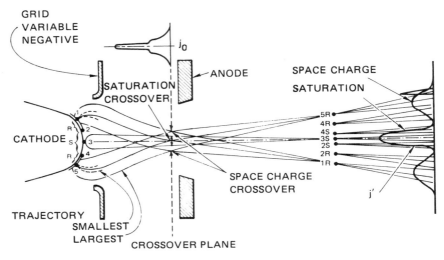

(a)   HOLLOW BEAM

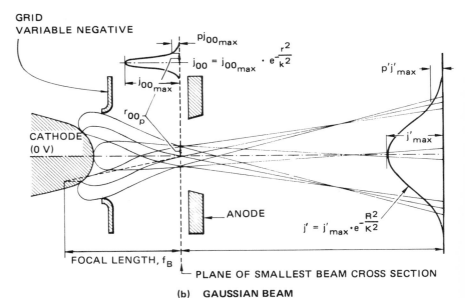

(b)   GAUSSIAN BEAM

FIGURE 2.9. Beam formation.

Figure 2.9(a) shows the formation of a hollow beam due to the transverse drift of electrons away from the laminar trajectory. Figure 2.9(b) depicts the Gaussian beam formation in the same triode electron gun without any space-charge effect. We see, therefore, that the action of space charge results in an increase in the crossover radius and a reduction of current in the crossover center. The quantitative theory of space-charge effects will be discussed briefly in connection with ion extraction and ion-beam formation (see Section 2.4.6).

2.3.5. Aberrations

The size and shape of the electron beam are also influenced by imperfections in the electron-optical elements that make up the electron gun. We briefly summarize these effects here; more complete information is available in the electron-optics literature.[20–24]

Spherical aberration of the immersion objective (first lens next to the cathode) influences the crossover formation. Its effect can be observed even at low beam currents when electrons are taken only from the central part of the cathode. As the current increases (by increasing the temperature) and electrons are taken from larger areas of the cathode surface, the conditions for paraxiality are violated, and other kinds of aberrations become noticeable. Hence an increase in beam current is generally accompanied by an increase in the size of the crossover.

The four types of electron-optical defects associated with the electron lenses used in electron-beam devices are listed below.

2.3.5.1. Spherical Aberration

Aberrations result in a minimum beam diameter, called the disk of least confusion, as shown in Figure 2.10(a). This minimum diameter d_s results from the crossing of several electron trajectories passing through the lens that do not come to a focus at the same axial position. The defect is caused by focusing fields that are stronger near the electrodes that produce them than at the paraxial focal position. The diameter of the disk of least confusion, d_s, is

$$d_s = 0.5 C_s \alpha^3 \qquad (2.36)$$

where C_s is the spherical aberration constant, related to the paraxial focal length f by $C_s = K_s f$, K_s is a constant dependent on the lens geometry, and

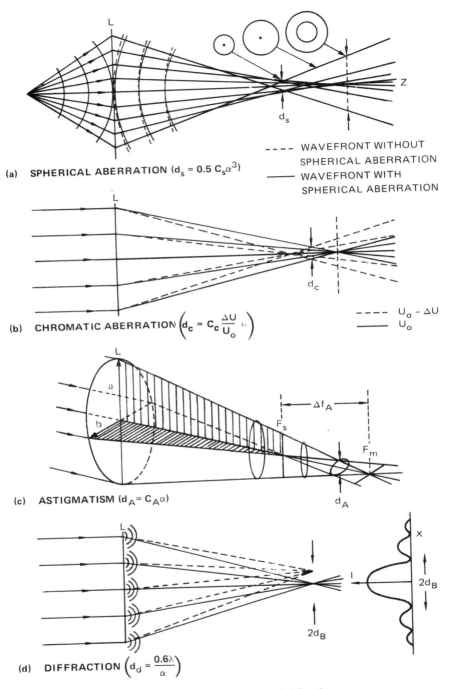

(a) SPHERICAL ABERRATION ($d_s = 0.5\,C_s\alpha^3$)

---- WAVEFRONT WITHOUT SPHERICAL ABERRATION
—— WAVEFRONT WITH SPHERICAL ABERRATION

(b) CHROMATIC ABERRATION $\left(d_c = C_c \dfrac{\Delta U}{U_o}\,\alpha\right)$

---- $U_o - \Delta U$
—— U_o

(c) ASTIGMATISM ($d_A = C_A\alpha$)

(d) DIFFRACTION $\left(d_d = \dfrac{0.6\lambda}{\alpha}\right)$

FIGURE 2.10. Aberrations and diffraction.

α is the convergence half-angle referenced to the image space. It can be related to the equivalent angle on the object side α_0 by the Lagrange–Helmholtz relation [Eq. (2.22)].

It is essentially impossible to reduce the spherical aberration introduced into a beam by a lens by any subsequent electron-optical action. It is therefore of great importance to design each electron optical element with a minimum C_s.

The effect of this aberration on the final image (spot) can be reduced, at the expense of beam current, by placing a limiting aperture in or before the lens. This aperture reduces the convergence angle α of the focused beam, and thereby the effect of spherical aberration. Such trade-offs between the spot diameter and the current density are part of the optimization of a system design.

The total aberration, resulting from two lenses in series spaced a distance L apart, with spherical aberration coefficient C_{s1} and C_{s2} and focal lengths f_1 and f_2, can be expressed in terms of $K_s = C_s/f$ as

$$K_s = \frac{K_{s1}f_2^2 + K_{s2}(L - f_1)^2}{(f_1 + f_2 - L)^2}$$

If the final lens is stronger, $f_1 \gg f_2$, and this expression reduces to

$$K_s = \left(\frac{f_2}{f_1}\right)^2 K_{s1} + K_{s2}$$

Thus the aberration of the stronger lens exerts a greater effect on the aberration of the system.

2.3.5.2. Chromatic Aberration

Chromatic aberration in a lens refers to the sensitivity of the focal properties to the energy with which the particles enter the lens. A particle with a higher energy will come to a focus farther from the lens than a particle with a lower energy. This effect is illustrated in Figure 2.10(b). It is seen that a disk of least confusion exists, and its diameter is

$$d_c = C_c \frac{\Delta U}{U_0} \alpha \tag{2.37}$$

where C_c is the chromatic aberration constant, frequently expressed as $C_c = K_c f$, which shows that chromatic aberration is also greater in lenses with longer focal lengths and ΔU is the total energy spread.

2.3.5.3. Astigmatism

Astigmatism can result if the apertures of the optical elements are not circular, are displaced, or are tilted relative to the optical axis. The disk of confusion is given by [see Figure 2.10(c)]

$$d_A = C_A \alpha \tag{2.38}$$

For a noncircular aperture C_A is given by δ, where δ is the difference between the axes of the ellipse that forms a limiting diaphragm, $\delta = a - b$. Astigmatism can be corrected by the use of a stigmator, which in its simplest form is a multipole element of opposite electric fields arranged around the beam.

2.3.5.4. Diffraction

If a particle beam is passing through a limiting aperture it will be diffracted to form a spot of diameter [see Figure 2.10(d)].

$$d_d = \frac{0.6\lambda}{\alpha} \tag{2.39}$$

where λ is the de Broglie wavelength of the particle and is given for electrons by

$$\lambda \,(\text{Å}) = \frac{12.26}{U_0^{1/2}} \qquad (U_0 \text{ in eV})$$

The crossover size, including all the various aberrations, is usually assumed to be given by the rms sum of the diameters of the disks of confusion; that is,

$$2r_{\text{eff}} = [(2r_c)^{1/2} + d_s^2 + d_c^2 + d_A^2 + d_d^2]^{1/2}$$

In practice only d_s is significant for the triode, the others (d_c, d_A, d_d) are usually small enough to be neglected.

2.3.5.5. Defocusing Due to Beam Energy Spread

There are two factors contributing to the energy spread in an electron beam: the temperature of the emitting cathode and the Coulomb interaction between electrons.[25] The thermal spread gives rise to a Maxwellian energy distribution, broadens significantly, and changes to wider Gaussian distribution as the beam current density increases. This effect, first reported by Boersch,[25] is now generally believed to be mainly a result of electron–electron interaction in the beam crossover, where the electrons are most tightly packed.

The total energy spread in the triode is given as

$$(\Delta U_{\text{triode}})^2 = (\Delta U_{\text{th}})^2 + (\Delta U_B)^2$$

and in the complete instrument by

$$|\Delta U_{\text{tot}}|^2 = |\Delta U_{\text{triode}}|^2 + |\Delta U_{\text{syst}}|^2$$

We can gain some physical insight into this effect by considering Figure 2.11, which describes the interaction in terms of geometrical parameters.[26] A reference electron is considered to travel along the optical axis; a second electron travels along a general trajectory. The closeness of approach is measured by an impact parameter b.

Three interactions are considered. Interactions parallel to b_z, that is, along the optic axis, lead to energy variations. Scattering along the direction of b_r leads to changes in trajectory displacement. Using Monte Carlo methods[27] the

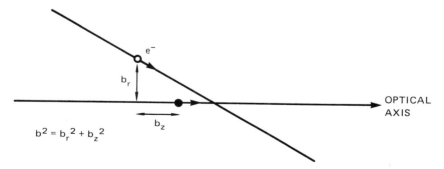

FIGURE 2.11. Geometry of the Coulomb interaction between two electrons.

expected mean values for the three interactions have been found to be as follows. The interaction energy change

$$\Delta U_i \approx |e^2/4\pi\epsilon_0| F_1(\lambda r_c)/\alpha_0 r_c \qquad (2.40)$$

the angular displacement

$$\overline{\Delta\alpha} \approx |e/4\pi\epsilon_0| F_2(\lambda r_c)/2 V_b\alpha_0 r_c$$

and the radial displacement

$$\overline{\Delta r} \approx |e/4\pi E_0| F_3(\lambda r_c)/2 V_b\alpha_0^2$$

where r_c is the radius of the crossover, α_0 is the beam half-angle subtended at the crossover, V_b is the beam voltage, and λ is given by

$$\lambda = |I_b/V_b^{1/2}|(e^3/m)^{-1/2}$$

where I_b is the beam current.

The functions F_1, F_2, and F_3 have the forms shown in Figure 2.12 when

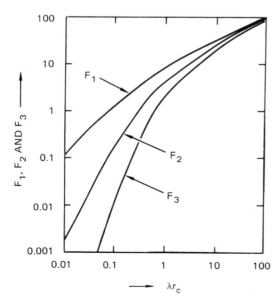

FIGURE 2.12. $F_1, F_2,$ and F_3 as functions of λr_c. Source: ref. 26.

plotted against λr_c. At high current levels all three functions tend to a $(I_b r_c / V_b)^{1/2}$ dependence. At lower currents the dependence is steeper and more variable. In the high-current regime we can use the relationship between the beam current and the brightness β to eliminate α_0 from Eq. (2.40) and to obtain

$$\overline{\Delta U} \approx (\beta r_c)^{1/2} V_b^{-1/4} \qquad (2.41)$$

This result has been confirmed experimentally.[28]

Since most beam paths in electron-optical columns can be considered as a sequence of crossovers, and since each crossover increases the energy spread, the number of these crossovers has to be minimized to achieve the smallest spot size. Coulombic interactions are "diffractionlike" in the sense that, unlike all other aberrations except for diffraction, they produce a loss of focusing ability that increases as the beam angle is decreased.

In microfabrication systems the ultimate in resolution is not usually required, so resolution (spot size) can be traded for increases in current. The limitations and trade-offs due to aberrations are shown in Figure 2.13 as a function of the beam half-angle. In general, system performance is limited by spherical aberration. To increase the spot current the angular aperture has to be increased to its maximum value, consistent with the required final image spot size (d_i), where

$$d_i^2 = d_0^2 + d_s^2 + d_c^2 + d_d^2 \qquad (2.42)$$

Here d_0 is the demagnified image of the gun source and d_s, d_c, and d_d are the aberration disk sizes. In this case the optimization is obtained by increasing the current in the final image spot (d_i) by increasing the angular aperture α.

2.3.6. Gun Structures

To extract electrons from the cathode and accelerate and shape them into the desired beam profile an arrangement of appropriately designed electrodes must be used. This electrode system must create the proper configuration of electric fields at the surface of the cathode and along the acceleration path and is often called the gun by itself. In this section the more common structures utilized in microelectronics applications are reviewed.

A general model[29] for a triode gun is shown in Figure 2.14. The electrons are emitted from a small region around the tip of the filament. In the immediate vicinity of the tip their trajectories are sharply curved, but the paths soon

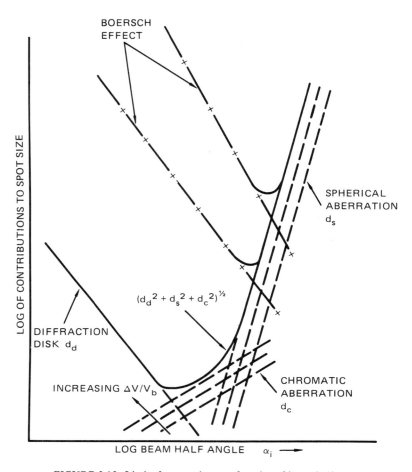

FIGURE 2.13. Limits for spot size as a function of beam half-angle.

straighten out and if we extrapolate back we find that the electrons appear to come from a virtual source, which, in general, does not coincide with the real filament tip. It is the size of this virtual source that we attempt to keep small in designing the triode.

The Wehnelt electrode grid and anode bring the accelerated electrons to a crossover or focus within the field of the electrodes; the size of this crossover is determined by the spherical aberration of the triode and the size of the virtual source, and is typically 10–50 μm in diameter.

In quantitative terms the radius of the virtual cathode is given by

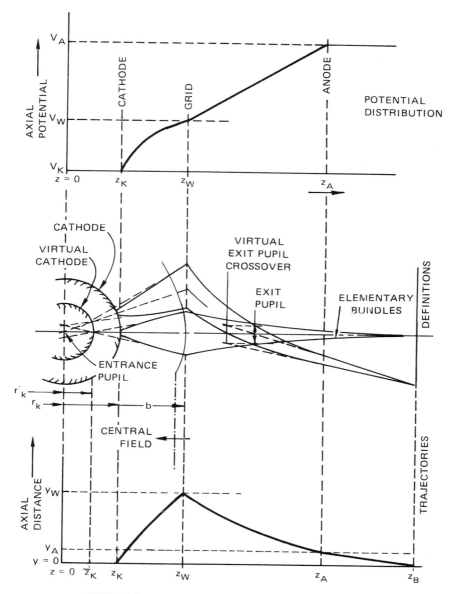

FIGURE 2.14. Lauer's model of the triode. Source: ref. 29.

$$r'_k \approx r_k \left(\frac{V_k}{V_w - V_k} \right)^{1/2} \tag{2.43}$$

where eV_k is the kinetic energy of the electrons at the cathode, r_k is the cathode radius, and V_w is the potential on the axis at the position of the Wehnelt cylinder. For a tungsten filament $eV_k = 0.25$ eV and the radius of the virtual cathode is

$$r'_k \approx r_k \left(\frac{0.25}{V_w} \right)^{1/2}$$

hence the crossover radius can be written as

$$r_c \approx r'_k \frac{1}{1 - (r_k + b)(V_A - V_w)/4(z_A - z_w)V_w} \tag{2.44}$$

where V_A is the anode potential and b, z_A, and z_w are the position coordinates depicted in Figure 2.14.

The telefocus gun [Figure 2.15(a)] is designed to produce a beam focused at a relatively large distance away from the anode. The long-focus effect results from the hollow shape and negative bias of the Wehnelt electrode. Near the cathode the electric field is diverging so that emerging electrons are given an outward radial velocity. Between the Wehnelt electrode and the anode the equipotentials first become flat and then converge toward the anode. The beam acquires a net radial velocity inward that is of smaller magnitude than the initial outward velocity because of the higher energy of the electrons. Thus the beam converges quite slowly and has a long focal length. An increase in the bias increases the focal length by increasing the curvature in the cathode region and starting the beam with more divergence, but it also has the effect of reducing the beam current. The spot size is further limited by the beam current owing to the space charge forces that act over an increased path length.

The gradient gun shown in Figure 2.15(b) has postacceleration. Its main advantage is that the beam current is independent of the final beam voltage V_5, since the current is controlled by the voltage V_1 on the first grid. Thus the total beam power may be varied over a wide range with only a small variation in spot size. On the other hand, relatively large voltages must be used. To take full advantage of the gun's capabilities, the final voltage V_5 must be a great deal larger than V_1, which in turn must be high enough to draw adequate emission from the cathode.

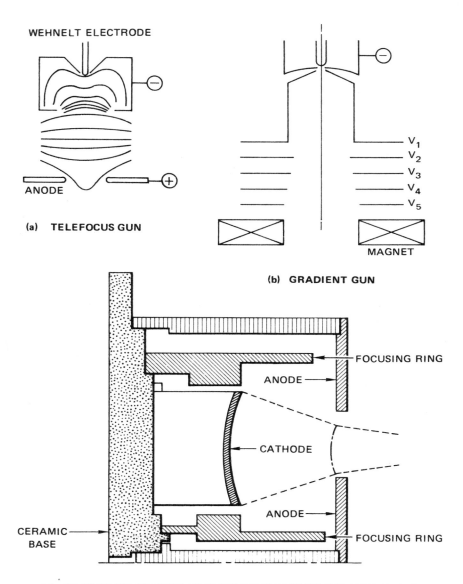

(a) **TELEFOCUS GUN**

(b) **GRADIENT GUN**

(c) **HALF CROSS SECTION OF MODIFIED PIERCE-TYPE TOROIDAL GUN**

FIGURE 2.15. Sample electron gun structures.

Perhaps the most significant advance in the design of space-charge-dominated electron guns was made by Pierce,[30] who evolved a theory and philosophy of design recording to which all high-current-density guns have been built in the past decades. He proposed to achieve a laminar flow in the emerging beam from the gun by forcing the electric field configuration external to the beam to be similar to that anticipated from an ideal diode. Pierce showed that the effect of space charge within a finite electron beam could be compensated for by a beam-forming electrode that causes the field in the region external to the beam to satisfy the proper boundary conditions at the beam edge. By forcing this boundary condition, the gun can be designed as if the electron flow were space-charge limited between the cathode and anode of an ideal diode of planar, spherical, or cylindrical configuration [see Figure 2.15(c)].

The current density that can be drawn with infinite-plane parallel electrodes under space-charge-limited emission conditions in a diode of spacing d with a potential difference V is given by Child's law as

$$J = \frac{4\sqrt{2}\epsilon_0}{9}\left(\frac{e}{m}\right)^{1/2}\frac{V^{3/2}}{d^2} = 0.0233\,\frac{V^{3/2}}{d^2}\,\text{A/cm}^2 \qquad (2.45)$$

where V is expressed in volts and d in centimeters.

The Pierce gun is designed to produce a uniform current density over the beam cross section. For converging guns the minimum beam diameter will lie well beyond the anode, and thereafter the beam will diverge uniformly. Aside from the design simplicity, the principal advantage of the Pierce gun is its high efficiency, which may range to 99.9% or more (that is, less than 0.1% of the cathode current will be lost to the gun electrodes) if reasonable care is exercised.

2.3.7. Gun Optics

The factors that limit the crossover size and the maximum achievable current in the crossover are source brightness, lens aberrations, diffraction, path changes due to space charge, and increased energy spread due to Coulomb interaction. In general, these factors cannot be eliminated. However, depending on the configuration and application, they may vary in significance from negligible to performance limiting.

As can be seen from Table 2, field-emission sources have a very high brightness as compared to thermionic sources. However, this high brightness can be achieved only over very small source sizes, resulting in a low total beam current. Hence the advantage of the high brightness of field-emission sources over conventional thermionic sources can be utilized only for small beam sizes (100–1000 Å) such as are usually employed in electron microscopy. The optical properties of point-cathode field-emission guns have been studied in some detail.[31] We note that in field-emission guns the object is usually not the crossover but the virtual cathode. Thus the final spot size (d) is set by the aberrations of the optical column and the allowed half-angle of the lens system. In other words, the small source size cannot be resolved by the lens system of the optical column.

In the case of a thermionic cathode with a resolvable source size (d_0) the final spot (d) is set by the effective crossover size ($d_i = 2r_c$) and the aberrations of the lenses in the optical column. By the use of Eq. (2.38) the crossover size is given as

$$d_i^2 = (2r_c)^2 = d_g^2 + d_s^2 + d_c^2 + d_d^2 \qquad (2.46)$$

where d_g^2 is the Gaussian spot size at the crossover (demagnified image of the source $d_g = md_0$). But from Eq. (2.34) $j = \pi\beta\alpha^2$, and the current contained within the Gaussian spot

$$\pi \frac{d_g^2}{4}$$

is

$$i = \beta \frac{\pi^2}{4} d_g^2 \alpha^2 \qquad (2.47)$$

and

$$d_g^2 = \left(\left| \frac{4i}{\pi^2\beta} \right| \right) \frac{1}{\alpha^2} = \frac{C_0^2}{\alpha^2} \qquad (2.48)$$

Then, using Eqs. (2.48) and (2.36)–(2.39), in Eq. (2.46) we get for the effective crossover size

$$d_{\text{eff}}^2 = \left(C_0^2 + (0, 6\lambda)^2 \right) \frac{1}{\alpha^2} + \tfrac{1}{4}C_s^2\alpha^6 + \left(C_c \frac{\Delta U}{U_0} \right)^2 \alpha^2 \qquad (2.49)$$

Now we can calculate the maximum beam current that can be achieved for a given source size d_0, a lens system (C_s, C_c), and desired operating characteristics [$\lambda(V_0)$, α].

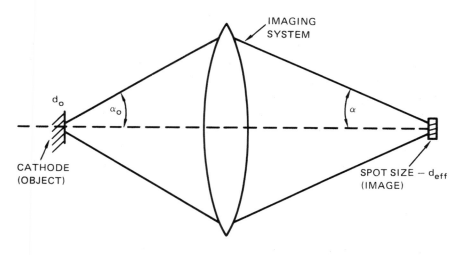

(a)   GUN OPTICAL SYSTEM

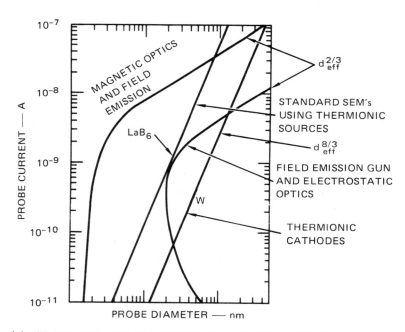

(b)   PROBE CURRENT AS A FUNCTION OF PROBE DIAMETER

FIGURE 2.16. Imaging system and current relationships for electron guns.

First we consider a resolvable thermionic source, and we will calculate the maximum current in the crossover with spot size d_{eff}. Figure 2.16(a) shows the simplified optical system of this gun, which images on an axis source of size d_0 to a final spot d_{eff} at the crossover plane. The imaging system is a single lens characterized by a linear magnification m and aberration coefficients C_c and C_s.

Since this type of optical system is usually limited by spherical aberration, the effective spot size is given [from Eq. (2.49)] by

$$d_{eff}^2 \approx \frac{C_0^2}{\alpha^2} + \tfrac{1}{4}C_s^2\alpha^6 \tag{2.50}$$

The spot size is minimized for a given current with respect to the beam half-angle if

$$\frac{\partial d_{eff}^2}{\partial \alpha} = -\frac{2C_0^2}{\alpha^3} + \tfrac{6}{4}C_s^2\alpha^5 = 0 \tag{2.51}$$

which gives the optimum half-angle for minimum spot size for a given current, or maximum current for a given spot size as

$$\alpha_{opt} = (\tfrac{4}{3})^{1/8} \left(\frac{C_0}{C_s}\right)^{1/4} \tag{2.52}$$

And with this

$$d_{eff,min} = (\tfrac{4}{3})^{3/4}(C_0^3 C_s)^{1/2} \tag{2.53}$$

Using Eq. (2.48) in (2.53), the maximum current in the beam is given by

$$i = \frac{3\pi^2}{16} \beta C_s^{-2/3} d_{eff}^{8/3} \tag{2.54}$$

as a function of the spot size d_{eff}. Figure 2.16(b) shows the maximum probe current as a function of the spot size for thermionic (W and LaB_6) and field-emission cathodes.[31]

For the extreme case of a nonresolvable source consider again the optical system shown in Figure 2.16(a) with our ideal point source ($d = 0$). For this condition Eq. (2.49) reduces to

$$d_{\text{eff}}^2 \approx d_g^2 + \tfrac{1}{4}C_s^2\alpha^6 = (md_0)^2 + \tfrac{1}{4}C_s^2\alpha^6 \approx \tfrac{1}{4}C_s^2\alpha^6 \qquad (2.55)$$

for a spherical-aberration-limited system. Because the source is just an ideal point, the usual definition of source brightness (β) (in A/cm$^2\cdot$sr^{-1} is meaningless, since the source current density (j_c) cannot be defined. In this case the source parameter of interest is the angular brightness, Ω (A/sr). This source parameter is also applicable to field-emission or thermal/field-emission sources that have a very small size (0.1–1 μm). For these sources the beam current delivered into a solid angle defined by the source half-angle α_0 can be expressed in terms of the angular brightness Ω as

$$i = \Omega\pi\alpha_0^2 \qquad (2.56)$$

Using Snell's law we can relate the half-angle in the object space α_0 to the half-angle in the image space by the equation

$$\alpha_0 = m\alpha\left(\frac{V}{V_0}\right)^{1/2} \qquad (2.57)$$

where V and V_0 are the beam potentials in the image and object spaces, respectively. Substituting Eq. (2.56) into Eq. (2.55), we obtain for the beam current

$$i = \Omega\pi m^2\left(\frac{V}{V_0}\right)\alpha^2 \qquad (2.58)$$

The beam half-angle α in the image space is defined in terms of the final effective spot size for a spherical-aberration-limited system from Eq. (2.55) as

$$\alpha = \left(\frac{2d_{\text{eff}}}{C_s}\right)^{1/3} \qquad (2.59)$$

and substituting Eq. (2.59) into (2.58) we get the maximum beam current in the imaged spot (d_{eff}),

$$i = \Omega\pi m^2\left(\frac{V}{V_0}\right)\frac{2d_{\text{eff}}^{2/3}}{C_s^{2/3}} \qquad (2.60)$$

A comparison of Eqs. (2.60) and (2.54) illustrates an important difference in the beam current capabilities of optical imaging systems with nonresolvable and resolvable sources. For the case of a resolvable source Eq. (2.54) shows that the beam current is proportional to the eight-thirds the power of the spot diameter (d_{eff}), but for a nonresolvable source the beam current is proportional to the two-thirds power of the final spot diameter (d_{eff}), as shown by Eq. (2.60).

In practice, this means that more current at small spot sizes can be achieved with field-emission sources that with conventional thermionic sources, whereas the converse is true for larger spot sizes, as shown in Figure 2.16.

In the case of small sources the current dependency on the spot size is generally given as

$$i = cd_{\text{eff}}^{2/3} \tag{2.61}$$

where the constant (c) depends strongly on the optical system constants (C_s) and on the electrode arrangement (diode, triode, tetrode) in the gun. With an electrostatic triode gun (Crewe gun[32]) a spot size of 100 Å has been achieved, which can be improved by the use of magnetic optical elements, as shown in Figure 2.16(b). Generally, the spot (probe) current (i) is a function of the spot size (d_{eff}), the type of electron source, and the type of optical elements (magnetic, electrostatic). However, the spot size is limited by the beam half-angle α for a given optical system (see Figure 2.13). Thus for system specifications two sets of data $d_{\text{eff}}(\alpha)$ and $i(d_{\text{eff}})$ are required, as shown in Figures 2.17(a) and (b).

For example, it can be seen from these figures that for a scanning electron microscope operating at a probe current in the range $i = 10^{-12}\text{–}10^{-11}$ A a probe size $d_{\text{eff}} = 100$ Å is practical with a thermionic cathode. The angular aperture of the optical system should be in the range $5 \times 10^{-3}\text{–}10^{-2}$ rad. With the use of α the limiting aperture sizes can be calculated for a given electrode geometry.

2.4. Ion Sources

2.4.1. Background

In this section we are concerned with the creation and extraction of ions. We will first discuss the different ways in which ions can be created and then the properties of ion sources used in microelectronic applications. This will be followed by a discussion of space-charge effects that play an important role in ion-beam formation used for ion-beam lithography, ion milling, ion implantation, and ion microscopy.[33,34]

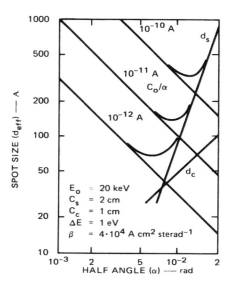

(a)   SPOT SIZE AS A FUNCTION OF BEAM HALF ANGLE

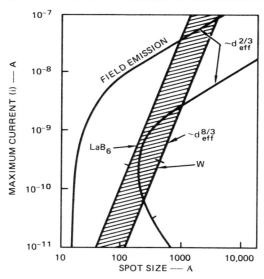

(b)   MAXIMUM CURRENT AS A FUNCTION OF SPOT SIZE

FIGURE 2.17. Limits for spot size and maximum current as functions of beam half angle for LaB₆, W, and field emission cathodes.

In general, each application of ion beams requires a somewhat different source and set of beam parameters. The source performance parameters of greatest importance are the following:

- The ion species produced
- The ion current produced by the source
- The brightness of the source
- The energy spread

There are other parameters of lesser importance (ionization efficiency, source life, energy conversion efficiency, etc.), and discussions of these can be found in the specialist literature.[35] To illustrate the different characteristics of the ion sources, three classes of ion sources are discussed: electron-impact, field-ion, and liquid-metal field-ion sources.

2.4.2. Electron-Impact Ionization Sources

Electron-impact ionization is the most used technique for generating ions for implantation and ion processing. In this ionization process the energy transferred to a molecule from an energetic electron exceeds the ionization energy eV_i, where V_i is the ionization potential for that molecule. Electrons for ionization can be created by thermionic or cold-cathode emission or can result from the discharge itself. These electrons are accelerated by the use of dc or rf fields and confined by the use of magnetic fields.

Energetic electrons can lose energy to gas particles by means of elastic collisions, the dissociation of gas particles, and the excitation of gas particles. The momentum transfer to the gas particles is usually negligible because of the large mass imbalance; thus the increase in kinetic energy of the gas particles is negligible. When an incident (primary) electron has an energy in excess of the ionization energy, this excess energy can appear as the kinetic energy of a scattered incident electron, as the kinetic energy of an ejected (secondary) electron, as the multiple ionization of a gas particle, the excitation of a gas particle, or any combination of these effects.

The number of ionizing collisions suffered by an electron in passing through a gas per unit path length per unit pressure is called the differential ionization coefficient S_e; the value of this coefficient depends on the electron energy U_e and the gas species. The differential electron-impact ionization coefficients are shown for several gas species in Figure 2.18. It is seen that electrons with energies much less than or much greater than the atomic and molecular ionization energies do not ionize these particles very effectively; electron ener-

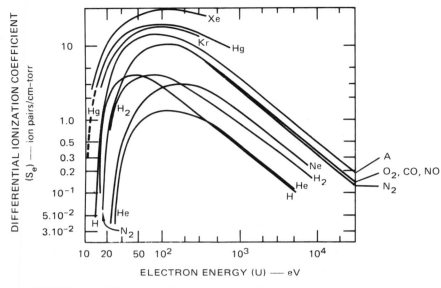

FIGURE 2.18. Differential ionization coefficients S_e as a function of electron energy U.

gies a few times the ionization energies are optimum for electron-impact ionization.

The number of ions per square centimeter per second ($\dot{n}_+$) produced by an electron current density j - (A/cm^2) passing a distance l (cm) through a gas of pressure p (Torr) is given by

$$\dot{n}_+ = \frac{j - plS_e}{e} \qquad (2.62)$$

As an energetic electron loses energy from ionizing and scattering collisions, its ionizing effectiveness changes. When its energy decreases below the lowest ionization energy of any atom in the volume, it no longer causes ionization and is called an ultimate electron. Energetic electrons can cause ionization efficiently only if they are maintained in the ionizing volume for a time long enough for them to lose their energy by ionizing collisions. A confining force resulting, for example, from a magnetic field or an oscillating rf field is often employed to increase path lengths and electron lifetimes. These concepts are illustrated in Figures 2.19(a)–(c).

In the hot- and cold-cathode ion sources the anode is maintained at a positive potential with respect to the two end plates so that electrons emitted by

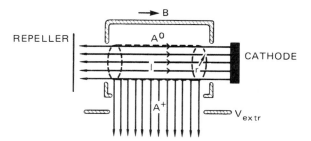

(a) OSCILLATING ELECTRON IMPACT-ION SOURCE

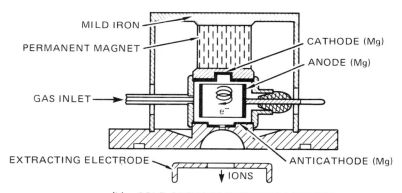

(b) COLD CATHODE "PENNING SOURCE"

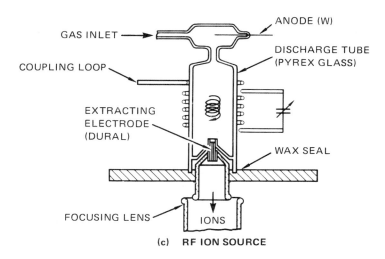

(c) RF ION SOURCE

FIGURE 2.19. Electron-impact ionization sources.

the cathode are injected through the plasma sheath in front of the cathode into a region where the potential varies with both the radial and axial distances. The application of an approximately axial magnetic field prevents the electrons from moving directly to the outer cylinder. In this crossed electric and magnetic field configuration the electrons spiral around the magnetic field lines and oscillate longitudinally. As a result of ionizing collisions between the electrons and the atoms of the source gas the plasma will be supplied continuously with ions and electrons. The primary electrons lose energy and diffuse across the magnetic field lines to the outer cylinder, where they are collected, causing an electric current to flow from the anode to the cathode. The plasma essentially fills the discharge chamber. In principle, this discharge mechanism can be used to ionize any gas.

In the rf ion source the ever-present electrons formed in the discharge tube fill the chamber and form a plasma sheath along the chamber wall and the extractor electrode. The wall potential is given by[4]

$$V_W = \frac{kT_e}{2e} \ln \frac{T_e m_i}{T_i m_e} \tag{2.63}$$

for an insulator surface, where T_e and T_i are the electron and ion temperatures and m_e and m_i are the electron and ion masses, respectively.[36] This wall potential causes the plasma boundary S_1 to move away from the wall surface S_2, as shown in Figure 2.20(a). The measure of the plasma withdrawal is given by λ_D, where λ_D is the Debye screening length

$$\lambda_D = \left(\frac{kT_e}{4\pi n_e e^2} \right)^{1/2} \tag{2.64}$$

The following are useful forms of Eq. (2.64):

$$\lambda_D = 6.9 (T/n)^{1/2} \text{ cm} \qquad (T \text{ in K}°)$$

and

$$\lambda_D = 740 (kT/n)^{1/2} \text{ cm} \qquad (kT \text{ in eV})$$

when $n_e = n_i = n$ and $T_e = T_i = T$.

If an aperture larger than $2\lambda_D$ is made in the chamber wall, the plasma will balloon out through this opening, as shown in Figure 2.20(a). In the open-

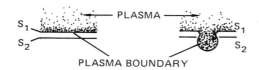

(a) **SEPARATION OF PLASMA BOUNDARY S₁ FROM THE WALL SURFACE S₂ CAUSED BY THE WALL POTENTIAL V_W**

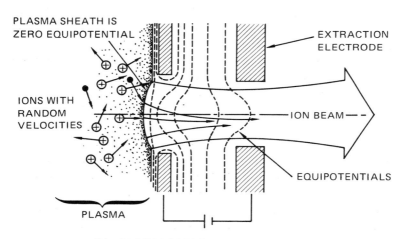

(b) **PLASMA BOUNDARY IN AN ION SOURCE**

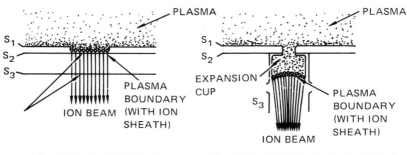

(c) **ION EXTRACTION BY THE USE OF GRIDS**

(d) **ION EXTRACTION BY THE USE OF AN "EXPANSION CUP"**

FIGURE 2.20. Beam extraction from plasma sources.

ing of the extractor electrode the value of the negative potential applied to this electrode determines the shape of the plasma surface [Figure 2.20(b)] because the plasma boundary must be an equipotential surface and the normal component of the electric field must be zero. This means that if the negative potential is increased, the plasma surface is forced back so that a curved surface will be the ion emitter forming a convergent beam of ions.

The shape and position of the ion-emitter surface of the plasma are thus dependent on the value of the negative potential applied to the extracting electrode. This means that the ion-optical system composed of the extracting electrode and the plasma surface can have different geometries determined by the value of the negative potential applied to the extracting electrode. This imposes limitations on the extracting voltage so that the beam intensity cannot be varied beyond a given value by varying the extracting voltage.

Under normal operating conditions the shape of the plasma boundary adjusts itself so that the ion extraction rate is equal to the supply rate from the plasma. In this case the extracted ion current density is given by the Child–Langmuir equation:

$$j_i = \frac{4\epsilon_0}{9} \left(\frac{2e}{m_i} \right)^{1/2} \frac{V_{ext}^{3/2}}{d^2} \tag{2.65}$$

where V_{ext} is the extractor potential, d is the extractor electrode–plasma boundary distance, and m_i is the ion mass.

The current limitation can be overcome by keeping the shape and position of the ion emitter surface fixed. This can be achieved by placing a grid of high transparency in the extracting aperture, which ensures both the constant shape and fixed position of the plasma boundary independently of the extractor voltage [Figure 2.20(c)].

For low-pressure (10^{-2}–10^{-3} Torr) and high-current operation it is important to restrict the extraction aperture size but enlarge the ion-emissive surface. This can be achieved by the use of an "expansion cup" as shown in Figure 2.20(d). These expansion cups are very useful devices in ion sources such as plasmatrons, where high-current-directed ion beams are required.

2.4.3. Plasmatrons

The plasmatron is an arc-discharge type of ion source that differs from that of the electron-impact ionization ion source in that higher gas pressures (3×10^{-3}–3×10^{-2} Torr) and higher electron currents are used. An arc discharge of a few hundred volts is maintained between a heated cathode and an anode. Electrons from the filament are accelerated toward the anode through the gas to be ionized. Near the anode the plasma is constricted by a conical intermediate electrode with an aperture (e.g., 5 mm) at a potential approximately halfway between those of the cathode and the anode (negative, ∽100 V) that creates a plasma bubble bounded by a charge double layer that focuses the electrons from the plasma on the cathode side in the region in front of the ion-extraction aperture. In the duoplasmatron shown in Figure 2.21(a) the plasma is also compressed by a nonuniform magnetic field created by a magnetic-pole lens. The poles are insulated from one another; one can also be used as the intermediate electrode and one can be the anode.

The magnetic field confines the electrons so as to limit intense ionization to a small region around the anode aperture.[37] The primary advantage of the duoplasmatron is the increased efficiency obtained (50–95%). One disadvantage in some applications is that in extracting large ion currents the ions are accelerated to 70 kV. Thus for low-energy applications a decelerating step is required, which can reduce the current deliverable to the target. Proton current densities of up to 65 A/cm² have been obtained from a pulsed duoplasmatron, corresponding to a current of about 1 A.

The shape of the extracted ion beam can be modified by the use of an expansion cup. If the axial magnetic field is very high in the plasma, a "halo beam" will be formed, as shown in Figure 2.21(b), but upon insertion of a conical element into the cup the beam becomes homogeneous again,[38] as seen in Figure 2.21(c).

2.4.4. Field-Ionization Sources

As the electric field surrounding a neutral atom or molecule is increased, the particle first becomes polarized and ultimately the valence electron can tunnel from within the particle to a region just outside of it. Once the particle is ionized, the intense electric field rapidly separates the ion and electron. The tunneling mechanism is similar in principle to the field emission of electrons from a metal crystal, but the detailed theory involves only the escape of a single electron from the ground state of the atom or molecule.[2]

The theory predicts the mean lifetime τ for free atoms or small molecules in an electric field E to be given by

$$\tau \approx 10^{-16} \exp\left(\frac{0.68\phi_i^{3/2}}{E}\right)$$

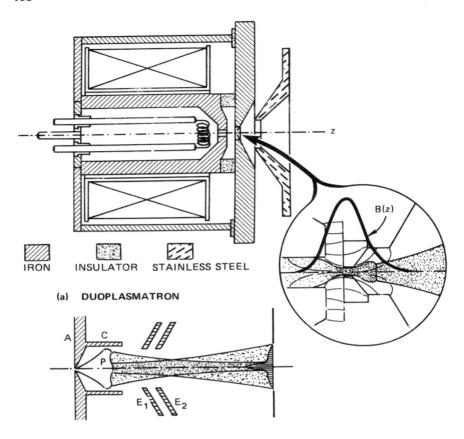

IRON   INSULATOR   STAINLESS STEEL

(a)   DUOPLASMATRON

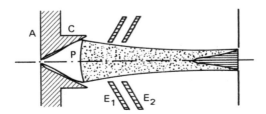

**(b)   FORMATION OF THE "HALO BEAM"
BY A HIGH AXIAL MAGNETIC FIELD**

**(c)   ELIMINATION OF HALO BEAM
BY USE OF CONICAL ELEMENT**

FIGURE 2.21. Beam extraction from plasmatrons.

where ϕ_i is the ionization energy. For larger molecules the calculations are more complex but are assumed to take on the same exponential form.

Since ionization energies of atoms and molecules are usually higher than the work functions of solid surfaces, the field required to extract the electron is usually an order of magnitude greater than for field emission from solid surfaces. Typically, fields of about 10^8 V/cm or 1 V/Å are required for field ionization. Such fields can be simply generated around sharp points of radius about 1000-Å. Since ionization occurs only an angstrom or so from the surface, the energy with which the electron strikes the tip is always small. The ions, on the other hand, are accelerated radially from a region close to the point, which forms a tiny virtual source. At very high field strengths the ion current i produced in a small volume element close to the tip is given by

$$i = e\dot{n}$$

where $\dot{n}$ is the number of particles arriving in the volume element. If A_0 is the area of the tip where the field is high, from kinetic theory we obtain

$$\dot{n}_0 = A_0 \frac{P}{(2\pi mkT)^{1/2}}$$

However, because of the high field gradient around the point, molecules are drawn toward it from a relatively large distance, where the fields are sufficient to polarize the molecule. This gives rise to a much larger effective area of capture A given by

$$A/A_0 = 1 - \frac{2}{3}\frac{V_0}{kT}$$

where $-V_0 = \mu E_0 + \frac{1}{2}\alpha E_0^2$ is the potential energy of the particles just outside the tip, of average permanent dipole moment μ in the field direction and of polarizability α, and E_0 is the field at the tip. The effective area of capture can exceed that of the tip by factors of 10–100, giving rise to correspondingly higher field-ionization currents.

For ion-beam lithography and microscopy, where a high brightness is required, the use of field-ionization (FI) sources is very attractive. With simple optical systems a resolution <1000 Å can be achieved with monatomic gases (H, He, Ar, Xe).[39]

An FI source can generate a total current of 10^{-8}–10^{-7} A, with a virtual source size of 10 Å and brightness

$$\beta \approx 10^8 \, \frac{\mathrm{A}}{\mathrm{cm}^2 \cdot \mathrm{sr}}$$

Figure 2.22 shows a comparison of currents as a function of beam diameter for two FI sources and a duoplasmatron source.[39] It can be seen from these curves that the FI source can provide more current in spots smaller than 2000 Å.

The source-extractor resolution for FI systems is limited by chromatic aberration due to the wide energy spectrum (4–15 V) of the extracted ions.

The ultimate limit of the emitted current density is set by the kinetic supply of ionizable gaseous particles to the high-field (>2.2 V/Å) region of the conically shaped emitter.[40] This limit can be removed if a liquid-film reservoir of the ionizable material is formed on the high-field region of the emitter.

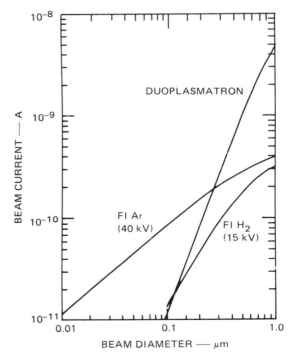

FIGURE 2.22. Limits on ion beam current as a function of beam diameter for field ionization (FI) and duoplasmatron sources.

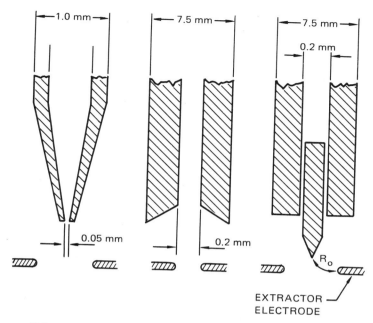

FIGURE 2.23. Liquid-metal ion source configurations. Source: ref. 44.

2.4.5. *Liquid Field-Ionization Sources*

A liquid-metal ion source[41-44] typically consists of a fine tungsten capillary (0.02–0.002 cm in diameter) or of a capillary with a tungsten needle projecting through it and an extraction electrode. As shown in Figure 2.23,[37,18] the operation of the source involves feeding liquid metal to the tip of the capillary tube and applying a voltage between the tip and the extractor electrode. The interaction of the electrostatic and surface-tension forces causes the liquid-metal meniscus to form a sharply peaked cone of small radius (Taylor cone).[45] The application of the critical Taylor voltage on a liquid-metal cone extracts ions or electrons, depending on the polarity. This critical cone-forming voltage is given by

$$V_c = 1.43 \times 10^3 \gamma^{1/2} R_0^{1/2} \tag{2.66}$$

where R_0 is the electrode spacing (cm) and γ is the surface tension (dyn/cm). To get some feeling for V_c let us assume that $R_0 \sim 0.1$ cm, and $\gamma = 700$ for liquid Ga; then we find that $V_c = 11$ kV.

It was shown by Taylor[45] that at the critical voltage, where the onset of cone formation occurs, the cone half-angle is 49.3°. As the cone forms the apex radius decreases sufficiently to enable field emission. If a negative field is applied to the emitter, field electron emission occurs when the local apex field reaches typical field-emission values (0.1–0.5 V/Å), while field desorption and ionization occur when the polarity is reversed and the field reaches values >1 V/Å. At higher voltages considerable heating takes place, resulting in the vaporization of the liquid metal followed by field ionization of the thermally evaporated atoms.[43]

Without a negative feedback mechanism the liquid-cone radius r_a (ion source size) would decrease without obvious limit with increasing apex field. The apex field (E_a) is given by[41]

$$E_a \approx 3322 \left(\frac{\gamma}{r_a} \right)^{1/2} \text{ V/cm} \qquad (2.67)$$

The formation of Ga ions via field desorption requires a field of 1.6×10^8 V/cm. With this E_a and $\gamma = 700$ dyn/cm Eq. (2.67) gives an r_a value of 30 Å.

The ion emission is extremely stable owing to space-charge effects in the emission process, even at low currents (<1 μA).[43] In the case of electron emission from liquid Ga cathode current pulses as high as 250 A with a 2- to 3-ns rise time have been observed at a repetition rate of 40–8000 pulses/s.[18] The mechanism involves the formation of a field-stabilized zone of the liquid cathode, which forms a sufficiently small apex radius (r_a) that a regenerative field electron current initiates an explosive emission process. Runaway is not observed for ion emission because there is always sufficient negative feedback present in all regimes of ion emission. This is provided by space charge and by the fact that a decrease in apex field leads to an increase in apex radius, hence a further decrease in the field and thus decreases in emission, which in turn reduce the space charge, thus leading to an increase in the field, and so on. Space charge is effective in ion emission but not in electron emission because the primary charge carriers in ion generation are massive and the counter-charge carriers (which reduce space charge) are light, whereas the opposite situation occurs in electron emission. Typical onset currents in liquid-ion sources are 10^{-5} A, which is a factor of 10^3 times as large as the equivalent current of the mass flow or the supply of liquid in the capillary.

Thus the conclusion is that, in contrast to the case of gas FI sources, for liquid-ion sources the atom supply is adequate for the formation of a space-charge-stabilized ion current. Probe diameters obtained from liquid-ion sources are limited by the chromatic aberration at the extractor lens. Spherical aberration is negligible in the region of half-angles between 1 and 6 mrad. Figure 2.24(a) shows[46] the beam and aberration disk diameter as functions of the lens half-angle, where $V_{ext} = 57$ kV, $C_c = 33$ mm, $C_s = 396$ mm, and $\Delta U = 10, 14,$ and 18 eV.

The liquid-metal source represents a major breakthrough in high-brightness ion-source technology. Its superior performance can be used to form focused ion beams with probe sizes and currents comparable to those used for scanning electron microscopy, as seen in Figure 2.24(b). The outlook for application to microcircuit fabrication will depend on the ability to generate other desired ion species (e.g., dopant ions) with liquid-metal sources. For diagnostics the impact will depend largely on the fundamental focusing limit. Extrapolating the probe size down to 100 Å or below by decreasing the lens half-angle is tempting, but perhaps a limiting virtual source size that is independent of α_0 will be reached. Ion-source performance parameters of greatest interest in microfabrication applications are compared in Table 3.

2.4.6. Space-Charge Effects

Whenever one attempts to obtain large currents (electron or ion) in a limited space, the mutual repulsion of the charged particles becomes evident and places definite limits on the particle densities obtainable. In diodes the space charge limits the current density that can be extracted with a given anode

TABLE 3. Comparison of Ion-Source Performance Parameters

Source	Total Current (A)	Current Density (A/cm²)	Energy Spread (eV)	Brightness (A/cm²·sr⁻¹)	Angular Brightness (A/sr)	Pressure (Torr)
uoplasmatron	10^{-3}–10^{-1}	10^{-2}–1	10	$<10^3$	—	10^{-2}–10^{-1}
f	10^{-4}–10^{-2}	10^{-3}–10^{-1}	30–500	—	—	10^{-4}–10^{-3}
enning (cold cathode)	10^{-5}–10^{-3}	10^{-4}–10^{-1}	50	—	—	10^{-1}–1
ot cathode, electron impact	10^{-3}	$\sim 10^{-2}$	10–50	—	—	10^{-4}–10^{-3}
eld ion	10^{-9}–10^{-8}	10^{-3}	2–5	10^5	1.5×10^{-6}	$\sim 10^{-8}$
quid-metal, field ion	10^{-8}–10^{-4}	$\sim 10^5$	5–4	10^7	10^{-4}	10^{-7}

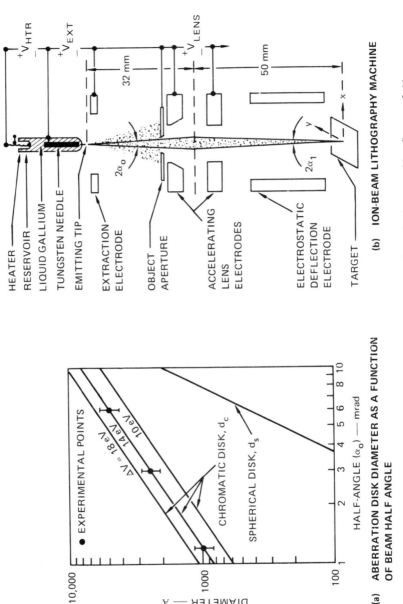

(a) **ABERRATION DISK DIAMETER AS A FUNCTION OF BEAM HALF ANGLE**

(b) **ION-BEAM LITHOGRAPHY MACHINE**

FIGURE 2.24. Limits on the ion-beam spot as a function of α_0 for an ion-beam machine. Source: ref. 46.

potential from both electron and ion sources and formed into beams. Space charge also limits the minimum spot size attainable in optical systems and the maximum current that can be passed through a drift tube. The best-known example of the influence of space charge is the limitation of the current density that can cross a simple infinite parallel-plane diode. In this case the maximum current density that can cross the diode with an electrode spacing d is given by Child's law:

$$j = \frac{4\epsilon_0}{9} \left(\frac{2\epsilon}{m} \right)^{1/2} \frac{V_a^{3/2}}{d^2} \tag{2.68}$$

where V_a is the potential difference between the anode and cathode. This solution, which comes from a direct integration of the one-dimensional Poisson equation, also gives the potential and the space-charge distributions as

$$V = V_a \, (x/d)^{4/3} \tag{2.69}$$

and

$$-\rho(x) = \frac{4\epsilon_0}{9} \left(\frac{V_a}{d^{4/3}} \right) x^{-2/3} \tag{2.70}$$

Equation (2.69) shows that the potential is lowered by the charge below what it would be (linear in x/d) were there no space charge. Near the cathode surface the potential depression is just sufficient to cancel any normal field there, clearly a stable self-limiting situation.

The $\frac{3}{2}$ power dependence of j on V_a [Eq. (2.68)] may be shown to be generally true for any geometry in which the off-cathode gradient vanishes because of space charge. The ratio $i/V_a^{3/2}$ (perveance) is a geometric factor roughly analogous to conductance. Similar results are available for coaxial and cylindrical diodes.[47] These solutions, along with the parallel-plane case, are especially important in electron- and ion-beam work as the starting point in Pierce's method of high-current electron and ion gun design.

In the infinite-plane parallel diode the potential lowering resulting from space charge is in the direction of electron flow. However, in general, a potential profile transverse to the direction of electron flow can exist. The most important case of this sort is the dense charged-particle beam inside the fixed-potential drift tube. Here the space charge results in a radial potential profile, with the beam axis at a lower potential than at the edges. In both the longitudinal and transversal cases we find that the space can support only a certain amount of current under given voltage conditions.

Consider a long charged-particle beam (electron or ion) carrying a current i, with the particles moving in the $+z$ direction with a constant velocity given by

$$\dot{z} = \left(\frac{2q}{m}\right)^{1/2} V \tag{2.71}$$

The radial force giving rise to the spreading in a nonrelativistic case ($\dot{z} \ll c$) comes from the charge enclosed in a cylindrical volume (see Figure 2.25). Using Gauss' theorem,

$$2\pi r \, l \, E_r + \pi r^2 \int_0^l \frac{\partial E_z}{\partial z} \, dz = \pi r^2 \, l \frac{\rho}{\epsilon_0} \tag{2.72}$$

where $\rho = nq$ is the space-charge density. Since in the drift space $E_z = 0$, the second term on the left-hand side of Eq. (2.72) is equal to zero. Then the space-charge density can be expressed as

$$\rho = \frac{j}{\dot{z}} = \frac{i}{\pi r_0^2 [(2q/m) V]^{1/2}} \tag{2.73}$$

and the radial component of the electric field intensity is

$$E_r = \frac{ir}{2\pi\epsilon_0 [(2q/m) V]^{1/2} r_0^2} \tag{2.74}$$

This expression shows that the radial component of the electric field produced by the space charge inside the beam is directly proportional to the radius r, that is, to the distance from the axis.

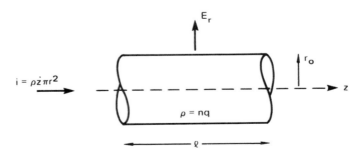

FIGURE 2.25. Charge enclosed in a cylindrical volume.

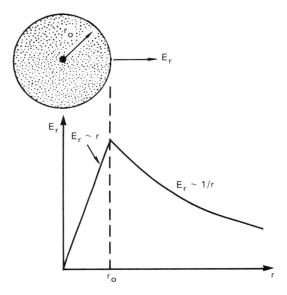

FIGURE 2.26. Currents and fields in space-charge-limited beams.

At the boundary of the beam ($r = r_0$) the radial component of the field is

$$E_r(r = r_0) = \frac{i}{2\pi\,\epsilon_0[(2q/m)\,V]^{1/2}r_0} \tag{2.75}$$

From this formula it is evident that at a constant current the radial field intensity increases with decreasing beam radius. In the case of laminar particle flow this means that the beam does not intersect the axis (no real crossover).

Equation (2.74) can also be used for the calculation of the field near the outer boundary of the beam ($r > r_0$) (see Figure 2.26). We get

$$E_r(r > r_0)\ \frac{i}{2\pi\,\epsilon_0[(2q/m)\,V]^{1/2}r} \tag{2.76}$$

Now the envelope of the beam can be found from the equation of radial motion, $m\ddot{r} = qE_r$. By using the radial space-charge field at the beam edge and $\eta = q/m$, the equation of motion becomes

$$\frac{d^2 r}{dt^2} = \eta \frac{i}{2\pi \epsilon_0 r(2\eta V)^{1/2}} \tag{2.77}$$

With the assumption that the axial velocity is constant, the time derivative may be changed to one with respect to z.

$$\frac{d^2 r}{dt^2} = \dot{z}^2 \frac{d^2 r}{dz^2}$$

which on substitution into Eq. (2.77) gives the final ray equation for this problem:

$$\frac{d^2 r}{dz^2} = \frac{1}{4\pi \epsilon_0(2\eta)^{1/2}} \frac{i}{V^{3/2}} \frac{1}{r} \tag{2.78}$$

The first term on the right-hand side is a constant coefficient, which for electrons is given by

$$\kappa_e = 1.52 \times 10^4$$

and for ions

$$\kappa_i = 6.5 \times 10^5 \, M$$

where M represents the atomic weight of the ions in the beam. The second term is the perveance P. Thus Eq. (2.78) can be written as

$$r'' - 1.52 \times 10^4 Pr^{-1} = 0 \qquad \text{for electrons} \tag{2.79}$$
$$r'' - 6.5 \times 10^5 \sqrt{M} \, Pr^{-1} = 0 \qquad \text{for ions}$$

where the primes denote the differentiation with respect to z. The solution of Eq. (2.79) will be simplified by the introduction of dimensionless variables R and Z,

$$R = \frac{r}{r_0}$$
$$Z = \frac{z}{r_0} \tag{2.80}$$

where r_0 is the initial radius of the beam. With Eq. (2.80), Eq. (2.79) reduces to

$$R'' - \frac{\kappa}{R} = 0 \qquad (2.81)$$

where $\kappa = \kappa_e P$ for electrons and $\kappa = \kappa_i PM$ for ion beams.

Equation (2.81) can be solved with the following initial conditions:

$$Z = 0$$
$$R = 1 \qquad (2.82)$$
$$R' = R'_0$$

that is, when a beam enters into a drift space with radius r_0 and it is either converging, diverging, or cylindrical ($R' = 0$). If we multiply Eq. (2.81) by $2R'$ we obtain

$$\frac{d}{dZ}(R')^2 = \frac{2kR'}{R} \qquad (2.83)$$

This equation can be integrated, taking into account the initial conditions in Eq. (2.82) as

$$R' = [2\kappa \ln R + (R'_0)^2]^{1/2} \qquad (2.84)$$

To find the entire beam envelope, $R = f(z)$, this equation must be integrated again.[47]

The beam, if initiated with a convergent slope, will decrease in radius until a certain minimum radius r_{min} is reached, beyond which it will expand again. The beam envelope will be symmetric around this minimum. At the position of minimum beam diameter $R' = 0$ in Eq. (2.84) and the minimum radius of the beam is

$$r_{min} = r_0 e^{-(R'_0)^2/2\kappa} \qquad (2.85)$$

Equation (2.85) shows that in order to achieve a small spot (i.e., a small minimum radius if the target is placed at the position $r = r_{min}$) a low-perveance beam with a high initial convergent slope should be used. However, a strongly convergent beam exhibits greater spherical aberration, which will limit the minimum spot size. Also, for any target location the smallest attainable radius, defined by the beam envelope, lies on the profile of a beam that has previously reached its waist and is expanding. For a minimum spot size, therefore, it is necessary to produce the beam waist a little before the target.[48,49]

In the case of electron beams the beam spread is also influenced by the residual gas pressure. The ionization of gas molecules by the beam produces positive ions that tend to accumulate along the beam axis.[50] This positive ion neutralization of the beam space charge reduces the spreading of the beam, but gives rise to other undesirable effects, such as beam attenuation from scattering, positive ion bombardment of the cathode, and beam oscillations.

In some applications (e.g., ion milling) neutralization of the ion beam is desired. In these machines electrons are injected into the ion beam for space-charge neutralization.

2.5. Components for Electron and Ion Optics
2.5.1. Electrostatic Lenses and Mirrors

The earlier discussion of Snell's law introduced the analogy between electron dynamics and geometrical optics. From that discussion we saw that the quantity analogous to the optical index of refraction is $\sqrt{V}$. Thus, considering the equipotential surfaces as refractive surfaces of an optical medium, we can find the trajectories of the electrons in an electric field by using the laws of geometrical optics. In the following paragraphs we will use electrons as test particles; but, by changing e/m, all of the conclusions are also valid for ions.

In a magnetic field the force acting on a charged particle depends on the magnitude and direction of the particle velocity. Using the analogy between geometrical optics and charged-particle motion, we observe that the presence of a magnetic field creates an anisotropic medium instead of an isotropic medium, as in the case of an electric field.

The refractive index in a magnetic field is expressed as[51-54]

$$n = \frac{e}{m} (\mathbf{A} \cdot \mathbf{s}) \tag{2.86}$$

where $\mathbf{A}$ is the vector potential and $\mathbf{s}$ is a unit vector tangent to the trajectory.

Although the analogy between light and charged-particle optics is close, it is important to bear in mind the distinctions between the propagation of light and the motion of charged particles. First, the energy of charged particles moving in an electric field varies continuously, but the energy of a photon does not change. Second, the refractive index in light optics can change discontinuously at the interface between two media with different refractive indices, but in electron optics the potential and therefore the refractive index varies continuously from point to point. Consequently, the path of a light ray usually consists of straight-line sections, while the trajectory of a charged particle is a smoothly varying curve. The third important difference between light and electron optics is the fact that in light optics the shape of refracting surfaces and the refractive index are not interconnected, but in electron optics the refractive index $(\sqrt{V})$ and the shape of the refracting (equipotential) surfaces are interdependent through the Laplace equation

$$\nabla^2 V = 0 \qquad (2.87)$$

Finally, charged particles interact through Coulomb forces, but there is no interaction between light particles.

On the other hand, despite these differences, the analogy between light and electron optics is very useful for a large set of the practical beam design problems we are concerned with.

A bundle of electron paths passing through a common point near the axis of an axially symmetric field system can be made to pass through another common point (or if the paths diverge, to appear to have passed through such a point) through the action of a relatively limited region of field variation. It seems appropriate to call the first common point the object, the second the image, and the region of fields bringing this about the electron lens. Imperfections in this process are naturally called aberrations. In electron optics the potentials often differ on opposite sides of the lens, corresponding to different media on opposite sides of a light lens. Indeed, there may be a varying potential on one side, corresponding to a medium of continuously varying index of refraction. Here we will first discuss purely electrostatic lenses and then consider the purely magnetic ones.[51,52]

In the following paragraphs we consider only the motion of bundles of electrons that are near the axis of an axially symmetrical system and whose paths make only small angles with the axis. The potential near the axis of a symmemtrical system cannot depend on the angle ϕ and, moreover, must be an even function of r. If we assume a power-series expansion in r for the potential, Laplace's equation enables us to determine the coefficients of the series in terms of the potential and its derivatives on the axis only. Thus we write

$$V(r, z) = V_0(z) + a_2(z)r^2 + a_4(z)r^4 + \cdots \qquad (2.88)$$

where $V_0(z)$ is the potential on the axis.

Laplace's equation in cylindrical coordinates, with no ϕ dependence, is

$$\frac{1}{r}\frac{\partial}{\partial r}\left(r\frac{\partial V}{\partial r}\right) + \frac{\partial^2 V}{\partial z^2} = 0 \tag{2.89}$$

Performing the indicated operations of the first term of the power series, we obtain

$$\frac{1}{r}\frac{\partial}{\partial r}\left(r\frac{\partial V}{\partial r}\right) = 2^2 a_2 + 4^2 a_4 r^2 + 6^2 a_6 r^4 + \cdots \tag{2.90}$$

Similarly, we can find the expression for $\partial^2 V/\partial z^2$, which must equal the expression in Eq. (2.90),

$$-\frac{\partial^2 V}{\partial z^2} = -(V_0'' + a_2'' r^2 + a_4'' r^4 + \cdots) \tag{2.91}$$

where the primes indicate the derivative with respect to z. Comparing these series term by term and solving successively for the coefficients, we finally find that

$$V(r, z) = V_0(z) - V_0''(z)\frac{r^2}{2^2} + V_0^{IV}(z)\frac{r^4}{2^2 4^2} - \cdots \tag{2.92}$$

This series converges quickly for small r and enables us to solve electron-optical problems after we have determined only the potential on the axis.

To take advantage of the convenience of Eq. (2.92) it is necessary to write the equations of motion in a particular form, which we will now derive. It is assumed, of course, that there is no motion in the ϕ direction; all electrons follow paths that lie in r–z planes. In this case the equations of motion are particularly simple:

$$\ddot{r} = -\frac{q}{m}\frac{\partial V}{\partial r} \tag{2.93a}$$

and

$$\ddot{z} = -\frac{q}{m}\frac{\partial V}{\partial z} \tag{2.93b}$$

where q is the charge and m is the mass of the particle. Time can be eliminated and the two equations combined into a single radial trajectory equation as follows:

$$\dot{r} = \frac{dr}{dz}\dot{z} \quad , \quad \ddot{r} = \frac{dr}{dz}\ddot{z} + \frac{d^2 r}{dz^2}\dot{z}^2 \tag{2.94}$$

and substituting from Eqs. (2.93b) and (2.94) into Eq. (2.93a) gives

$$-\frac{q}{m}\frac{\partial V}{\partial r} = -\frac{q}{m}\frac{dr}{dz}\frac{\partial V}{\partial z} + \frac{d^2 r}{dz^2}(\dot{z})^2 \tag{2.95}$$

But by using the energy conservation $\frac{1}{2}m\,(\dot{r}^2 + \dot{z}^2) = -qV$ and with Eq. (2.94), we can write

$$\dot{z}^2 = \frac{-2(q/m)\,V}{1 + (dr/dz)^2} \tag{2.96}$$

Using Eq. (2.96) in Eq. (2.95) and rearranging gives us the general ray equation

$$2V\frac{d^2 r}{dz^2} = \left(\frac{\partial V}{\partial r} - \frac{dr}{dz}\frac{\partial V}{\partial z}\right)\left[1 + \left(\frac{dr}{dz}\right)^2\right] \tag{2.97}$$

This is a nonlinear differential equation for r as a function of z, but it can be made linear by assuming that the paths never make large angles with the axis (paraxial rays). In this case $(dr/dz)^2 \ll 1$, and the equation, now called the paraxial ray equation, becomes

$$2V\frac{d^2 r}{dz^2} = \frac{\partial V}{\partial r} - \frac{dr}{dz}\frac{\partial V}{\partial z} \tag{2.98}$$

It is now possible to substitute the series expansion [Eq. (2.92)] into this equation. Since it has already been assumed that r and dr/dz are small, it is reasonable to approximate V, $\partial V/\partial r$, and $\partial V/\partial z$ by the first terms of only their respective series. Thus

$$V \approx V_0(z)$$
$$\frac{\partial V}{\partial r} \approx -V_0''(z)\frac{r}{2} \tag{2.99}$$
$$\frac{\partial V}{\partial z} \approx V_0'(z)$$

These substitutions lead to the final form of the paraxial ray equation:

$$\frac{d^2 r}{dz^2} + \frac{dr}{dz}\left(\frac{V_0'}{2V_0}\right) + \frac{r}{4}\frac{V_0''}{V_0} = 0 \qquad (2.100)$$

This equation has several significant features. First, it is independent of the ratio q/m. Second, the derivatives V_0' and V_0'' are normalized with respect to V_0; hence it is the field distribution or shape but not its intensity that governs the trajectories. Finally, the equation is unchanged in form if a scale factor is applied to r. It is this fact that allows us to talk of electron lenses, since it implies the same focus ($r = 0$) for all trajectories parallel to the axis and independent of the initial radius. Of course, this fortunate result is the outcome of neglecting all but the first terms of the potential and its axial derivatives. If an extra term is included in the analysis, the various aberrations of such lenses are revealed. These are discussed in Section 2.5.4. The paraxial ray equation can be rewritten in the form

$$\frac{d}{dz}\left(\sqrt{V_0}\frac{dr}{dz}\right) = -\frac{r}{4}\frac{V_0''}{\sqrt{V_0}} \qquad (2.101)$$

And from this, after integration,

$$\sqrt{V_0}\frac{dr}{dz}\bigg|_1^2 = -\frac{1}{4}\int_1^2 \frac{rV_0''}{\sqrt{V_0}}\, dz \qquad (2.102)$$

where 1 and 2 are as yet only arbitrary limit points. The integral will receive contributions to its value only in regions where V_0'' has some finite value. In many cases of interest only a limited range of z contains such values of V_0''. Elsewhere the field $-V_0'$, or the potential V_0 are constant, making $V_0'' = 0$ there. The limits 1 and 2 are chosen at values of z that completely enclose this active lens region, as shown schematically in Figure 2.27.

Equation (2.102) demands only that the active region of the lens be limited between points 1 and 2, but if this requirement is made even more stringent, the equation is easier to solve. The lens must be thin (or weak); that is, its active region is short compared to its focal length, so r cannot change appreciably between points 1 and 2 (although dr/dz does). Now r can be taken out of the integral and we obtain

$$\sqrt{V_2}\left(\frac{dr}{dz}\right)_2 - \sqrt{V_1}\left(\frac{dr}{dz}\right)_1 = -\frac{r}{4}\int_1^2 \frac{V_0''}{\sqrt{V_0}}\, dz \qquad (2.103)$$

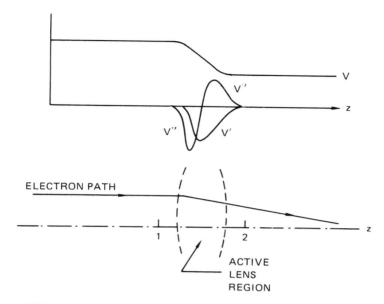

FIGURE 2.27. Electron trajectories and fields in electrostatic lens elements.

If $V_0(z)$ is known explicitly, the integral can be evaluated either analytically or graphically. The focal lengths f_1 for electrons moving parallel to the axis and to the right in region 1, and f_2 for those moving parallel to the axis and to the left in region 2, are found by setting $(dr/dz)_1 = 0$ and $(dr/dz)_2 = 0$ separately (see Figure 2.28). This procedure results in

$$\frac{1}{f_2} = \frac{-(dr/dz)_2}{r_1} = \frac{1}{4\sqrt{V_2}} \int_1^2 \frac{V_0''}{V_0} \, dz \qquad (2.104)$$

and

$$\frac{1}{f_1} = \frac{-(dr/dz)_1}{r_2} = -\frac{1}{4\sqrt{V_1}} \int_1^2 \frac{V_0''}{\sqrt{V_0}} \, dz \qquad (2.105)$$

Thus

$$\frac{f_2}{f_1} = -\frac{\sqrt{V_2}}{\sqrt{V_1}} \qquad (2.106)$$

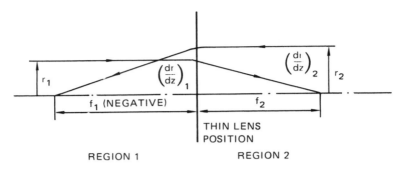

FIGURE 2.28. Definition of the focal points for a thin lens.

which is again analogous to the corresponding law in light optics for a lens between two different media.

If f_2 is positive, then f_1 is negative, but both conditions imply convergence. The sign of the lens depends on the sign of V_0'', which determines the integral, since all other factors are positive. Thus wherever the axial potential curves concavely upward, it is a converging region, and wherever it curves concavely downward, it is diverging.

To obtain more physical insight it is instructive to write $1/f_2$ in a different form. If the integral of Eq. (2.104) is integrated by parts, we obtain two terms contributing to $1/f_2$:

$$\frac{1}{f_2} = \frac{1}{4\sqrt{V_2}}\left(\frac{V_0'(z_2)}{\sqrt{V_2}} - \frac{V_0'(z_1)}{\sqrt{V_1}}\right) + \frac{1}{8\sqrt{V_2}}\int_1^2 \frac{(V_0')^2}{V_0^{3/2}}\, dz \quad (2.107)$$

If the axial fields on both sides of the lens are zero, then only the integral remains. It is clear from the squared factor in the integral that this must indicate a convergent lens. We are thus led to an important and useful result: *All electric lenses that lie between field-free regions are convergent.* Such lenses are formed by the field between the ends of two long coaxial cylinders, as shown in Figure 2.29(a). These lenses, although always converging, are asymmetrical because of the differing potentials on each side. To explain the converging nature of these lenses consider the influence of the longitudinal field component on the particle motion in the case of a symmetrical two-electrode (bipotential) lens, shown in Figure 2.29(a).

In passing from the first half of the lens to the second the sense of the radial field is reversed; if the field had a radial component only, it would have no overall effect. There is, however, a longitudinal component that accelerates

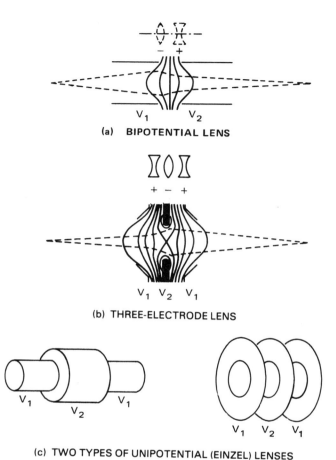

(a) **BIPOTENTIAL LENS**

(b) **THREE-ELECTRODE LENS**

(c) **TWO TYPES OF UNIPOTENTIAL (EINZEL) LENSES**

FIGURE 2.29. Examples of electrostatic lens configurations.

the electrons as they pass through the lens, which as a result are moving faster in the second half of the lens than in the first half. In the first half, therefore, the converging power is increased, and in the second the diverging power is diminished; that is, the whole lens is convergent. In the three-electrode lens the action can be analyzed similarly [see Figure 2.29(b)] into an assembly consisting of a convergent unit between two divergent units.

Two forms of these symmetrical "einzel" or unipotential lenses are shown in Figure 2.29(c). These lenses have the virtue of leaving the beam energy unchanged and may be "inserted" anywhere convenient along the beam.

Regardless of whether $V_2 > V_1$ or $V_2 < V_1$, the lens is convergent and its power is conveniently adjusted by varying V_2. In practice, V_2 is usually less than V_1 because a greater lens power can be achieved with smaller changes in voltage.

When either of the gradients $V_0'(z_2)$ or $V_0'(z_2)$ are not zero, the first term of Eq. (2.103) dominates; then, if the lens is thin, $V_1 \sim V_2$:

$$\frac{1}{f_2} \approx \frac{1}{4V_2} [V_0'(z_2) - V_0'(z_1)] \tag{2.108}$$

This well-known expression applies to the lens action of a round aperture in a plate with different longitudinal fields on either side. Such apertured plates, or their equivalent, are virtually unavoidable in electron guns in which electrons are accelerated to an anode through which they must pass. The lens action of these anodes must be taken into account by a suitable application of Eq. (2.108).

In the following paragraphs we summarize the optical properties of three types of electrostatic lenses: aperture lens, the bipotential (immersion) lens, and the unipotential lens.

The aperture lens [Figure 2.30(a)] is formed by a disk-shaped electrode with a circular aperture at a potential V_d. Fields with constant but different potentials, E_1 and E_2, act on both sides of the electrode. In some particular cases, one of these fields may be absent (e.g., E_1 or E_2 is along the axis and $V_0'' \neq 0$).

The optical power of an aperture lens can be easily calculated with some approximation. Since the potential near the aperture varies insignificantly, we may in a first approximation transfer the term V_0 in Eq. (2.104) from under the integral sign. In this case the aperture lens power is calculated by the following approximate formula:

$$\frac{1}{f} \approx \frac{1}{4V_0} [V']_1^2 = \frac{1}{4V_d} (V_2' - V_1') \tag{2.109}$$

where V_1' and V_2' are potential gradients on both sides of the aperture lens. Taking into account that $V' = -E_2$, Eq. (2.109) may be conveniently rewritten in the form

$$\frac{1}{f} = \frac{|E_2| - |E_1|}{4V_d} \tag{2.110}$$

where E_1 and E_2 are the field intensities on the two sides of the aperture.

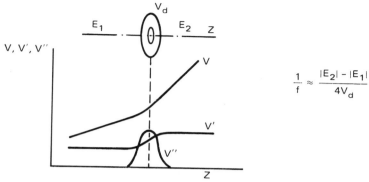

$$\frac{1}{f} \approx \frac{|E_2| - |E_1|}{4V_d}$$

(a) APERTURE LENS

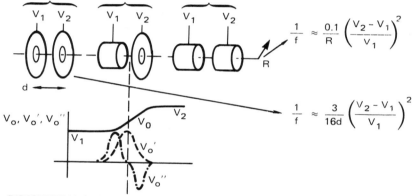

$$\frac{1}{f} \approx \frac{0.1}{R}\left(\frac{V_2 - V_1}{V_1}\right)^2$$

$$\frac{1}{f} \approx \frac{3}{16d}\left(\frac{V_2 - V_1}{V_1}\right)^2$$

(b) BIPOTENTIAL LENSES

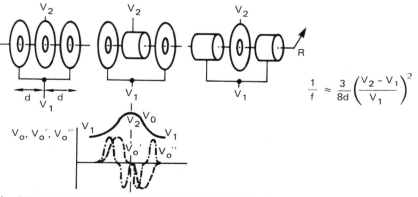

$$\frac{1}{f} \approx \frac{3}{8d}\left(\frac{V_2 - V_1}{V_1}\right)^2$$

(c) SYMMETRIC UNIPOTENTIAL LENS CONFIGURATION

FIGURE 2.30. Lens and field configurations.

Depending on the absolute values of the field intensities, the aperture lens can be converging (positive) or diverging (negative). In the case where an electron passes from the domain with a lower intensity to the domain of a higher intensity of the field, the lens is converging. The direction of the force acting on the electron in the lens domain is determined by the sign of the second derivative V''. The aperture lens has a limited application as a self-contained focusing device, but it is often used as a component in electron-optical systems.

The bipotential lens is usually formed by two coaxial apertures, an aperture and a cylinder, or two coaxial cylinders at different potentials [Figure 2.30(b)]. The diagrams of the variations of the potentials as well as their first and second derivatives are shown in Figure 2.30(b). When the electrons pass through a bipotential lens, their velocities vary; at $V_2 > V_1$ the lens is accelerating and at $V_1 > V_2$ it is decelerating.

From Figure 2.30(b) it is clear that the bipotential lens has zones with a positive and a negative second derivative of the potential $V_0''(z)$; that is, it has converging and diverging portions of the field. However, the particle velocity determined by the potential value is higher in that zone where $V_0''(z) < 0$; that is, the diverging region is passed by the electron in a shorter time, so that an impulse imposed upon this electron radially towards the axis turns out to be greater than that directed from the axis. Thus the converging action is predominant.

The focal lengths of the bipotential lenses cannot be calculated from the paraxial equation for every geometry; however, the focal length of the weak immersion lens formed by two apertures is given by the expression

$$\frac{1}{f} \approx \frac{3}{16d} \left(\frac{V_2 - V_1}{V_1} \right)^2 \tag{2.111}$$

where d is the distance between the apertures. This expression is valid when

$$\frac{V_2 - V_1}{V_1} \leq 0.2$$

From Eq. (2.111) it is evident that the bipotential aperture lens is always converging, regardless of the sign of the potential difference $V_2 - V_1$.

The approximate expression for the focal length of a bipotential lens formed by two cylinders of the same radius R is given as

$$\frac{1}{f} \approx \frac{k}{R} \left(\frac{V_2 - V_1}{V_1} \right)^2 \tag{2.112}$$

where $k \sim 0.1$.

The given Eqs. (2.111) and (2.112) show that the focal length of the bipotential lenses depends on the potential ratio V_2/V_1 of the electrodes. The optical properties of the lenses are also affected by the geometrical configurations, particularly the radius of the cylinder. A bipotential lens in which the object (usually the electron emitter) is positioned in the field of the lens (objective lens) is widely used in electron-beam machines (electron-beam lithography, electron microscopy).

The unipotential lens [Figure 2.30(c)] is formed by three coaxial electrodes (apertures or cylinders), and outer electrodes being at a common potential V_1 and the center electrode at a different (lower or higher) potential V_2. If the outer electrodes are of the same shape and are located at the same distance from the center electrode, then the field of the lens is symmetrical relative to the midplane of the lens. Such a unipotential lens is called symmetrical. The electrode systems shown in Figure 2.30(c) form symmetrical unipotential lenses. When the outer electrodes are not identical or are unequally spaced from the center electrode, the unipotential lens is asymmetrical. The unipotential lens is characterized by the equality of the potentials on the outer electrodes. As a result the energy of electrons passing through this lens does not change, it is the velocity vector direction that changes. The equality of the potentials on the outer electrodes results in the electric fields on either side of the midplane (plane of symmetry) of the lens acting in opposite directions. For a weak symmetrical unipotential lens

$$\frac{V_2 - V_1}{V_1} \ll 1$$

The approximate expression for the focal length can be written as

$$\frac{1}{f} \approx \frac{3}{8d}\left(\frac{V_2 - V_1}{V_1}\right)^2 \tag{2.113}$$

In the case of asymmetrical unipotential lenses the focal lengths have to be calculated numerically, since analytical expressions are not known.

The unipotential lenses can be used with both $V_1 > V_2$ and $V_2 > V_1$. However, the selection of $V_1 > V_2$ is more practical because the lens has lower aberrations and simpler voltage sources.

Unipotential electrostatic lenses with electron-transparent foils as outer electrodes have interesting optical properties, including negative values for both their focal lengths and a third-order spherical aberration coefficient.[55]

More complicated electrostatic electron-optical systems formed by several electrodes can be regarded as a combination of aperture lenses and bipotential

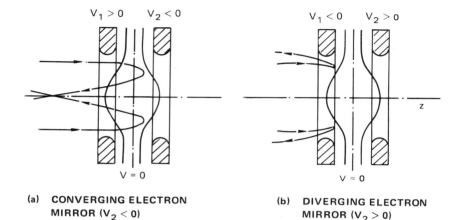

FIGURE 2.31. Bipotential lenses used as electron mirrors.

or unipotential lenses. The total optical power of a complex electron-optical system can be approximated by the light-optical formula

$$\frac{1}{f_{\Sigma}} = \frac{1}{f_1} + \frac{1}{f_2} + \cdots + \frac{1}{f_n} \qquad (2.114)$$

where $f_1, f_2, \ldots, f_n$ are the focal lengths of the lenses forming the electron-optical system.

The bipotential and unipotential electrostatic lenses with their electrodes at certain potential differences can be used for reflecting the electron flow; that is, they can be converted into electron mirrors [Figures 2.31(a) and (b)]. Depending on the sign of curvature of the equipotential surfaces in the reflecting area, the mirror can be either converging (concave) or diverging (convex). Shown in Figure 2.31(a) is a bipotential lens with the right-hand electrode at a potential $V_2 < 0$. Of course, the electron beam near the reflecting surface cannot be regarded as paraxial, because the angles formed by the electron trajectories relative to the axis in the reflection area cannot be considered small, even as a rough approximation. Therefore, when determining the optical parameters of electron mirrors, the equations of paraxial optics may be used only for a preliminary estimation. Furthermore, electron mirrors have considerable chromatic aberration.

A unipotential lens with the central electrode potential at $V_2 < 0$ can also be used as an electron mirror. Varying this potential makes it possible to change the sign of the optical power of the mirror (to obtain converging and diverging mirrors) and control the optical power within a wide range. When the central electrode potential of a unipotential lens is reduced, the area having a potential below zero and reflecting the electrons is first formed near the electrode itself, while the potential along the lens axis remains positive. In this case the lens serves as a lens for near-axis electrons and changes into a mirror for peripheral electrons. This property of a unipotential lens can be used to control the intensity of the electron flow passing through the lens. By varying the potential of the central electrode from zero to a negative value such that the equipotential $V = 0$ lies at the saddle point of the field, it is possible to smoothly vary the electron-beam current from its maximum to zero. Under these conditions the unipotential lens is similar to the "iris" diaphragm widely used in optical devices.

2.5.2. Magnetic Lenses

In addition to electrostatic lenses, electron-beam devices widely employ magnetic lenses. These lenses consist of a limited region of axially symmetric magnetic fields, usually produced by ring-shaped magnets, as shown in Figure 2.32. As the following brief analysis shows, this arrangement has useful and attractive lens characteristics. The equation of motion for a charged particle of

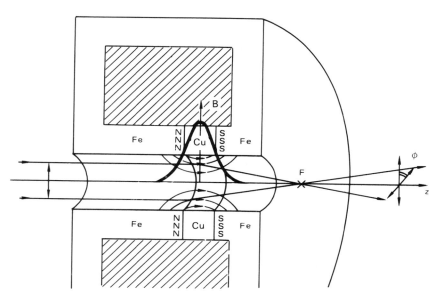

FIGURE 2.32. Magnetic lens configuration.

mass m and charge q in cylindrical coordinates in an axially symmetric magnetic field are

$$\ddot{r} = r\dot{\phi}^2 - \frac{q}{m}(B_\phi \dot{z} - B_z r\dot{\phi})$$

$$\ddot{z} = -\frac{q}{m}(B_r r\dot{\phi} - B_\phi \dot{r}) \tag{2.115}$$

$$r\ddot{\phi} + 2\dot{r}\dot{\phi} = \frac{q}{m}(\dot{r}B_z - \dot{z}B_r)$$

where the dot notation refers to total time derivatives. The last equation may be rewritten as

$$\frac{1}{r}\frac{d}{dt}(\dot{\phi}r^2) = \frac{q}{m}(\dot{r}B_z - \dot{z}B_r) \tag{2.116}$$

The difference between the flux linking a circular ring element of radius r and of width dr (ϕ_2) and the radial flux leaving through the curved surface is given by $\psi_2 - \psi_1 = 2\pi r(B_z\,dr - B_r\,dz)$. If this change of flux occurs in a time dt as a result of the electron motion in a magnetic field of axial symmetry from position 1 to position 2, then

$$\frac{d\psi}{dt} = 2\pi r(\dot{r}B_z - \dot{z}B_r)$$

Substituting into Eq. (2.116) gives

$$\frac{1}{r}\frac{d}{dt}(\dot{\phi}r^2) = \frac{q}{2\pi rm}\dot{\psi}$$

or

$$\frac{d}{dt}\left(\dot{\phi}r^2 - \frac{q}{2\pi m}\psi\right) = 0$$

Integration with respect to time yields

$$m\dot{\phi}r^2 - \frac{q}{2\pi}\psi = \text{const.}$$

or

$$\dot{\phi} = -\frac{q}{m}\left(\frac{\psi_c - \psi}{2\pi r^2}\right) \tag{2.117}$$

where ψ_c is the flux when $\dot{\phi} = 0$. If the magnetic field is confined to the lens region only, then $\psi_c = 0$, and

$$\dot{\phi} = -\frac{q}{m}\frac{\psi}{2\pi r^2} \tag{2.118}$$

This equation (Busch's theorem) tells us that the position of the electron in the initial and final radial planes are different because in the lens $\dot{\phi}$ had a finite value; that is, the image will be rotated away from the object in the ϕ direction.

As in the electrostatic case, we can simplify the equations of motion by using an approximate expression for the magnetic field. Since the axial component of magnetic field B_z must satisfy Laplace's equation, it may be expanded in the same series as was the electrostatic potential, that is,

$$B(z) = B_0(z) - B_0''\frac{r^2}{2^2} + B_0''''\frac{r^4}{2^2 4^2} - \cdots \tag{2.119}$$

But the magnetic flux density B is a vector and it is necessary to include its other components. We assume that $B_\phi = 0$, and B_r may be determined in terms of B_z by means of the divergence equation, that is,

$$\frac{1}{r}\frac{\partial}{\partial r}(rB_r) + \frac{\partial B_z}{\partial z} = 0 \tag{2.120}$$

Substitution of Eq. (2.119) into Eq. (2.120) and the solution for B_r give

$$B_r = -B_0'\frac{r}{2} + B_0'''\frac{r^3}{2^2 \cdot 4} - V_0''''\frac{r^5}{2^2 \cdot 4^2 \cdot 6} + \cdots \tag{2.121}$$

Thus the complete magnetic field may be represented in terms of the axial component and its axial derivatives.

Neglecting all terms higher than the first power in r in Eqs. (2.119) and (2.121) therefore gives

$$B_z \approx B_0(z) \tag{2.122}$$

$$B_r \approx -B_0'(z)\frac{r}{2}$$

Putting these values into the equations of motion [Eq. (2.115)] Gives

$$\ddot{r} = r\dot{\phi}^2 + \frac{q}{m} B_0 r\dot{\phi}$$

$$\ddot{z} = \frac{q}{m} B_0' \frac{r}{2} r\dot{\phi} \tag{2.123}$$

$$\dot{\phi} = -\frac{q}{m} \frac{B_0}{2}$$

Substituting the latter expression for $\dot{\phi}$ in the first two equations, we obtain

$$\ddot{r} = -\left(\frac{q}{m}\right)^2 \frac{B_0^2}{4} r \tag{2.124}$$

For paraxial rays r is small so that $\ddot{z} \cong 0$. Note that the radial acceleration is linear in r, a necessary condition for image formation. Now

$$\ddot{r} = \frac{d^2 r}{dz^2} \dot{z}^2 = -\left(\frac{q}{m}\right)^2 \frac{B_0^2}{4} r$$

[from Eq. (2.124)] and

$$\dot{z}^2 = 2\left(\frac{q}{m}\right) V$$

where V is the electric potential energy referred to the cathode. Thus

$$\frac{d^2 r}{dz^2} = \frac{(q/m) r B_0^2}{8V} \tag{2.125}$$

This is the paraxial ray equation for the axially symmetric magnetic fields. In contrast to its counterpart for electric fields, this paraxial equation does contain the ratio q/m, so the magnetic-lens effect does depend on the particles involved. If B_0 is reversed in sign, the equation is unchanged; thus these lenses are symmetrical. This is consistent with the constant potential throughout the system, which implies the same index of refraction on the two sides of the lens. The important feature of Eq. (2.125), however, is its invariance to the change of scale in r, which is an essential property for image formation. Again, this invariance appears only because we could approximate the fields near the axis to make the right-hand side linear in r.

The focal length is found as before by integrating once with respect to z between points 1 and 2, completely enclosing the active lens region:

$$\left(\frac{dr}{dz}\right)_2 - \left(\frac{dr}{dz}\right)_1 = \int_1^2 \frac{q}{m}\frac{B_0^2 r}{8V}\,dz \qquad (2.126)$$

If the lens is sufficiently thin, r may be taken outside the integral. The focal length f_2 is then found by setting $(dr/dz)_1 = 0$. Thus

$$\frac{1}{f_2} = \frac{-(dr/dz)_2}{r_1}\bigg|_{|dr/dz|_1=0} = \frac{-q/m}{8V}\int_1^2 B_0^2\,dz \qquad (2.127)$$

Clearly f_2 is always positive, so the lens is always converging; and because of the symmetry already noted, f_1 equals $-f_2$. From Eq. (2.123)

$$\dot{\phi} = -\frac{q}{m}\frac{B_0}{2}$$

hence

$$\frac{d\phi}{dz} = -\frac{q}{m}\frac{B_0}{2\dot{z}} = -\frac{q}{2m}\frac{B_0}{[2(q/m)V]^{1/2}}$$

Thus the image rotation is given by

$$\phi_2 - \phi_1 = \left(\frac{q}{8mV}\right)^{1/2}\int_1^2 B_0\,dz \qquad (2.128)$$

The axial magnetic field of a short circular coil carrying a current i and having n turns and a mean radius R_m is given by

$$B(z) = \frac{\mu_0 R_m ni}{2(R_m^2 + z^2)^{3/2}} \qquad (2.129)$$

Thus, using equation Eq. (2.127), the focal length of a thin magnetic lens is given by

$$f \approx 98\frac{VR_m}{(ni)^2}\text{ cm} \qquad (2.130)$$

and the angle of image turn is

$$\Delta\phi = \phi_2 - \phi_1 = 10.7 \frac{ni}{\sqrt{V}} \text{ degrees} \qquad (2.131)$$

The values of V, i, and R in Eqs. (2.130) and (2.131) should be in volts, amperes, and centimeters, respectively, to obtain f in centimeters and $\Delta\phi$ in angular degrees.

2.5.3. Energy-Analyzing and Selecting Devices

Energy analysis of charged-particle beams is useful in two classes of instruments used in manufacturing, namely, electron microscopes and surface-analysis equipment [Auger spectroscopy and low energy electron diffraction (LEED)].

In energy-analyzing microscopes the image is formed by electrons with a selected energy. The spectral width (0.1–1 eV) is given by the product of the dispersion of the analyzer and the width of the selecting aperture. The energy-selecting microscope produces the electron-optical analogue of a monochromatic photograph.

The following analyzers are used in energy-selecting electron microscopes:[56]

- The cylindrical electrostatic analyzer
- The cylindrical magnetic analyzer
- The electrostatic mirror magnetic prism

The cross-sectional view of a cylindrical electrostatic analyzer is shown in Figure 2.33(a). A fine slit S a few microns wide and about 1 cm long apertures the incoming charged-particle beam. The apertured particles pass through regions of high-energy dispersion near the two electrodes C, which are biased at a potential near the cathode (or ion source). The energy spectrum is recorded on the exit side of the lens. At high accelerating voltages the problem of electrical breakdown limits the use of the electrostatic analyzer. A purely magnetic system [shown in Figure 2.33(b)] avoids insulation problems. The magnetic analyzer is the magnetic analogue of the cylindrical electrostatic system discussed above.

The electron-optical properties of a double magnetic prism and an electrostatic mirror are shown in Figure 2.34. Here the prism employs a uniform magnetic field and a concave electrostatic mirror that is biased to the source (cathode) potential.

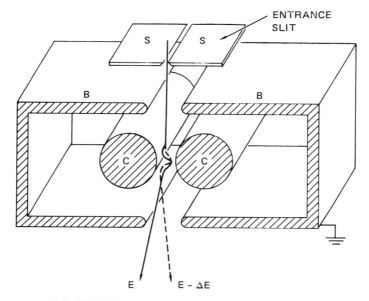

(a) CYLINDRICAL ELECTROSTATIC ENERGY ANALYZER

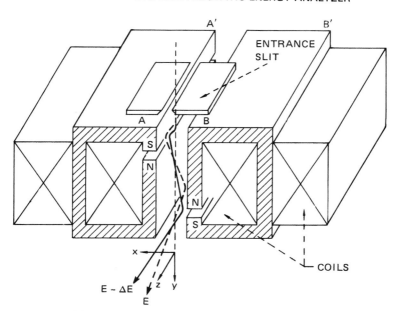

(b) CYLINDRICAL MAGNETIC ENERGY ANALYZER

FIGURE 2.33. Electrostatic and magnetic energy analyzers. Source: ref. 56.

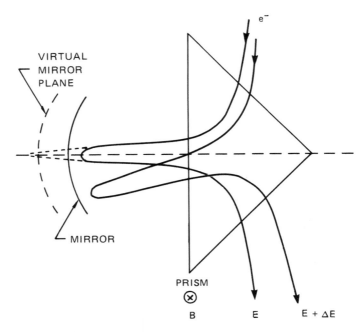

FIGURE 2.34. Behavior of charged particles of energies E and $E + \Delta E$ in traversing a double magnetic prism and electrostatic mirror.

In Auger electron spectroscopy it is useful to measure the derivative of the secondary-electron energy distribution to display the effects more vividly. Energy analysis is obtained by using a retarding-field or cylindrical-mirror analyzer, shown in Figure 2.35.

In a retarding-field analyzer the secondary electrons are collected by a fluorescent screen after passing the grid system [Figure 2.35(a)]. A retarding potential is applied to the two central grids, with a small modulating voltage added. The outer and inner grids are at ground potential. The modulated collector current that contains the information is amplified and demodulated using lock-in amplifier techniques.

The collector current is given by

$$I_{\text{coll}}(E) = I_p \int_E^{E_p} N(E) \, dE$$

where $N(E)$ is the secondary emission distribution, E_p and E are the primary energy and the retarding potential, respectively, and I_p is the primary current.

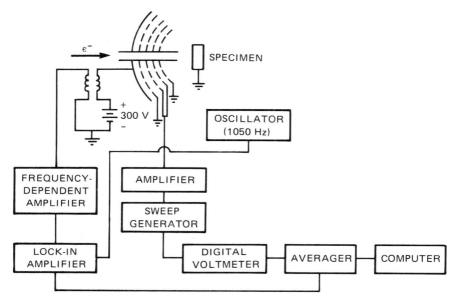

(a)   **RETARDING FIELD ANALYZER**

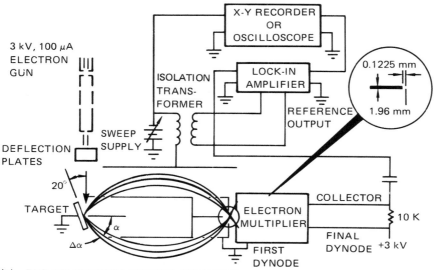

(b)   **THE CYLINDRICAL MIRROR ANALYZER**

FIGURE 2.35. Analyzers for recording the derivative Auger spectra. Source: ref. 57.

It can easily be shown that for a small modulating voltage the first harmonic coefficient in the Taylor expansion of the current is proportional to $N(E)$, the second harmonic coefficient to $dN(E)/dE$, and so on. In the second harmonic mode, the dc component is eliminated and electronic amplification can be considerably increased. The cylindrical-mirror analyzer is a deflection bandpass instrument with high transmission (10%) for moderate resolution (0.3%) [Figure 2.35(b)].[57] Here the first harmonic coefficient is proportional to $E[dN(E)/dE]$ for small modulation voltages. If true energy distribution curves are desired, an integration has to be performed.

2.5.4. Image Aberrations

2.5.4.1. Geometric Aberrations

In the previous derivation of the basic equations of electron (ion) optics we only considered the rays close to the optical axis, that is, paraxial beams. The condition of paraxiality was expressed in the approximation that we used, by taking into account only the first terms of the field expansion series [Eqs. (2.92) and (2.121)]. However, neglecting the subsequent terms in the expansion can result in considerable error. As seen from the radial motion equation in electric fields [Eq. (2.100)] and magnetic fields [Eq. (2.125)], one condition for obtaining ideal images is linearity in r.

The use of subsequent terms in the field expansion results in terms proportional to the third, fifth, and so on, degrees of r or dr/dz in the equation of motion. If only the third-order terms (r^3, $r^2 \, dr/dz$, $r \, d^2 r/dz^2$, and $d^3 r/dz^3$) are taken into account in the equations of motion the errors in image formation are called third-order aberrations.

In image formation both r, the distance between the trajectory and the axis, and dr/dz, the inclination angle relative to the optical axis, can be expressed by r_a, the distance between the object point and the axis, and r_d, the radius of the limiting aperture for the lens (see Figure 2.36). To classify the aberrations for wide beams we assume that a narrow paraxial beam nn is emitted from the point a of an object and creates an undistorted image of the point a in the plane z_b. For the wide beam mm emitted from the point a a certain aberration figure is formed in the plane z_b rather than a point.

Since the aberration figures in the image plane are not circles, it is expedient to use Cartesian coordinates in the object and image planes to estimate the distortion of the image. The origin of these coordinates coincide with the point where the planes intersect the axis, and the aberrations are characterized by the values Δx and Δy of the deviation of the points of the aberration figure from the point of the image formed by the paraxial beam.

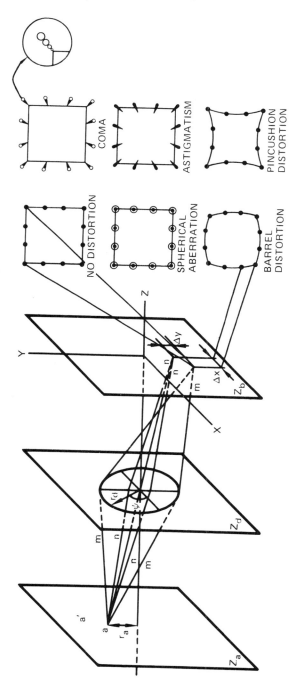

FIGURE 2.36. Image formation and examples of common aberrations.

Since the aperture limiting the beam must be circular, the coordinates of
the aperture edges are preferably expressed in polar coordinates:

$$x_d = r_d \cos \psi, \qquad y_d = r_d \sin \psi$$

With these coordinates the deviations in the image plane from the point of the
image formed by the paraxial beam can be expressed as[52]

$$\Delta x = Br_d^3 \cos \psi + F(2 + \cos 2\psi)\, r_d^2\, x_a + (2C + D) \cos \psi r_d x_a^2 + Ex_a^3$$
$$\Delta y = Br_d^3 \sin \psi + F \sin 2\psi r_d^2 x_a + D \sin \psi r_d x_a^2$$

$$(2.132)$$

where the coefficients B, C, D, E, and F are functions of the axial potential
and its derivatives of the first, second, third, and fourth orders. These coefficients characterize the five geometrical aberrations of the third order.

From Eq. (2.132) it is clear that for points of the object lying on the system axis ($x_a = 0$) all the terms except the first vanish and there remains only
one aberration characterized by the coefficient B. This error is known as the
spherical aberration and is very important for electron-beam devices. In the
presence of spherical aberration the image of a point on the axis is a circle with
radius Δr centered on the axis and given by

$$\Delta r = [(\Delta x)^2 + (\Delta y)^2]^{1/2} = Br_d^3 \qquad (2.133)$$

Since the radius of the circle for spherical aberration is proportional to
r_d^3, this error can be reduced by decreasing the aperture radius, that is, by using
narrower beams. However, a decrease in the aperture radius also decreases the
beam current.

The radius of the limiting aperture is proportional to the tangent of the
aperture angle α. Since for small angles $\tan \alpha \cong \alpha$, Eq. (2.133) is usually
written in the form

$$\Delta r = C_s \alpha^3 \qquad (2.134)$$

where C_s is the spherical aberration coefficient of the lens. C_s can be approximated for unipotential lenses by the empirical formula

$$C_s \approx k \frac{f^3}{R^2} \qquad (2.135)$$

where f is the focal length of the lens, R is the radius of the electrode, and k = 2.5–5 is an empirical constant.

The coefficient F in Eq. (2.132) characterizes the image error called the coma. The aberration figure produced by this error also has the form of a circle but with a center that does not coincide with the point of the image formed by the paraxial beam. Putting all the coefficients in Eq. (2.132) except for F equal to zero and squaring and summing these equations gives

$$(\Delta x + 2Fx_a r_d^2)^2 + (\Delta y)^2 = F^2 x_a^2 r_d^4 \qquad (2.136)$$

Equation (2.136) is an equation of a circle with radius $\Delta r = Fx_a r_d^2$ and with the center at a distance $2Fx_a r_d^2$ from the image point.

Since the aberration figures are formed by the electrons passing through the entire aperture and not only by those close to the electrodes, we have in the image plane a combination of superimposed circles with gradually increasing radii and with centers located at a greater distance from the point of the image created by the paraxial beam.

The coefficients C and D characterize the image error known as astigmatism. Combining the terms with the coefficients C and D in Eq. (2.132), we obtain the ellipse equation

$$\frac{(\Delta x)^2}{[(2C + D)\, r_d x_a^2]^2} + \frac{(\Delta y)^2}{(D r_d x_a^2)^2} = 1 \qquad (2.137)$$

whose center coincides with the point of undistorted image and with the axis values proportional to $r_d x_a^2$. With $D = 0$ the ellipse degenerates into a straight line of length $4C r_d x_a^2$. With $C = 0$ the aberration figure transforms into a circle.

The final geometric error determined by the coefficient E in Eq. (2.132) is called distortion. This aberration is a result of nonlinear scaling, which results in a system magnification that depends on the distance of the object from the axis, $\Delta x = Ex_a^3$. This distortion for larger than Gaussian magnification is called barrel distortion, and for smaller than first-order magnifications, pincushion distortion. The aberration figures corresponding to the geometric image errors for electrostatic lenses are shown in Figure 2.36.

Aberrations are also inherent in magnetic lenses. However, because of the "twisting" action of the magnetic field, the magnetic lenses have additional "anisotropic" aberrations. This group of aberrations includes anisotropic coma, anisotropic astigmatism, and anisotropic distortion.

2.5.4.2. Electronic Aberrations

Geometrical aberrations depend on the fields forming the electron lenses and are independent of the electron-beam parameters. So far we have assumed that all the electrons at a given point in the system have the same energy that is uniquely determined by the potential of this point. This assumption is valid only for electrons entering the systems with zero initial velocity. Real electron beams enter the system with a finite energy distribution. Thus an electron at potential $V(r_1, z)$ will have a kinetic energy given by

$$\frac{mv^2}{2} = e[V(r, z) + \Delta V] \qquad (2.138)$$

where $e\,\Delta V$ is initial kinetic energy of the electron.

When attempting to focus electrons having different initial kinetic energies, there arise additional errors called chromatic aberrations, by analogy to light optics. From the physical point of view chromatic aberrations results from the dependence of the optical parameters of the electron lenses on the electron energy in the beam being focused. This dependency follows directly from Eqs. (2.104) and (2.127).

Consider the trajectories of two electrons, one of which has an initial energy $e\,\Delta V$ and the other zero. If these electrons are emitted from point z_a (see Figure 2.37) and pass through a lens, they will intersect the axis at $z_b + \Delta z$ and z_b. For electrons with larger initial energies the value of Δz is greater. Generally, electrons with nonzero initial energies will create a circle of confusion in the image plane at z_b. The radius of this chromatic aberration circle is equal to

$$r_c = \Delta z \tan \gamma_2 \approx r_d \frac{\Delta z}{z_b} \qquad (2.139)$$

where r_d is the radius of the aperture. Then for the two electrons the least and most energetic in the beam we can use the basic lens laws

$$\frac{1}{z_a} + \frac{1}{z_b} = \frac{1}{f}$$

$$\frac{1}{z_a} + \frac{1}{z_b + \Delta z} = \frac{1}{f + \Delta f} \qquad (2.140)$$

where Δf is the increase in focal length corresponding to the most energetic electron in the beam.

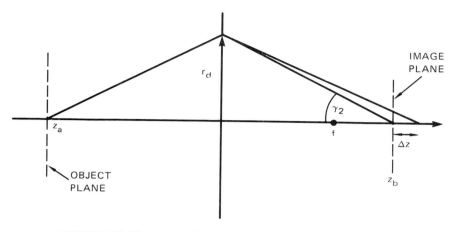

FIGURE 2.37. Electron imaging in a thin lens with a finite energy spread.

Subtracting the first equation of system (2.140) from the second and neglecting Δf as compared to f and Δf as compared z_b, we get

$$\frac{\Delta z}{z_b} = \frac{z_b}{f}\frac{\Delta f}{f} \tag{2.141}$$

From Eqs. (2.141) and (2.139) it is clear that the radius of the confusion circle (disk) for chromatic aberration is proportional to the ratio $\Delta f/f$. Using the formula for the electrostatic lenses we can calculate the focal distances as

$$\frac{1}{f} = \frac{1}{4\sqrt{V}}\int_{-\infty}^{+\infty}\frac{V_0''}{\sqrt{V_0}}\,dz$$

$$\frac{1}{f+\Delta f} = -\frac{1}{4(V+\Delta V)^{1/2}}\int_{-\infty}^{+\infty}\frac{V_0''}{\sqrt{V_0}}\,dz \tag{2.142}$$

Dividing the first equation by the second, we obtain

$$\frac{\Delta f}{f} = \frac{1}{2}\frac{\Delta V}{V} \tag{2.143}$$

Thus, the ratio $\Delta f/f$ is proportional to the ratio of the initial energy to the final energy. Finally, from Eqs. (2.139), (2.141), and (2.143) the radius of the chromatic aberration disk can be expressed as follows:

$$r_c = r_d \frac{z_b}{f} \frac{1}{2} \frac{\Delta V}{V} = C_c \frac{\Delta V}{V} \gamma_2 \qquad (2.144)$$

This expression shows that the chromatic aberration is proportional to the aperture radius r_d. To reduce this aberration the diameter of the aperture should be reduced and the beam energy increased.

Other types of electronic aberrations include space-charge errors in intense electron beams and diffraction errors, which are the consequence of the wavelike properties of electron beams.

2.5.5. Multiple-Focusing Elements

In axially symmetric lenses the electron trajectories are approximately parallel to the field lines; hence these lens elements are referred to as longitudinal systems.[53] In longitudinal systems electron beams are focused by a small (compared to the longitudinal) transversal component of the field. Transversal systems in which the lines of force of the field are directed across the beam are more effective. Transversal electron-optical systems have gained wide use, particularly for focusing high-energy particles (strong focusing). Transversal systems are interesting in that they can be used to design aberration-free focusing systems.

Transverse focusing fields are usually created by four electrodes or four

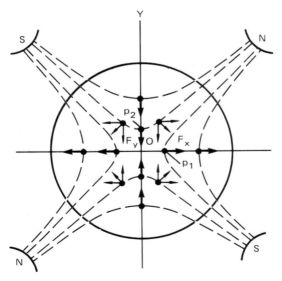

FIGURE 2.38. Pole configuration for a magnetic quadrupole.

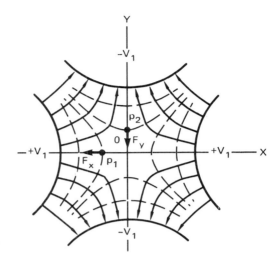

FIGURE 2.39. Electrode configuration for an electrostatic quadrupole lens.

magnetic coils located about the axis of the system, the diametrically opposite electrodes or magnets being of similar polarity and the adjacent ones of different polarity (Figures 2.38 and 2.39). Such four-pole systems having two symmetrical planes are known as quadrupole lenses. A specific feature of the quadrupole lenses is that the axial component of the field is equal to zero.

As an example, consider a quadrupole electrostatic lens formed by four identical electrodes in the form of hyperbolic cylinders located symmetrically and equally spaced from the axis of potentials $\pm V_1$ (Figure 2.39). The potential distribution in such a system is described by the equation

$$V_1(x, y) = -\tfrac{1}{2}k(x^2 - y^2) \tag{2.145}$$

where $k = 2V_1/a^2$ and a is the distance from the optical axis to the electrode surface. With the potential distribution described by Eq. (2.145), the field intensity components E_x and E_y are linear functions of the coordinates:

$$E_x = -\frac{\partial V_1}{\partial x} = -kx$$

$$E_y = -\frac{\partial V_1}{\partial y} = ky \tag{2.146}$$

$$(E_z = 0)$$

that is, the gradients of the field intensity are constant. For this reason the quadrupole lens with hyperbolic electrodes is called a constant-gradient lens.

Consider the motion of an electron entering such a lens parallel to the axis $0Z$. One can easily see that at the point p_1 (Figure 2.39) the electron is acted on by the force $F_x = -eE_x = ekx$ directed from the lens axis, while at the point p_2 it is under the action of the force $F_y = -eE_y = eky$ directed towards the axis. Thus in the plane $X0Z$ the lens is diverging, whereas in the plane $Y0Z$ it is converging. In any point of the field not lying in the symmetry planes the electron is acted on by a force having a component forcing the electron towards the plane $X0Z$ and away from the plane $Y0Z$. The equations of motion in this quadrupole field can be written as

$$m\frac{d^2 x}{dt^2} = -eE_x = \frac{2eV_1}{a^2}x$$

$$m\frac{d^2 y}{dt^2} = -eE_y = -\frac{2eV_1}{a^2}y$$

(2.147)

Since there is no field on the lens axis, it is permissible to consider the longitudinal component of the electron velocity as constant for the region near the axis (paraxial approximation); hence

$$m\left(\frac{2e}{m}V_0\right)^{1/2}\frac{d}{dz}\left[\left(\frac{2e}{m}V_0\right)^{1/2}\frac{dx}{dz}\right] = \frac{2eV_1}{a^2}x$$

$$m\left(\frac{2e}{m}V_0\right)^{1/2}\frac{d}{dz}\left[\left(\frac{2e}{m}V_0\right)^{1/2}\frac{dy}{dz}\right] = -\frac{2eV_1}{a^2}y$$

(2.148)

that is,

$$x'' - k_E^2 x = 0$$

$$y'' + k_E^2 y = 0$$

where $k_E = V_1/a^2 V_0$ and the primes indicate differentiation with respect to z.

For the symmetrical magnetic quadrupole lens (Figure 2.38) we assume that the electrons enter the lens as a beam parallel to the z axis with velocity

$$v_z = \left(\frac{2e}{m}V_0\right)^{1/2}$$

At the point P_1 the electron is affected by a force $F_x = ev_z B_y$ directed from the axis, while at the point P_2 the electron is affected by the force $F_y =$

$-ev_zB_x$ directed towards the lens axis. At any point not lying in the symmetry planes the electron is acted on by a force having a component forcing the electron towards the plane $X0Z$ and from the plane $Y0Z$. Thus the magnetic quadrupole lens will be converging in the plane $Y0Z$ and diverging in the plane $X0Z$.

The equations of motion in a quadrupole magnetic lens are

$$m\frac{d^2 x}{dt^2} = ev_zB_y$$
$$m\frac{d^2 y}{dt^2} = -ev_zB_x$$

(2.149)

The magnetic field components are similar to the axially symmetrical case; that is,

$$B_x(x, y, z) = F(z)y - \tfrac{1}{12}F''(z)(3x^2y + y^3) + \cdots$$
$$B_y(x, y, z) = F(z)x - \tfrac{1}{12}F''(z)(x^3 + 3xy^2) + \cdots$$

(2.150)

where

$$F(z) = \left(\frac{\partial B_x}{\partial y}\right)_0 = \left(\frac{\partial B_y}{\partial x}\right)_0$$

Within the limits of paraxial approximation we use only first terms in Eq. (2.150), and the equations of motion become

$$\frac{d^2 x}{dt^2} = \frac{e}{m}v_z\left(\frac{\partial B_y}{\partial x}\right)_0 x$$
$$\frac{d^2 y}{dt^2} = -\frac{e}{m}v_z\left(\frac{\partial B_y}{\partial x}\right)_0 y$$

(2.151)

Assuming that in the paraxial region

$$v_z = \left(\frac{2e}{m}V_0\right)^{1/2} = \text{const.}$$

we can change the differentiation with respect to t in Eq. (2.151) to differentiation with respect to z, giving

$$\frac{d^2 x}{dz^2} - \left(\frac{e}{2mV_0}\right)^{1/2} \left(\frac{\partial B_y}{\partial x}\right)_0 x = 0$$

$$\frac{d^2 y}{dz^2} + \left(\frac{e}{2mV_0}\right)^{1/2} \left(\frac{\partial B_y}{\partial x}\right)_0 y = 0 \tag{2.152}$$

or

$$x'' - k_M^2 x = 0 \tag{2.153}$$
$$y'' + k_M^2 y = 0$$

where

$$k_M^2 = \left(\frac{e}{2mV_0}\right)^{1/2} \left(\frac{\partial B_y}{\partial x}\right)_0$$

Equations (2.153) are similar in structure to those describing the motion of electrons in an electrostatic quadrupole lens. Therefore in the general case the equations of motion for electrons near the axial region of a symmetrical electrostatic or magnetic lens can be presented in the form

$$x'' - k^2 x = 0 \tag{2.154}$$
$$y'' + k^2 y = 0$$

where $k = k_E$ for the electrostatic lens and $k = k_M$ for the magnetic lens.

If the lens has a length L, the lens field is concentrated between the planes $z = +L/2$ and $z = -L/2$ and abruptly drops down to zero outside this region. In this case we can consider the field independent of z and the motion equations (2.154) have solutions of the form

$$x = A \cosh kz + B \sinh kz \tag{2.155}$$
$$y = C \cos kz + D \sin kz$$

Here the constants A, B, C, and D are determined by the initial conditions. For example, if in the plane $z = -L/2$ (in the "entrance" plane of the lens) $x = x_0$, $y = y_0$, $x' = x_0'$, and $y' = y_0'$, then

$$x = x_0 \cosh k \left(z + \frac{L}{2} \right) + \frac{x_0'}{k} \sinh k \left(z + \frac{L}{2} \right)$$

$$y = y_0 \cos k \left(z + \frac{L}{2} \right) + \frac{y_0'}{k} \sin k \left(z + \frac{L}{2} \right) \tag{2.156}$$

To explore the optical characteristics of the quadrupole lens we consider the path of an electron entering the lens parallel to the axis $0Z$. In this case

$$x_0' = y_0' = 0$$

$$x = x_0 \cosh k \left(z + \frac{L}{2} \right) \tag{2.157}$$

$$y = y_0 \cos k \left(z + \frac{L}{2} \right)$$

The focal lengths are equal to

$$f_x = -\frac{x_0}{x'} \quad \text{and} \quad f_y = -\frac{y_0}{y'} \tag{2.158}$$

If we differentiate Eq. (2.157) and determine x' and y' when $z - L/2$ (in the "exit" plane of the lens), we obtain

$$x' \left(\frac{L}{2} \right) = x_0 k \sinh kL$$

$$y' \left(\frac{L}{2} \right) = -y_0 k \sin kL \tag{2.159}$$

Substituting Eq. (2.159) into Eq. (2.158) then gives

$$f_x = -\frac{1}{k \sinh kL}$$

$$f_y = \frac{1}{k \sin kL} \tag{2.160}$$

The expressions for the focal lengths show that, indeed, in the plane $Y0Z$ the lens is converging ($f_y > 0$, real focus) and in the plane $X0Z$ the lens is diverging ($f_x < 0$, imaginary focus). Thus, upon passing through the quadrupole lens, the electron beam parallel to the axis is deformed in such a way that a line focus is produced in the focal plane of the image space.

In the case of a thin, weak quadrupole lens, $kL \ll 1$, so that $\sin kL \simeq \sinh kL \simeq kL$, and the expressions for the focal lengths are simplified to

$$f_x \approx -\frac{1}{k^2 L}$$

$$f_y \approx \frac{1}{k^2 L}$$

(2.161)

that is, for the thin, weak lenses the focal lengths are equal in magnitude and the principal planes coincide with the center plane of the lens. These equations are valid only for lenses with a constant field gradient. This condition is satisfied when using lens electrodes or pole pieces in the form of hyperbolic cylinders positioned symmetrically around the lens axis. In practice, electrodes and pole pieces having the shape of circular cylinders are usually used, since they are easier to make. To achieve an effectively constant gradient field a configuration in which the radius of the circular cylinder R is in the range $1.1a$–$1.15a$ for the electrostatic lens is required, where a is the distance from the electrode axis. With magnetic lenses the optimum value for R is $1.15a$.

Similarly, the assumption of an abrupt drop of the field at the boundary

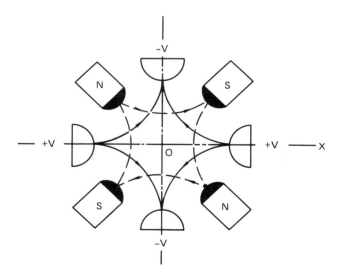

FIGURE 2.40. Combined electric and magnetic quadrupole lens.

of the electrodes or pole pieces is not a realistic approximation. Satisfactory results, however, may be obtained if the value L is regarded as an "effective length" including the stray fields. It has been found experimentally that good agreement of the design data with the experimental data is obtained when the effective length $L = l + 1.1a$, is used, where l is the length of the electrodes or pole pieces and a is the distance from the surface of the electrode (pole) to the axis.

Two quadrupole lenses located along an axis, called a doublet, produce a point image of a point object. Each lens of a doublet produces a linear image of a point object, but a doublet is not identical to an axially symmetric lens because the quadrupole lenses do not have the same magnification in the $0X$ and $0Y$ directions.

Quadrupole lenses have the same axially-symmetric aberrations as the axially symmetric lenses, but in contrast to the other lenses compensation for spherical aberration can be made. It is also possible to build an achromatic combination quadrupole lens with electric and magnetic fields superimposed in one lens (Figure 2.40).

2.5.6. Deflection Systems

In this section we are concerned with the aberrations caused when charged-particle beams are deflected. In many applications (electron microprobes, microscopes) deflection aberrations are of somewhat secondary importance because they limit the field size but not the basic probe size and beam current. Deflection aberrations distort the spot from its ideal round shape and distort the raster scan lines into curves. These subjects are treated in detail in the literature, and a brief discussion is given here.[58,59]

The deflection of a round beam of charged particles by a pair of electrostatic plates is illustrated in Figure 2.41, where the distortions of the scan pattern and of the spot shape are shown. The displacement of the spot on the screen can be considered to be made up of linear terms x and y and error terms Δx and Δy, the latter being functions of the deflection angle, the entrance angle, and the deflection field configuration. The linear deflection term and some insight into the causes of beam distortion can be derived from the first-order deflection equation

$$m \frac{d^2 y}{dt^2} = (-e) E_y$$

$$m \frac{d^2 z}{dt^2} = 0$$

$$(2.162)$$

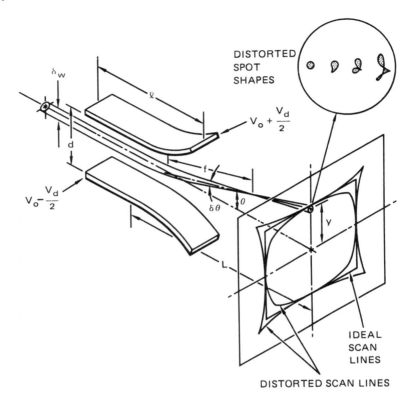

FIGURE 2.41. Deflection distortions from electrostatic plates. Source: ref. 33.

where $E_y = \partial V/\partial y$ and the initial conditions are given by

$$\left(\frac{d_y}{dt}\right)_0 = 0 \quad , \quad \left(\frac{dz}{dt}\right)_0 = v = \left(\frac{2eV_0}{m}\right)^{1/2}$$

$$y_0 = 0 \qquad z_0 = 0$$

By integrating twice Eq. (2.162) we obtain

$$y = \frac{-e}{2m} E_y t^2 \tag{2.163}$$

$$z = vt$$

Elimination of time t from Eq. (2.163) yields the equation of the particle trajectory:

$$y = \frac{-e}{2m} E_y \left(\frac{z}{v}\right)^2 \tag{2.164}$$

This is the equation of a parabola. Now substituting $(2eV_0/m)^{1/2}$ for v we get

$$y = \frac{z^2}{V_0} \frac{V_d}{4d} \tag{2.165}$$

where d is the plate separation distance. The angle θ of deflection is obtained by differentiating Eq. (2.165), which yields

$$\tan \theta = \frac{dy}{dz} = \frac{z}{V_0} \frac{V_d}{2d} = \frac{l}{V_0} \frac{V_d}{2d} \tag{2.166}$$

This is the standard electrostatic deflection equation for charged particles (electrons here) passing between ideal plates. The deflection on a screen a distance L away will be

$$y \simeq L\theta = \frac{V_d}{V_0} \frac{l}{2} \frac{L}{d} \tag{2.167}$$

This simple expression does not show the several optical defects present in a deflected beam. For example, the particles in a beam of finite thickness δw will attain different axial velocities in their transit through the plates. Depending on their transverse position, this effect results in different values of the deflection angle θ. This effect can be evaluated by differentiating Eq. (2.166):

$$\delta\theta = -\theta \frac{\delta V_0}{V_0} \tag{2.168}$$

where

$$\delta V_0 = \frac{V_d}{d} \delta w$$

thus

$$\delta\theta = -\frac{2\theta^2}{l} \delta w \tag{2.169}$$

The electrons closest to the positive electrodes will move faster and will be deflected less; the two rays will therefore cross at a distance f away; that is, the electrostatic deflection process acts as a convergent lens, f, and the focal length is given by

$$f = \frac{\delta w}{\delta \theta} = -\frac{l}{2\theta^2} \tag{2.170}$$

It is seen that the spot diameter at the target will vary with the deflection angle (unless the target is curved). To minimize this effect, the "focal length" f should be made equal to the distance from the deflection plate to the target and the beam diameter should be as small as possible. The deflection plates should always be operated in a "push-pull" mode so that the beam axis remains an equipotential independent of the deflecting voltage. Other types of deflection errors include distortion of the spot shape and of the scan line.

The simplest deflection or scanning system is a set of electrostatic deflection plates as illustrated in Figure 2.42. This type of deflection does not depend

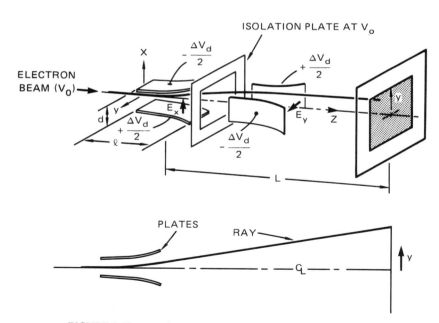

FIGURE 2.42. x–y electrostatic deflection system. Source: ref. 33.

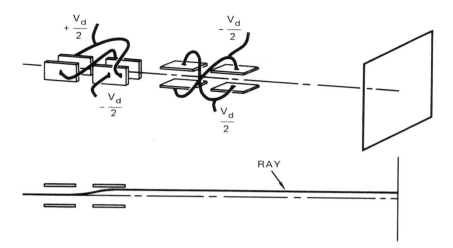

FIGURE 2.43. Double $x-y$ electrostatic deflection. Source: ref. 33.

on e/m and therefore can be used for electrons or ions. In the case of magnetic deflection systems, because the deflection is proportional to the particle velocity (and therefore to e/m), a very large magnetic field is required for ion deflection.

If a constant impingement angle is required on the target plane, a double-deflection method can be used (see Figure 2.43). An interesting cylindrical octupole deflector configuration[60] is shown in Figure 2.44(a). In this system y deflection is obtained by applying the voltages $\pm V_y$ to the upper and lower pairs of plates, which can be called the "main" plates for this operation and by applying $\pm aV_y$ to the other or "auxiliary" plates in the pattern shown in Figure 2.44(b). The value of the parameter a is a constant less than unity that determines the degree of deflection-field shaping provided by the auxiliary plates. For x deflection, the roles of the main and auxiliary plates are interchanged. In principle, combined x and y deflection can be obtained by summation [Figure 2.44(c)].

The aberrations resulting from the deflection of two sets of electrostatic or magnetic deflections have been studied in the literature.[59-63] The equations of motion for electrons (or any other charged particles) in electrostatic and magnetostatic fields, characterized by $V(x, y, z)$ and the vector potential $\overline{A}$ (x, y, z), can be derived from the variation of the action integral

$$\delta \int n \, ds = 0 \qquad (2.171)$$

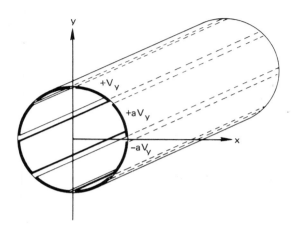

(a) OCTUPOLE DEFLECTION STRUCTURE

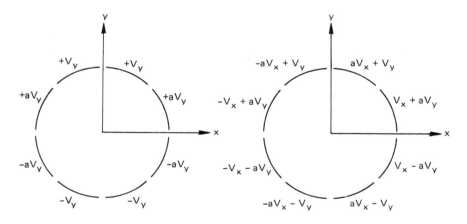

**(b) POTENTIALS ON THE
ELECTRODES FOR
y-DEFLECTION**

**(c) POTENTIALS ON THE
ELECTRODES FOR
x-y DEFLECTION**

FIGURE 2.44. Cylindrical octupole deflection system.

where, using rectangular coordinates,

$n(x, y, x', y')$

$$= \sqrt{V(x, y, z)} - \left(\frac{q}{2m}\right)^{1/2}\left(A_x\frac{dx}{ds} + A_y\frac{dy}{ds} + A_z\frac{dz}{ds}\right)$$

and

$$ds = dz(1 + x'^2 + y'^2)^{1/2} \tag{2.172}$$
$$x' = \frac{dx}{dz}, \quad y' = \frac{dy}{dz}$$

The Langrange equation associated with Eq. (2.171) can be written in the form

$$\frac{\partial n}{\partial x} - \frac{d}{dz}\left(\frac{\partial n}{\partial x'}\right) = 0$$
$$\frac{\partial n}{\partial y} - \frac{d}{dz}\left(\frac{\partial n}{\partial y'}\right) = 0 \tag{2.173}$$

Then, with Eq. (2.172) we get the general trajectory equations

$$\frac{1}{2\sqrt{V}}(1 + x'^2 + y'^2)^{1/2}\frac{\partial V}{\partial x} = \frac{d}{dz}\left[\frac{\sqrt{V}\,x'}{(1 + x'^2 + y'^2)^{1/2}} - \left(\frac{q}{2m}\right)^{1/2}A_x\right]$$

$$\times\left(\frac{q}{2m}\right)^{1/2}\left[x'\frac{\partial A_x}{\partial x} + y'\frac{\partial A_y}{\partial x} + z'\frac{\partial A_z}{\partial x}\right]$$

$$\frac{1}{2\sqrt{V}}(1 + x'^2 + y'^2)^{1/2}\frac{\partial V}{\partial y} = \frac{d}{dz}\left[\frac{\sqrt{V}\,y'}{(1 + x'^2 + y'^2)^{1/2}} - \left(\frac{q}{2m}\right)^{1/2}A_y\right]$$

$$\times\left(\frac{q}{2m}\right)^{1/2}\left[x'\frac{\partial A_x}{\partial y} + y'\frac{\partial A_y}{\partial y} + z'\frac{\partial A_z}{\partial y}\right]$$

$$\tag{2.174}$$

The aberrations of a two-dimensional electrostatic or magnetic system can be derived by integrating these motion equations [see Eq. (2.174)]. The general technique used is to develop expansions for the potential [$V(x, y, z)$] and the fields (E_x, E_y, H_x, H_y), using the symmetry of the deflection system. The elimination of various cross terms is achieved by making the expansions consistent with Laplace's equation. Direct integration of the trajectory equations is then possible using a number of approximations.[61,62] The aberrations or deviations from first-order (Gaussian) deflection theory then appear as terms having two parts: an integral over some component or derivative of the deflection field, and a multiplier showing the dependence on Gaussian deflection and beam entry conditions (diameter and angle). The aberration integrals are dependent only on the deflection-field distribution.

2.6. Electron Interactions

2.6.1. Electron–Solid Interactions

2.6.1.1. Introduction

In the previous sections we discussed the physics of beam generation and focusing into a small spot. The following sections provide a brief overview of the surface interactions of electron, ion, and photon beams with solid-state targets.

The surface of a solid constitutes a region where the geometrical arrangement of atoms, the electronic structure, and the chemical composition may be quite different from the bulk. Under atmospheric conditions all surfaces are contaminated by adsorbed layers. If an atomically clean surface is generated (by sputtering or fracture of a crystal) in a vacuum of 10^{-6} Torr, it will be covered by a complete monolayer (10^{15} atom/cm^2) from the gas phase in about 1 s; therefore an ultrahigh vacuum (10^{-10} Torr) is required to study atomically clean surfaces. In such an environment one can characterize a surface region by specifying the following[64]:

- The atomic and molecular species on the surface region (chemical characterization)
- The arrangements of these atoms and molecules on the surface
- The valence electron distribution on the surface

For chemical characterization, electron-, ion-, and photon-beam excitation processes are used. In this section the basic excitation processes will be discussed; applications to electron- and ion-beam microscopies will be treated in Chapter 6.

Most of our knowledge of surface geometry has been obtained by electron diffraction and field-ion microscopy.[65,66] From these experimental studies we learned that the surface of a crystal can differ from the bulk structure in a number of ways (see Figure 2.45). Foreign atoms may be chemisorbed into an overlayer [Figure 2.45(a)] or the adsorption of reactive species may form a

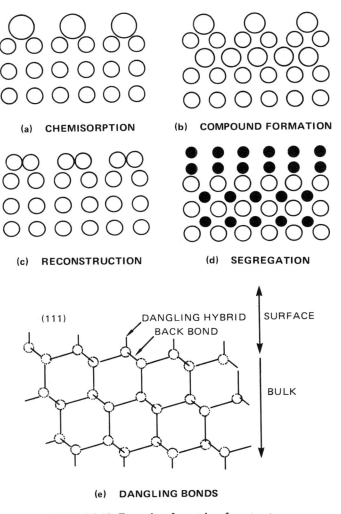

FIGURE 2.45. Examples of crystal surface structures.

three-dimensional surface-layer compound [Figure 2.45(b)]. The surface atoms may be displaced, forming a new lateral periodicity [Figure 2.45(c)], or at the surface of an alloy segregation of one of the components may occur, as shown in Figure 2.45(d).

The chemical composition of a surface is generally not the same as the bulk. For example, all metals have a native oxide on the surface that can vary in thickness from 1 to 10 nm. These oxide layers affect the surface mechanical and chemical properties if they are not removed by etching or chemical polishing.

Surface contamination normally consists of absorbed or adsorbed gases, organic materials, water, and inorganic ions.

The electronic structure at the surface of a crystal is usually different from that of the bulk. Some of the orbitals that are forming all the bonds in the crystal remain unbonded at the surface. These orbitals, directed out of the surface, are called "dangling bonds," as shown in Figure 2.45(e). This is the case for many semiconductors (Si, Ge, GaAs, etc.) in which the bonds in the bulk are each made up of two electrons shared by adjacent atoms, but at the surface dangling bonds are left, each with only one electron. Thus both filled and empty electron surface-state energy levels, which are different from the bulk energy levels, occur for these "dangling bonds." Such surface states play an important role in band bending and work-function-related effects. The differences between the bulk and surface properties of solids are important owing to their effects in surface bombardment by electron, ion, and photon beams.

2.6.1.2. Electron Transmission Through Thin Films

If a beam of electrons with kinetic energy E_0 impinges upon a thin film sample (i.e., one whose thickness is much smaller than the range of the electrons in the target material), it is convenient to classify the transmitted electrons into three categories[67] (see Figure 2.46):

- The unscattered beam
- The elastically scattered beam
- The inelastically scattered beam

Unscattered electrons pass through the sample without being deflected or suffering any energy loss.

2.6.1.3. Elastic Scattering

Elastic collisions are primarily due to the screened nuclear field and can result in large angle ($> 90°$) deflections, although most are smaller angle deflections. Some of the electrons are scattered on phonons, suffer small (1–10 meV) energy losses and are indistinguishable from elastically scattered electrons.

The angular distribution of elastically scattered electrons can be calculated as Rutherford scattering screened by the electron cloud of the atom.[68]

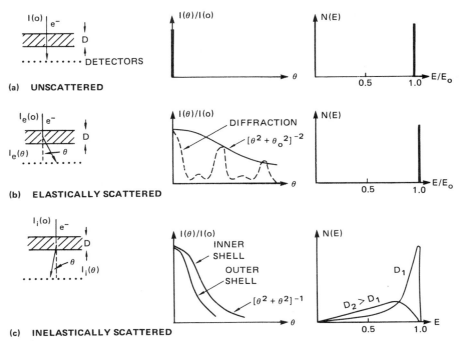

FIGURE 2.46. Angular and energy distributions of the electrons transmitted through thin films (arbitrary units).

The incoming $I_e(0)$ and scattered $I_e(\theta)$ current ratios can be related to the angular distribution[69] by

$$I_e(\theta)/I_e(0) = (\theta^2 + \theta_0^2)^{-2}$$

where $\theta_0 = \lambda/2\pi a$ is the characteristic screening angle and a is the atomic radius given by

$$a = 0.9\, a_0 Z^{-1/4}$$

a_0 is the Bohr radius, Z is the atomic number, and λ is the incident electron wavelength.

This angular distribution formula is approximately correct for amorphous samples, but in the case of crystalline films the angular distribution is modulated by diffraction effects, and peaks appear at the angles where the Bragg condition $\theta \approx \lambda/d$ is satisfied (see Figure 2.46).

2.6.1.4. Beam Spreading in Thin Films

The spatial distribution of electrons as a function of vertical penetration z and radial distance from the center of a fine (100-Å) incoming beam due to scattering has been calculated.[70] For film thicknesses of several thousand angstroms the incident electrons suffer many collisions and the cumulative deflections can be determined statistically by using the Boltzmann equation.

Using Fermi's solution,[70] the spacial probability distribution density of the scattered electrons is given as

$$H(r, z) = \frac{3\lambda}{4\pi z^3} \exp\left[-\frac{3}{4}\left(\frac{\lambda r^2}{z^3}\right)\right]$$

$$= \frac{1}{\pi r_0^2} \exp\left|-\frac{r^2}{r_0^2}\right| \tag{2.175}$$

where r is the radial distance measured from the axis of incoming electrons, z is the vertical distance measured along the axis of electron penetration, and λ is the scattering mean free path.

The normalization condition

$$\left(\frac{3\lambda}{4\pi z^3}\right) \int_{-\infty}^{\infty} \int_{-\infty}^{\infty} \exp\left(-\frac{3}{4}\frac{\lambda(x^2 + y^2)}{z^3}\right) dx\, dy = 1$$

assures that at every thickness the spatial distribution is Gaussian. For nonrelativistic electrons

$$\lambda(\text{Å}) = \frac{5.12 \times 10^{-3}\ U^2\ A}{\rho Z^2 \ln(0.725\ U^{1/2}/Z^{1/3})} \tag{2.176}$$

where A is the mass number, U is the electron energy (eV), ρ is the film density (g/cm³), and Z is the atomic number. From these formulas [Eqs. (2.175) and (2.176)] the response density at any radius r_0 from the center of the Gaussian at penetration depth z_0 can be calculated. The effective beam radius $\bar{r}'$ after passing through the film of thickness z is related to the Gaussian radius of the incoming beam ($\bar{r}_i$) by

$$\bar{r}'^2 = \bar{r}_i^2 + \frac{4 z_0^3}{3\lambda} \tag{2.177}$$

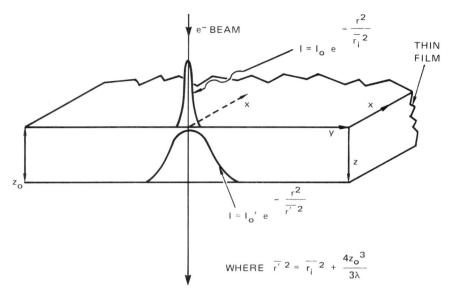

FIGURE 2.47. Spatial distribution of electrons as a function of vertical penetration and radial distance from the beam center.

where the Gaussian radius $\bar{r}_i$ is for the incident beam at $z = 0$. The second term gives the spread in radius, as illustrated in Figure 2.47. This calculation assumes no loss of energy of electrons in passing through the material, only elastic scattering.

2.6.1.5. Inelastic Scattering

In the case of inelastically scattered electrons the angle of scattering θ depends on the energy lost by the incident electron. The angular distribution[67] in this case is given by

$$I_i(\theta)/I_i(0) = (\theta^2 + \theta_E^2)^{-1} \qquad (2.178)$$

where $\theta_E = E/pv$ and E is the energy lost by an incident electron of velocity v and momentum p.

The energy loss in thin and thick targets resulting from inelastic scattering occurs as discrete events, with the creation of secondary electrons of low energy (up to 50 eV), excitation of density oscillations in metal electron plasmas, inner-shell ionizations that lead to X-ray emission and Auger electron emission, creation of electron–hole pairs followed by photon emission (cathodoluminescence), transition radiation,[71] and lattice vibrations (phonon excitation) (see Figure 2.48).

The energy losses from all inelastic processes are conveniently described by the Bethe continuous energy loss relation[72]:

$$\frac{dE}{ds} = \left(\frac{N_A\, e^4}{2\pi\, \epsilon_0^2}\right)\left(\frac{Z\rho}{A}\right)\left(\frac{1}{E}\ln\frac{1.66E}{J}\right)$$
$$= -7.85 \times 10^4 \frac{\rho Z}{AE}\ln\frac{1.66E}{J}\left[\frac{\text{keV}}{\text{cm}}\right] \tag{2.179}$$

where dE is the change in energy of the electron of energy E (keV) while transversing a path of length ds (cm), Z is the atomic number, A is the atomic weight, ρ is the density, and J is the mean ionization potential. J is an increasing function of atomic number and can be approximated as

$$J = 1.15 \times 10^{-2}\, Z \text{ (keV)} \tag{2.180}$$

The rate of energy loss with distance traveled is thus dependent on the atomic number because of the dependence on Z/A and on density: Thus dE/ds generally increases with increasing Z. The rate of energy loss also increases with

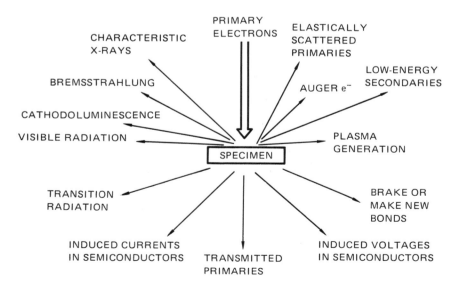

FIGURE 2.48. Basic electron–target interactions used in microfabrication.

decreasing energy, since the E term in the denominator of Eq. (2.179) dominates the $\ln E$ term in the numerator.

2.6.2. Energy-Loss Mechanisms

A schematic spectrum of electrons emitted from a thin film under electron bombardment is shown in Figure 2.49. Electrons of different origin are observed in the four energy ranges (I–IV in Figure 2.49). The relative intensity of the various contributions to the spectrum depends on the primary energy E_0 (5–50 keV), the angle of incidence, and the angle of emission.[73] An actual elastic energy spectrum obtained under particular experimental conditions may therefore differ significantly in shape from Figure 2.49. The narrow peak at the incident energy E_0 is made up of elastically scattered electrons plus diffracted electrons if the thin film is crystalline. Energy losses in range I due to photon excitations (see insert (a) in Figure 2.49) can be resolved only by the most sophisticated energy analyzers.

In metals and semiconductors most of the energy lost in range II (due to electronic excitations and ionization losses) is transferred to the conduction or valence electrons through individual or collective excitations. In metals the free electrons can be excited in unison by the incident electron beam, resulting in a collective or "plasma" oscillation of the valence electrons. Typical energy losses for plasma excitation are around 10–20 eV (see insert (b) in Figure 2.49).

Above 50-eV energy losses (range III) the energy spectrum is shaped by various atomic excitation mechanisms. The most striking is the appearance of sharp edges corresponding to the excitation of inner-shell atomic electrons into the vacuum continuum (see insert (d) in Figure 2.49). These edges correspond to the energy necessary to ionize the atoms, and the energy loss at which they occur is related to the atomic number. In these processes a core hole is generated by a primary or secondary electron and filled by another core or valence electron. In the Auger process an electron drops down to fill the hole and a second electron (Auger electron) is emitted (see insert (c) in Figure 2.49). In the case of X-ray emission (X-ray fluorescence), instead of the second electron a photon is emitted. These processes (Auger, X-ray emission) are shown in Figure 2.50. Finally, range IV (0–50 eV) contains the "true" secondary electrons, which are the result of cascade collision processes of the primary and target electrons. The secondary-electron distribution (dashed line in Figure 2.49) shows a most probable energy at 3–5 eV. One cannot find a clear upper energy limit for secondary electrons, but the use of 50 eV is the present convention.

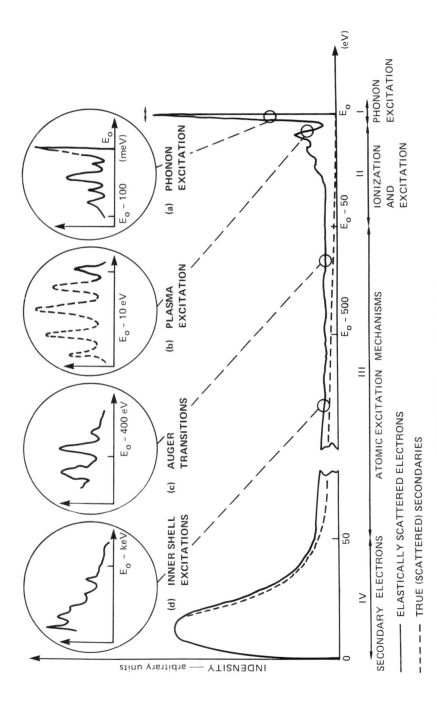

FIGURE 2.49. Spectrum of inelastic electrons.

2.6.2.1. Electron Range

The penetration depth of electrons in a solid and the distribution of energy loss as a function of penetration depth are of great importance in microfabrication.

The rate of change of the electron kinetic energy E along the path s is described by the Bethe energy loss formula [Eq. (2.179)]

$$\frac{dE}{ds} = \left(\frac{N_A e^4}{2\pi \epsilon_0^2} \right) \left(\frac{Z\rho}{A} \right) \left(\frac{1}{E} \ln \frac{1.66 E}{J} \right) \tag{2.181}$$

which consists of three terms: The first is a constant, the second depends only on the material, and the third depends on both the energy and the material [$J = 1.15 \times 10^{-2} Z$ (keV)]. Since Z/A is nearly constant, the second term varies with the density, which changes with the position of the element in the periodic table. However, if the distance is measured in terms of ρs (g/cm^2), the change with density is accounted for.

To calculate the electron range one has to integrate Eq. (2.181), starting with $E = E_0$ at $s = 0$ (electron entrance):

$$R^B = \int_{E_0}^{0} \frac{dE}{dE/d(\rho s)} \tag{2.182}$$

The maximum value of ρs for complete slowing down ($E = 0$) is called the Bethe range R_B. The Bethe range R_B can be larger than the practical range R because of elastic scattering of the electrons in the target.

The practical electron range R, obtained by measuring the electron transmission of thin films, can be approximated by

$$R = 10 E_0^{1.43} \left[\frac{\mu g}{cm} \right] \tag{2.183}$$

where E_0 is in keV and R is in μg/cm. This formula is useful in the ranges of $E_0 = 1$–100 keV.

For low-Z material the practical range R is on the same order as the Bethe range R_B. The probability for large-angle elastic scattering is small owing to Z^2 in the Rutherford formula. More specifically, the inelastic/elastic total scattering cross sections can be expressed as

$$\frac{\sigma_i}{\sigma_e} = \frac{1}{Z} \ln \left(\frac{4E_0}{\bar{E}} \right) \tag{2.184}$$

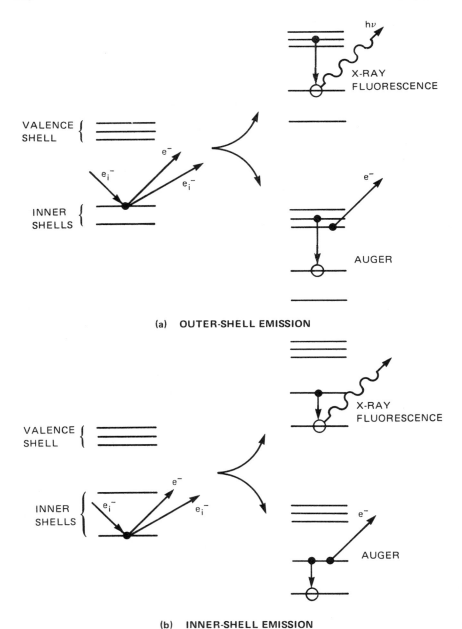

FIGURE 2.50. Two-step Auger and X-ray fluorescence processes.

where $\overline{E} \approx 5Z$, and

$$\sigma_e = \frac{1.5 \times 10^{-4}}{(v/c)^2} Z^{3/2} \text{ (in Å}^2\text{)} \qquad (2.185)$$

where v is the velocity of the electron.[67] From Eq. (2.184) it is clear that the ratio σ_i/σ_e increases slowly with increasing voltage and with decreasing Z.

In low-Z materials most of the electrons pass through the target with multiple inelastic small-angle scattering events. For example,

$$\sigma_i \cong 3\sigma_e \qquad \text{for carbon}$$

and

$$\sigma_i \cong 0.4\sigma_e \qquad \text{for platinum}$$

In high-Z targets the electron trajectories are bent more strongly and no electron can penetrete to a depth of R_B.

Many theoretical models[74-76] for electron penetration have been built on variations of Eq. (2.182). In these models the electrons penetrate to a depth z_d into the material, losing energy according to the Bethe law and scattering according to a screened Coulomb cross section. At depth z_d, owing to multiple scattering, the electron paths become isotropic, traveling in all directions until they reach their total range (R_B) in the target material. The "depth of complete diffusion"[75] is given by

$$z_d \approx R_B \left(\frac{40}{7Z} \right) \qquad (2.186)$$

Based on this equation one can characterize the shape of the electron diffusion cloud with the shape of a fruit. For low-Z material the diffusion cloud resembles the shape of a pear, and for high-Z targets that of an apple.

The energy dissipated $dE/d(\rho z)$ by an electron beam is used mainly for X-ray production, secondary emission, and electron–hole pair production. A typical plot of $dE/d(\rho z)$ as a function of ρz is shown in Figure 2.51. The extrapolation of the linear portion of the curve defines the so-called Grün range $R_G(E)$, which can be expressed for low-Z targets[74] by

$$R_G = 4.0 \, E_0^{1.75} \qquad (2.187)$$

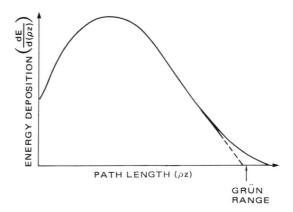

FIGURE 2.51. Energy deposition as a function of path length.

where E_0 is the beam energy in keV and the range is expressed in $\mu g/cm^2$ (see Figure 2.52). Knowledge of the energy dissipation $dE/d(\rho z)$ profile as a function of distance ρz ["depth-dose distribution" $\phi(z)$] from the point of entry of the primary-electron beam is necessary for calculating secondary yields.[74,75] Figure 2.53 shows the depth-dose curves[75] for carbon and gold calculated by the Monte Carlo method.

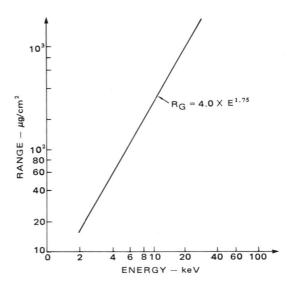

FIGURE 2.52. The Grün range as a function of the electron energy.

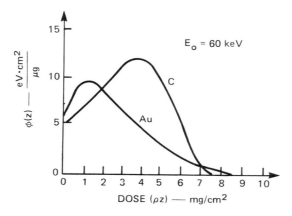

FIGURE 2.53. Depth-dose distribution $\phi(z)$ generated by 60-keV electrons in carbon and gold.

2.6.2.2. Backscattered Electrons

When an electron beam impinges on a plane target, the electrons emerging from the target may be divided into two groups: backscattered electrons and "true" secondary electrons. Backscattered electrons leave the target with energies greater than 50 eV. These can be subdivided into the following:

- Elastically reflected primaries, which leave the target with only a small loss of energy
- Diffused electrons, which leave the target with a greater loss of energy
- Auger electrons, which have an energy related to the target material

True secondary electrons typically have energies of only a few volts. However, the secondary-emission coefficient δ is customarily defined as the ratio of the total electron current leaving the target to the primary-electron current striking the target. Thus δ has two components: η, the ratio of reflected-electron current leaving the target to the primary current striking the target, and δ_s, the ratio of the secondary-electron current leaving the target to the primary-electron current striking the target; that is,

$$\delta = \eta + \delta_s$$

From the integration of the Rutherford scattering for large scattering angles[77] one can define η as

$$\eta = \frac{a - 1 + 0.5^a}{a + 1} \tag{2.188}$$

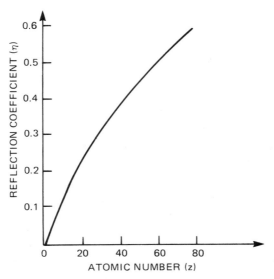

FIGURE 2.54. Reflection coefficient η as a function of atomic number Z.

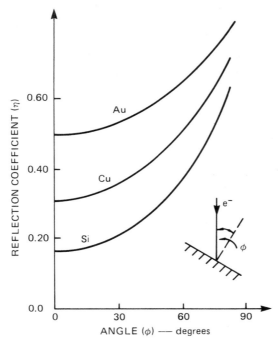

FIGURE 2.55. Reflection coefficient η as a function of the tilt angle ϕ.

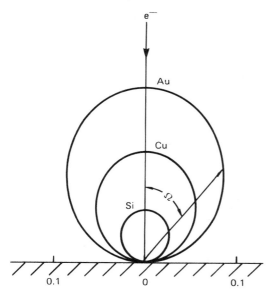

$$\frac{d\eta}{d\Omega} — \text{arbitrary units}$$

FIGURE 2.56. Angular distribution of backscattered electrons at normal incidence.

where $a = 0.045 Z$. Figure 2.54 shows η as a function of Z. This expression predicts that η is independent of the primary-electron energy and depends only on the atomic number of the target material and is in good agreement with the experimental data. Note that η increases with increasing tilt angle ϕ,[78] as shown in Figure 2.55. The angular distribution of the backscattered electrons at normal incidence ($\phi = 0$) can be approximated by a sin Ω law[79] where Ω is the take-off angle for backscattered electrons (see Figure 2.56). The exit depth of the backscattered electrons is on the order of half the electron range for normal incidence.[80]

2.6.2.3. Secondary Electrons

The energy distribution of the secondary electrons shows (Figure 2.49) that the most probable energy is between 2 and 5 eV.

The secondary-electron yield δ_s, defined as the average number of emitted secondaries per incident primary, does not show a monotonic increase with increasing Z as does the backscattered electron yield, but it is influenced by the variation of the work function and surface conditions.[81] δ_s shows a maximum larger than unity at primary energies in the range $E_0 = 300–800$ eV.

The observed behavior of the secondary emission can be explained in terms of the production and escape mechanisms.[82] Let $n(z) \, \Delta z$ be the number of secondary electrons produced between z and $(z + \Delta z)$; let $g(z)$ be the average probability of escape of a secondary electron produced at a depth z. Then δ_s is given by

$$\delta_s = \int_0^{z_0} n(z) g(z) \, dz \tag{2.189}$$

where z_0 is the range of the primary electrons in the target material ($z_0 \sim R_G$). At very small primary energies the penetration depth is small so that $g(z) = g_0$ (see Figure 2.57) and

$$\delta_s = g_0 \int_0^{z_0} n(z) \, dz = g_0 N \tag{2.190}$$

where N is the total number of secondary electrons produced at a given primary energy E_0. If on the average an energy E_s is needed to produce a secondary electron, then $N = E_0 / E_s$ and

$$\delta_s = g_0 \frac{E_0}{E_s} \tag{2.191}$$

that is, δ_s is roughly proportional to E_0 at low energies.

At high primary energies ($R_G > z_e$) the energy loss [Eq. (2.181)] can be approximated by

$$\frac{dE_0}{dz} = - \frac{a}{E_0^\alpha} \tag{2.192}$$

where α is a very slow function of E_0, starting at $\alpha < 1$ and approaching unity for large primary energies. Equation (2.192) can be integrated if we treat α as a constant, with the condition $E_0 = E_{00}$ at $z = 0$, obtaining the extended Whiddington law as

$$E_0^{1+\alpha} = E_{00}^{1+\alpha} - (1 + \alpha) az \tag{2.193}$$

From this the range $z_0 \sim R_G$ is found by setting $E_0 = 0$. This yields, from Eq. (2.193),

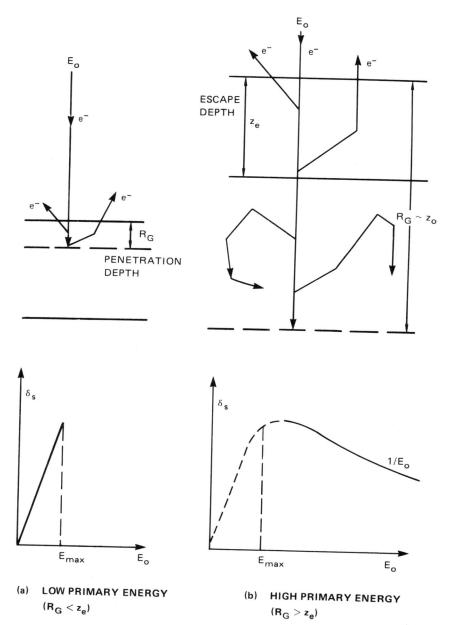

FIGURE 2.57. Secondary-electron yield (δ_s) as a function of the energy of primary electrons.

$$z_0 = \frac{E_{00}^{1+\alpha}}{(1 + \alpha) a} \tag{2.194}$$

Now if we assume that $n(z)$, the number of secondary electrons produced per unit length is proportional to $-dE_0/dz$. That is, if b is a proportionality factor,

$$n(z) = \frac{ba}{E_0^{\alpha}} \tag{2.195}$$

and the escape function has the form

$$g(z) = g_0 \exp(-\beta z) \tag{2.196}$$

Substituting Eqs. (2.195) and (2.196) into Eq. (2.189) yields

$$\delta_s = bg_0 \int_0^{z_0} \frac{a}{E_p^{\alpha}} \exp(-\beta z) \, dz \tag{2.197}$$

Then using Eqs. (2.192) and (2.194) one obtains

$$\delta_s = bg_0 \int_0^{E_{00}} \exp\left[-\frac{\beta}{(1 + \alpha) a} (E_{00}^{1+\alpha} - E_0^{1+\alpha}) \right] dE_0 \tag{2.198}$$

This integral goes through a maximum of $E_{00} = E_{\max}$ and decreases for $E_{00} > E_{\max}$. If E_{00} is larger than a few times $E_{\max}$, then the integral yields

$$\delta_s = \frac{bg_0 a}{\beta \, E_{00}^{\alpha}} = \frac{\text{const.}}{E_{00}^{\alpha}} \tag{2.199}$$

This result agrees with measurements above 10 kV, where α approaches unity.

The exit depth (z_e) for secondary electrons is on the order of 1–10 nm.[83] The variation of the secondary-electron yield with tilt angle ϕ can be approximated by

$$\delta_s(\phi) = \frac{\delta_s(0)}{\cos \phi} \tag{2.200}$$

2.6.2.4. X-Ray Emission

When electrons excite atoms in a solid, the excitation energy can be transferred to other electrons in an Auger process (see Figure 2.48) or it can be emitted in the form of characteristic photons (X-ray fluorescence). At low energies the transfer rate is much greater for the electron process than for photon emission. The photon process generally becomes more probable with increasing energy.[84] The interaction of an electron with a solid from the standpoint of X-ray emission can be divided into two distinct modes: emission resulting from the radiative capture of energetic electrons (normal bremsstrahlung) and emission resulting from inelastic collisions with core electrons (Auger electrons or characteristic X rays). The fraction of incident-electron intensity that is converted into bremsstrahlung is found experimentally to vary linearly with both the Z of the material and the incident-electron energy:

$$f_B \approx 7 \times 10^{-10} \, ZE_0 \tag{2.201}$$

where Z is the atomic number of the target and E_0 is given in eV.[85]

The separation of channels producing X rays and electrons from an electron-bombarded target is shown in Figure 2.58. The channel on the right side involves the dissipation of energy by nuclear and valence band electron–electron scattering, and the left channel includes the energy dissipation involving atomic core-level excitation.[86] Here again, if one wants to calculate the number of emitted X-ray quanta from the solid, one has to calculate the number of ionization processes along the electron trajectory, using the Bethe formula.

The number of emitted X-ray quanta per path element $d(\rho s)$ per unit of solid angle for an element concentration c_a can be written[87] as

$$dN_K = \frac{1}{4\pi} \omega_K P_K \frac{c_a N_A}{A} \frac{\sigma_K}{dE/d(\rho s)} \, dE \tag{2.202}$$

where ω_K is the fluorescence yield, $\omega_K \sim 2 \times 10^{-2}$ for low Z and ~ 0.97 for gold, P_K is the ratio of intensity of the observed characteristic line to all lines of the K series, and σ_K is the total cross section of K shell ionization.[88]

The total number of X-ray quanta generated from the slowdown of an electron from energy E_0 to the average K shell ionization energy E_1 can be obtained from

$$N_K = \frac{1}{4\pi} \omega_K P_K \frac{c_a N_a}{A} \int_{E_K}^{E_0} \frac{\sigma_K}{dE/d(\rho s)} \, dE \tag{2.203}$$

Analogous to the depth distribution for ionization, there is a depth distribution of X-ray emission $\phi_x(z)$ with a similar shape but reduced dimensions.

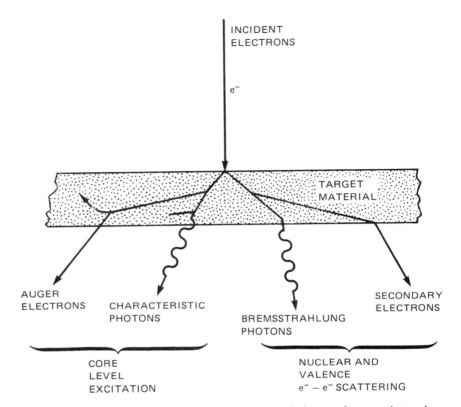

FIGURE 2.58. Separation channels producing X rays and electrons from an electron-bombarded target.

Now we can summarize (Figure 2.59) the wide range of depths from which the secondary particles (e^-, X ray) are originated. Figure 2.59 illustrates the range and spatial resolution of various secondaries that are generated when an electron beam strikes a solid target.[89]

2.6.3. Physical Effects in the Target

2.6.3.1. Polymerization: Bond Breaking

When energetic (1–100 keV) electrons strike a target made of organic molecules, the molecules may either break down to smaller fragments or link together to form larger molecules. Usually both types of change occur simultaneously, but for a given material one of these processes dominates and determines the net effect upon the material. For polymers it is known that if every carbon atom in the main chain is directly linked to at least one hydrogen atom, larger molecules are generally formed upon irradiation by the process known as cross-linking.

If the polymer is cross-linked, it forms a network and becomes insoluble and infusible. This type of polymer forms the basis of negative electron resist,

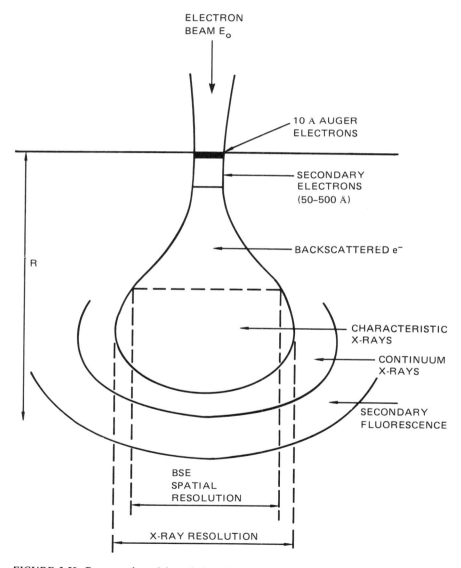

FIGURE 2.59. Range and spatial resolution of secondaries generated when an electron beam strikes a solid target. Source: ref. 89.

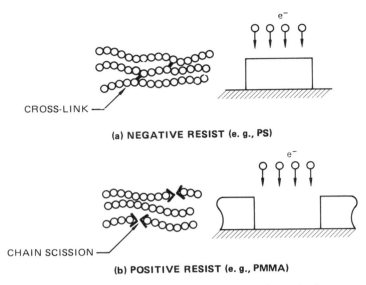

FIGURE 2.60. Schematic representation of basic resist mechanisms.

which can be dissolved when it is not irradiated but cannot be removed when irradiated. Thus a residual film pattern remains on the irradiated areas after development.

Another group of polymers is known in which irradiation causes a break in the main polymer chain, resulting in a lower average molecular weight in the irradiated material than in the unirradiated polymer. Since the lower-molecular-weight material can be dissolved selectively, the residual film pattern forms a positive working electron resist. Polymethyl methacrylate (PMMA) is the best-known high-resolution long-chain polymer resist. Figure 2.60 is a schematic representation of the basic mechanisms of resists, showing the cross-linking of negative resists and the chain scission of positive resists.

2.6.3.2. Thermal Effects

Electron beams have many special properties that make them particularly well suited for thermal processing and machining.[90]

The first of these is a very high resolution. The second feature is the large energy densities that can be obtained ($\leq 10^{10} \text{ W/cm}^2$). Thirdly, the dissipation of this power in surfaces can be used to alloy and anneal thin films (thermal lithography), catalyze chemical reactions, and selectively remove materials. Finally, electron beams are easy to control.

The distribution of the input electron-beam power P_e striking a target can be described by the following equation:

$$P_e = P_1 + P_2 + P_3 + P_4 \qquad (2.204)$$

where P_1 is the power utilized in vaporizing the material, P_2 is the average power lost as a result of heat conduction, P_3 is the average power lost as a result of radiation, and P_4 is the power lost to ionization and excitation of the evaporated atoms in the path of the beam.

P_3 and P_4 are usually negligible. The efficiency of the evaporation is very low; much less than 1% of the beam power is used for evaporation.

Thus the major effect of the beam is to cause the temperature to rise in the target area. The temperature rise caused by the beam is given as

$$\Delta T = \frac{3}{2} \frac{V_0 I_0}{\pi \lambda R} \qquad (2.205)$$

where V_0 is the beam voltage, I_0 the beam current, λ the heat conduction coefficient ($J \cdot m^{-1} \cdot s^{-1} \cdot degree^{-1}$), and R the range of the electrons in the material.

For metals ΔT is very small, but in plastics and insulators with low heat conductivities the temperature rise can cause the melting of the material. The heating effect of an electron beam will be discussed in more detail in connection with electron-beam annealing (Section 5.4.1).

2.6.3.3. Current Injection

If an electron beam impinges on a semiconductor target, collisions with the valence electrons produce hole–electron pairs. The energy necessary to create one hole–electron pair, E_p, is the same order of magnitude as the band-gap energy E_g. $E_p = 3.6$ eV in silicon and 2.8 eV in Ge. The number of electron–hole pairs created by one electron with energy E_0 is

$$N = \frac{E_0}{E_p} (1 - k\eta) \qquad (2.206)$$

where η is the backscattering coefficient and $k = E_{BS}/E_0$, where E_{BS} is the mean energy of backscattered electrons.

To calculate the number of electron–hole pairs as a function of the penetration depth one has to assume that the carrier generation is proportional to the main energy loss.[74] Measurement of the electron–hole generation profile (conductivity modulation) forms the physical basis for many election-beam techniques used for device characterization.

2.7. Ion Interactions

2.7.1. Introduction

In this section we consider the fate of energetic ions (1–100 keV) incident on a solid surface. The 10 ways in which ions can interact with a surface are illustrated in Figure 2.61. An incoming ion can be backscattered by an atom or group of atoms in the bombarded sample [1]. The backscattering process generally results in a deflection of the ion's incident path to a new trajectory after the encounter and an exchange of energy between the ion and the target atom. The energy exchange can be either elastic or inelastic, depending on the constituent particles and the energy of the ions. The momentum of an ion can be sufficient to dislodge a surface atom from a weakly bound position on the sample lattice and cause its relocation on the surface in a more strongly bound position [2]. This process is called atomic dislocation. Ions with greater energies can cause internal dislocations in the bulk of the sample [3]. Physical sputtering [4] results when ions strike the sample surface and transfer enough momentum to entirely free one or more atom. Ions can penetrate into the lattice and become trapped as their energy is expended (ion implantation) [5]. Chemical reaction of the ions with the surface atoms can occur, and, as a result, new compounds can be formed on the sample surface or the outermost layer of atoms may leave as a gas (chemical sputtering) [6]. The bombarding positive ion can gain an electron from the surface by Auger neutralization and be reflected as a neutral atom [7]. Ions can become bound to the surfaces of the sample as adsorbed ions [8]. Secondary electron emission [9] occurs under suitable conditions of ion bombardment of metal surfaces. Secondary ion emission [10] results when surface atoms are excited to ionized states and emitted from the sample.

In slowing down, the incident ion transfers energy to the solid. In analyzing the energy loss it is convenient to distinguish between two major energy loss processes: electronic collisions and nuclear collisions.

The first process is the interaction of the fast ion with the lattice electrons, resulting in excitation and ionization. Since the density of the electrons in the target is high and the collisions are so numerous, the process, similar to the electron energy loss, can be regarded as continuous.[91]

The nuclear loss results from collisions between the screened nuclear charges of the ion and the target atoms. The frequency of these collisions is lower, so they may be described by elastic two-body collisions. At high energies they are accurately described by Rutherford scattering, at medium energies by screened Coulomb scattering, while at low energies the interaction becomes even more complex.[92]

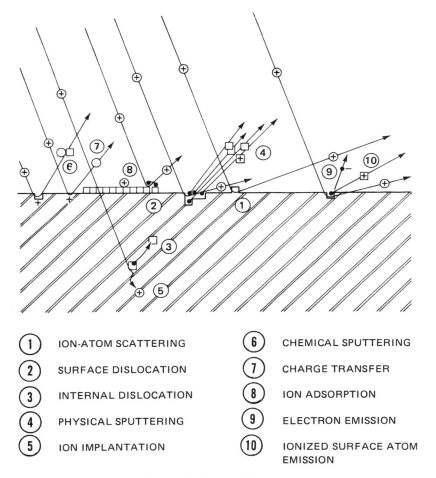

①	ION-ATOM SCATTERING	⑥	CHEMICAL SPUTTERING
②	SURFACE DISLOCATION	⑦	CHARGE TRANSFER
③	INTERNAL DISLOCATION	⑧	ION ADSORPTION
④	PHYSICAL SPUTTERING	⑨	ELECTRON EMISSION
⑤	ION IMPLANTATION	⑩	IONIZED SURFACE ATOM EMISSION

FIGURE 2.61. Ion–solid interactions.

In addition, there is a contribution to the energy loss resulting from the charge exchange between the moving ion and target atom [93]; this energy loss reaches a maximum when the relative velocity is comparable to the Bohr electronic velocity ($2 \times 10^6 \text{ ms}^{-1}$).

The total energy loss $-dE/dz$ can thus be regarded as the sum of three components, nuclear, electronic, and charge exchange,

$$\frac{dE}{dz} = \left(\frac{dE}{dz}\right)_n + \left(\frac{dE}{dz}\right)_e + \left(\frac{dE}{dz}\right)_{ch} \qquad (2.207)$$

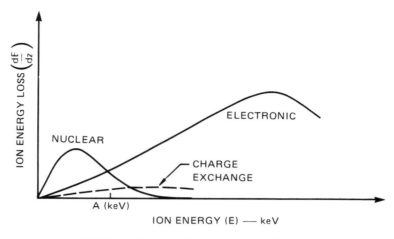

FIGURE 2.62. Variation of ion energy loss with ion energy.

At low ion energies nuclear stopping dominates and is responsible for most of the angular dispersion of an ion beam. At high energies electronic collisions are more important. A useful rule of thumb is that the energy lost to the lattice by nuclear collisions becomes predominant at energies less than A keV, where A is the atomic weight of the incident ion. At intermediate energy regions the charge-exchange contribution may rise to roughly 10% of the total. The variation of the energy loss with ion energy is shown in Figure 2.62.

Inelastic interactions with the target electrons lead to secondary electron emission, emission of characteristic X rays, and optical photon emission; elastic

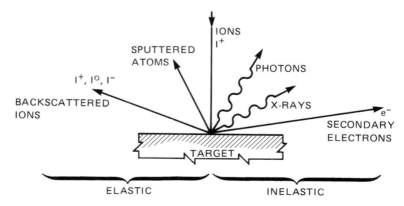

FIGURE 2.63. Schematic representation of ion–target interactions.

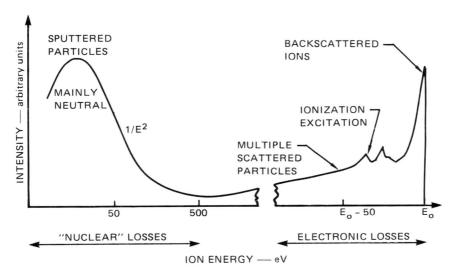

FIGURE 2.64. Energy spectrum of ions scattered from a solid target.

interactions lead to the displacement of lattice atoms, the formation of defects, and surface sputtering, as illustrated schematically in Figure 2.63.

The energy spectrum of ions[94] with initial energy E_0 scattered from a solid target is represented schematically in Figure 2.64. There is a broad low-energy hump (10–30 eV) corresponding to the emitted neutral atoms (sputtered atoms) and a high-energy hump distributed about the incident ion energy E_0 (elastically scattered ions).

2.7.2. Energy-Loss Mechanisms

2.7.2.1. Nuclear Stopping

The calculation of elastic stopping requires a screened nuclear (Thomas–Fermi) potential,[95] but if a simple inverse-square potential is used, a good approximation[96] for the energy loss can be obtained:

$$\left(\frac{dE}{dz}\right)_n = 27.8 \frac{Z_1 Z_2}{(Z_1^{2/3} + Z_2^{2/3})^{1/2}} \frac{M_1}{M_1 + M_2} N \ (\text{eV} \cdot \text{Å}^{-1}) \quad (2.208)$$

Here Z_1 and M_1 and Z_2 and M_2 are the atomic numbers and masses of the ion and lattice atoms, respectively, and N is the atomic density (particles/Å^3).

This expression gives the magnitude of the energy loss rate in the region of 10 to 100 eV per angstrom for the majority of ion–target atom combinations.

The energy E of an ion of incident energy E_0 after scattering through a given angle θ in any given system is determined by the laws of conservation of energy and momentum[97] and can readily be shown to be

$$E = \frac{E_0 [M_1 \cos \theta \pm (M_2^2 - M_1^2 \sin^2 \theta)^{1/2}]^2}{(M_1 + M_2)^2} \qquad (2.209)$$

This expression is valid for $M_2 > M_1$ only with the positive sign, but for $M_2 \leq M_1$ both signs are valid, provided that $\sin \theta \leq M_2 / M_1$. The maximum energy that can be transferred to a target atom in a single collision is, from Eq. (2.209), for $\theta = \pi$:

$$E_{max} = \frac{4 M_1 M_2}{(M_1 + M_2)^2} E_0 \qquad (2.210)$$

If $M_1 < M_2$, backscattering can occur in a single collision, whereas if $M_1 > M_2$, the scattering must be forward.

Since the stopping power in Eq. (2.208) is independent of energy, the ion range R_n is

$$R_n = \frac{E_0}{(dE/ dz)_n} = 2 k E_0 \qquad (2.211)$$

where

$$k = \frac{0.018}{N} \frac{(Z_1^{2/3} + Z_2^{2/3})^{1/2}}{Z_1 Z_2} \frac{M_1 + M_2}{M_1} \text{Å} \cdot \text{eV}^{-1}) \qquad (2.212)$$

As expected, the range decreases when the atomic number Z or the density N increases.

2.7.2.2. Electronic Stopping

At ion velocities greater than the electron velocity in the K shell, an ion will have a high probability of being fully stripped of its electrons. The energy loss under these circumstances will be described by the Bethe formula, Eq. (2.181). However, for low ion velocities and high atomic numbers (Z_1, Z_2) the Bethe formula is invalid because it does not take into account the charge fluctuations, excitation of plasma resonance, and the charge-exchange effect.

An alternative approach has been formulated[98,99] that accounts for the electronic stopping behavior at low and intermediate energies for heavy ions. According to the Lindhart, Scharff, and Schiott (LSS) model,[95,98] the energy loss is proportional to the velocity of the ion,

$$\left(\frac{dE}{dz} \right)_e = k' E^{1/2} \qquad (2.213)$$

The value of k' depends on both the ion being used and the target material, as given[99] by

$$k' = 0.328 (Z_1 + Z_2) M_1^{-1/2} N \text{ Å}^{-1} \cdot \text{eV}^{1/2} \qquad (2.214)$$

2.7.2.3. Range

To calculate the range[100] it is customary to assume that the nuclear and electronic losses are independent of each other, as follows:

$$R = \int_0^R dz = \int_0^{E_0} \frac{dE}{(dE/dz)_n + (dE/dz)_e}$$

$$= \int_0^{E_0} \frac{dE}{1/2k + k' E^{1/2}} = \frac{2 E_0^{1/2}}{k'} - \frac{1}{kk'^2} \ln (1 + 2kk' E_0^{1/2})$$

$$(2.215)$$

And using the approximation

$$\ln (1 + x) \approx x - (\tfrac{1}{2}x^2) + \cdots$$

we get

$$R \cong 2 k E_0 (1 - \tfrac{4}{3} kk' E_0) \qquad (2.216)$$

It is convenient to define a stopping cross section (stopping power) S as

$$S = \left(\frac{1}{N} \right) \frac{dE}{dz} = \frac{1}{N} [S_n(E) + S_e(E)] \qquad (2.217)$$

Generally, when $S_n(E)$ and $S_e(E)$ are known, Eq. (2.217) can be integrated to give the total distance R that an ion with energy E_0 will travel before coming to rest:

$$R = \int_0^R dz = \frac{1}{N} \int_0^{E_0} \frac{dE}{S_n(E) + S_n(e)} \tag{2.218}$$

In amorphous targets R is called the average total range.

2.7.2.4. Channeling

In crystalline materials the target atoms are arranged symmetrically in space so that an incident energetic ion may have a correlated collision with these atoms. These correlated deflections (collisions) along a crystallographic direction can channel the energetic ions (see Figure 2.65).

The channeling property of the crystalline substrate depends on the angle of entrance ψ of the individual ions and on the ion and substrate characteristics. If the entrance angle ψ is large, then the amplitude of oscillation will be large and the ion will not be channeled (trajectory a in Figure 2.65). There will be a maximum permissible amplitude at which the ion will remain channeled (trajectory b in Figure 2.65); ψ_c is the critical angle that produces this maximum amplitude trajectory.

This critical angle for channeling can be approximated[101] by

$$\psi_c = \left(\frac{2 Z_1 Z_2 e^2}{4 \pi \epsilon_0 E d} \right)^{1/2} \tag{2.219}$$

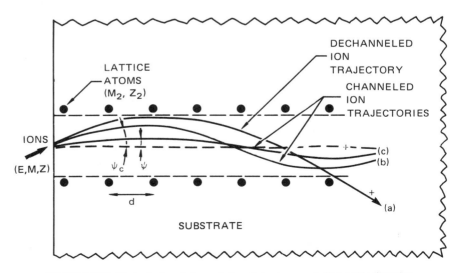

FIGURE 2.65. Trajectories of channeled particles in a crystallographic direction.

It is clear from this formula that as the ion energy E is increased, ϕ_c decreases and it becomes more difficult to remain in a stable trajectory. On the other hand, for closely packed atomic rows d decreases and channeling is easier.

2.7.2.5. Secondary Processes

2.7.2.5.1. Sputtering In the sputtering process an energetic ion incident on the target surface creates a cascade of atomic collisions in the bulk material. Some atoms in this cascade may be in the surface and escape. The sputtering yield S is defined as the number of atoms (mostly neutral) ejected per incident ion. The principal features of the sputtering of metals were explained by the first experimental investigators.[97] They showed that the sputtering yield was a function of the atomic number and that the yield showed a periodicity that was closely correlated with that of the periodic table.

At low ion energies there is a threshold for sputtering to occur. Above the threshold the sputtering yield rises to a maximum and eventually, at very high energies, decreases again as the ion energy is deposited so far into the solid that it cannot reach the surface.

A comprehensive and detailed theory of sputtering has been formulated in ref. 102. This model is based on random collision processes, applying Boltzmann's equation and general transport theory. When the energy of the primary ion E is sufficient to transfer an energy greater than the displacement energy of a lattice atom, then a collision cascade is initiated. If this cascade intersects the surface with atoms whose energy is greater than the surface-binding energy (V_0), then sputtering takes place.

For low energies up to 1 keV the expression for the sputtering yield is given by[102]

$$S(E) = \frac{3}{4\pi^2} \alpha \frac{4 M_1 M_2}{(M_1 + M_2)^2} \frac{E}{U_0} \qquad (2.220)$$

where α depends on M_2/M_1, given in Table 4 and plotted in the original paper, and U_0 is the surface binding energy. For keV energies and heavy-to-medium mass ions

$$S(E) = 0.0420 \, \alpha \frac{4\pi Z_1 Z_2 e^2 a}{U_0} \left(\frac{M_1}{M_1 + M_2} \right) S_n(\epsilon) \qquad (2.221)$$

where $\epsilon = [M_2 E/(M_1 + M_2)] (Z_1 Z_2 e^2/a)$ and $a = 0.8853 \, a_0$ $(Z_1^{2/3} Z_2^{2/3})^{-1.2}$. a_0 is the Bohr radius and $S_n(\epsilon)$ is a universal function tabulated in Table 5 and discussed in more detail in the original paper.

TABLE 4. α as a Function of Mass Ratio		TABLE 5. S_n (ϵ) as a Function of Ion Parameters (E, M_1, M_2, Z_1, Z_2)	
M_2/M_1	α	ϵ	$S_n(\epsilon)$
0.01	0.17	0.01	0.211
0.5	0.20	0.1	0.372
1.0	0.23	1.0	0.356
5.0	0.98	10.0	0.128
10.0	5.0	40.0	0.0493

A statistical model of sputtering[103] predicts yields as a function of ion energy, mass, and incident angle. According to this model, the probability of displacing an atom in the target depends only on the distance of that atom from the point of ion impact.

This model assumes that the yield is proportional to the area defined by the intersection of the "cascade volume" with the target surface (see Figure 2.66). At low energies the yield varies as

$$S(E, \theta) \approx E^{2/3}/\cos \theta \qquad (2.222)$$

The sputtering processes widely used in microfabrication can be divided into the following application areas:

- Ion-beam machining or etching
- Production of atomically clean surfaces in vacuum
- Surface analysis
- Deposition of thin films

2.7.2.5.2. Inelastic Processes The interaction of ions with atoms has been discussed in the previous sections on scattering and sputtering. The interactions with electrons do not result in any appreciable scattering of the incident ion, since momentum transfer is very small, but causes excitation and ionization of the electron shells of both the incident and target atoms. When this excitation occurs near the surface, it results in the emission of electrons, photons, and X rays.[94,97]

2.7.2.5.3. Electron Emission The ion striking the target surface has both potential energy, because of its ionized state, and kinetic energy, because of its relative velocity to the target. Electron emission can occur as a result of the rearrangement of either potential or kinetic energies in the ion–metal system. The secondary-electron yield (γ) is proportional to the energy loss of the incident ion and is observed, experimentally, to be the same for all metals.

2.7.2.5.4. X-Ray Emission X-ray emission results from the production of an inner electron shell vacancy as a result of a collision by the incident ion with a target atom. Once the inner shell vacancy is produced, Auger electron emission and X-ray emission compete to deexcite the atom. Both have energies characteristic of the excited atom.

2.8. Photon Sources and Interactions

2.8.1. X-Ray Sources for Lithography

There are several different types of photon sources used in microfabrication processes. X-ray sources are used for the mask replication of submicron features and for the fabrication of microdevices (optical elements, X-ray gratings, and microscopies). (See Sections 4.6 and 4.7.)

The requirements for an ideal X-ray source for lithography can be described as follows: The source should be intense for short exposure times (< 1 min) and small (< 1 mm) in order to cast sharp shadows. The X-ray energy should exceed 1 keV in order to be able to penetrate the mask support but should be absorbed in the resist (< 10 keV). Using these selection rules, the following X-ray sources can be considered:

- Electron bombardment sources
- Plasma sources
- Synchrotron radiation

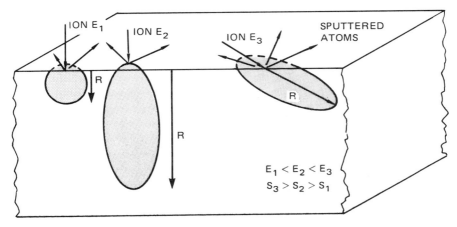

FIGURE 2.66. Sputtering model predicting yield as a function of ion energy, mass, and incident angle. Source: ref. 100.

2.8.1.1. Electron Bombardment Sources

X rays have traditionally been generated by accelerating electrons and allowing them to hit a solid target (W, Au, Pt, Cu, Al). The efficiency of this method is quite low; typically $<10^{-3}$ of the electron energy is converted to X rays. It is even lower for soft X rays (≤ 1 keV) because of the absorption in the target. An attraction of these sources is that the strong characteristic emission lines from the target atoms can be isolated by simple filters. The Al K (0.83 nm), Cu L (1.33 nm), and C K (4.48 nm) radiation sources have been specifically developed[104] for use in X-ray microlithography.

For the more powerful of such sources the source size is typically 1 mm or larger because of space-charge problems (especially at low-voltage operation) and heat load. The X rays are emitted in all directions and only a small fraction can be collected on the target. (See Section 4.6.2.)

2.8.1.2. Plasma Sources

Hot plasmas generated by high-power lasers or electrical discharges (exploded wire, vacuum spark, plasma focus) produce soft X rays.[105] These sources, although as of yet untapped, may play a future role in microlithography. In these sources the X-ray emission comes in a very intense pulse of short (10^{-9}s) duration from a small region (0.1–1 mm). (See Section 4.6.3.)

2.8.1.3. Synchrotron Radiation

Because of its intensity, tunability, small source size, and small divergence, synchrotron radiation comes close to being the ideal universal source.[104,105] Beam lines for X-ray lithography were recently designed for electron storage rings and synchrotrons. The spectral and temporal characteristics of the various X-ray sources are shown in Figure 2.67. (See Section 4.7.)

2.8.2. Light Sources

2.8.2.1. Laser Sources for Annealing

In annealing ion-implant damage in semiconductors by means of pulsed lasers the defect layer, generally amorphous, must be melted by the laser. The melt depth must extend slightly into the single-crystal layer to achieve optimum annealing and substitution for the implanted ions in the lattice structure.

For Si implanted with As, B, or P, a Q-switched Nd:glass, Nd:YAG, or ruby laser can be used with a pulse range of 10- to 100-ns duration. The energy

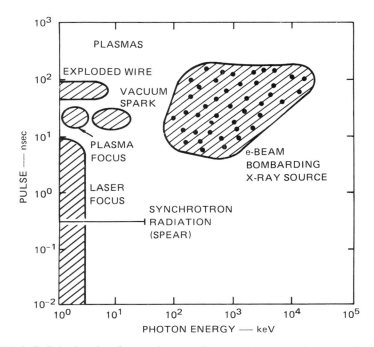

FIGURE 2.67. Pulse length and spectral energy of electron-impact, synchrotron-radiation, and high-temperature plasma flash X-ray sources. Source: ref. 105.

density of irradiation on the material is in the range of 0.5–10 J/cm². While pulsed laser annealing restores the electrical characteristics of the ion-implanted semiconductors, it causes the dopant to redistribute itself deeper into the host material. Another approach to laser annealing is the use of continuous wave (cw) lasers, such as Ar-ion lasers, the Kr-ion laser, or the CO_2 laser. The scan rate of the cw beam is usually about 0.5–10 cm/s, giving a dwell time of about 10–160 ms. The typical fluence is about 200 J/cm².

2.8.2.2. Arc Lamps for Photolithography

High-pressure mercury-arc lamps in the 200–600 nm range are used almost universally as light sources in optical lithography. Figure 2.68 shows the spectal output $S(\lambda)$ of a high-pressure mercury-arc lamp with the spectral sensitivity curves $R(\lambda)$ for several positive and negative resists.[107] For printers using reflective optics the light spectrum of the wafer is essentially $S(\lambda)$. Printers employing refractive optics corrected for printing at one or two wavelengths ("g" and "h") must employ sharp cutoff filters to remove other emissions dur-

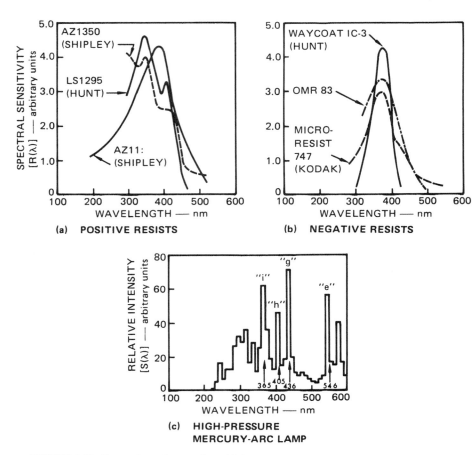

FIGURE 2.68. Comparison of spectral sensitivity curves for several positive and negative resists with spectral output of a high-pressure mercury-arc lamp. Source: ref. 107.

ing exposure. If $F(\lambda)$ is the transmission characteristic of the printer, then the light spectrum reaching the photoresist-coated wafer is $S(\lambda) F(\lambda)$. As far as the photochemical reactions in the resist are concerned, the effective light intensity at the resist surface in the wavelength range $\Delta\lambda$ is proportional to $S(\lambda) F(\lambda) R(\lambda) \Delta\lambda$.

2.8.3. Photon Interactions

X-ray photons interact with matter chiefly via three processes: the photoelectronic process, the Compton effect, and pair production. A complete treatment of the three processes is rather complicated and requires the tools of quantum electrodynamics. The essential facts, however, are simple. In the photoelectric effect the photon is absorbed by an atom and an electron from one of the shells is ejected. In the Compton effect the photon scatters from an atomic electron. In pair production the photon is converted into an electron–positron pair. This process is impossible in free space because energy and momentum cannot be conserved simultaneously when a photon decays into two massive particles; however, it can occur in the Coulomb field of a nucleus, where energy and momentum can be balanced.

The energy dependence of these processes are very different. At low energies below a few keV the photoelectric effect dominates, the Compton effect is small, and pair production is energetically impossible. At an energy of 1 MeV pair production becomes possible, and it soon dominates completely. Two of the three processes, the photoelectric effect and pair production, eliminate the photons undergoing interaction. In Compton scatterings the scattered photon is degraded in energy.

The characteristics of the transmitted beam are described by an exponential law:

$$N(z) = N(0)\, e^{-\mu z}$$

where the absorption coefficent μ is the sum of three terms,

$$\mu = \mu_{\text{photo}} + \mu_{\text{Compton}} + \mu_{\text{pair}}$$

The number of transmitted particles decreases exponentially. No range can be defined, but the average distance traveled by a particle before undergoing collision is called the mean free path and it is equal to $1/\mu$.

Figure 2.69 shows the photon absorption coefficent μ (in cm^{-1}) for single-crystal silicon as a function of photon energy from 10^{-2} to 10^8 eV. Optical photons at the lowest energies can be absorbed by lattice phonons; since the silicon lattice is not infrared active, this absorption requires multiphonon processes. Absorption due to impurities can also occur in the energy range below 1 eV, but such absorption is not indicated in Figure 2.69. Photoelectric absorption begins at the energy corresponding to excitation across the forbidden gap (E_g). For energies near E_g the photoexcited electron remains in the solid, and this process is called the internal photoeffect, while at higher energies (at approximately the ionization energy of the free atom) the photoelectron can leave the solid, and this process is called the external photoeffect. For energies near E_g (1.2 eV) the magnitude of the absorption cross section depends sensitively on the band structure of the solid. At higher photon energies photoelectron absorption is the dominant process, until Compton scattering and pair production take over.

TABLE 6. *Characteristics of Particle Beams Used in Microfabrication*

Beam Type	Wave Length $\lambda(M)^a$	Range^b	Deflection Properties	Interactions with Matter Used in Microfabrication
Photon ($h\nu$)	2000–4000 Å	$I = I_0 e^{-\mu z}$	Optical elements (lenses, mirrors)	Bond breakage and polymerization; photoelectron production; local heating (annealing); plasma generation; geometrical alignment; diffusion
X ray ($h\nu$)	2–20 Å	$R \sim$ Range of photoelectrons	Crystals	Bond breakage and polymerization; material analysis; alignment
Electron (e^-, m_0)	~0.1 Å	$R\left[\dfrac{\mu g}{cm}\right] = 10\,E_0^{1.43}\;E_0$ in keV	E, H fields	Bond breakage and polymerization; secondary-e production; bremsstrahlung (X-ray production); plasma generation; induced current, voltage production; alignment diagnostics (electron microscopies); annealing
Ion (e^+, M)	0.01 Å	$R \simeq 2kE_0(1 - \frac{4}{3}kk'E_0)$	E, H fields	Sputtering; deposition; etching; ion implantation (surface treatment); diagnostics (microscopies)
Atom (A, Z)	0.01 Å	Eq. (2.216)	Gradient fields when μ, $p \neq 0$	Sputtering; crystal growing; molecular beam epitaxy

$^a \dfrac{12.26\,A}{M/m_0 V}$ (eV)

$^b R = -\displaystyle\int_{E_0}^0 \dfrac{dE}{dE/dz}$

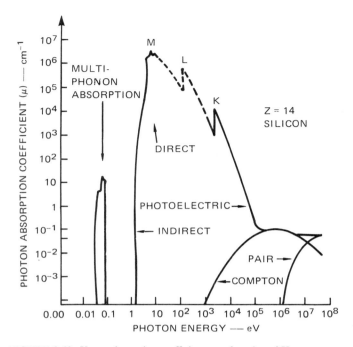

FIGURE 2.69. X-ray absorption coefficient as a function of X-ray energy.

Recently laser-beam technologies[108,109] based on the thermal interaction of high-power (10^8-W/cm^2) lasers have offered solutions for materials processing. These include the following:

- Laser-assisted diffusion
- Laser annealing
- Laser cutting and drilling
- Surface alloying and heat treating

Laser-assisted diffusion and laser annealing will be discussed in more detail in the following chapters.

2.9. Summary

In this chapter we discussed the source properties for generating particle beams (electron, ion, photon), the formation of microbeams, and beam–target interactions. Table 6 summarizes the most important characteristics of these beams and those interactions that are important in microscience.

Thin Films

3

3.1. Introduction

Growing and utilizing thin films has attracted scientists for many generations, particularly since Faraday first observed film deposition from an exploding wire in 1857.[1] The subject matter is extraordinarily varied and complex, as a glance at the recent specialist textbooks devoted to thin films soon reveals.[2-6]

Thin films are important in fabricating microdevices because by depositing a thin film at least one dimension of the device is relatively straightforward to obtain and control with high precision, from single molecular layers (monolayers) a few angstroms thick to a film several light wavelengths thick (microns). Of equal consequence are the unique physical properties exhibited by certain thin-films because of the influence of the surfaces, the uniqueness of the structures of thin films, the crystal arrangements obtained in them, and the chemical compositions that may be deposited.

Our objective in this chapter is to motivate the reader to tackle the larger body of literature on thin films, and arm him with some basic physical concepts and a knowledge of the terminology used rather than to give specific methods.

3.2. Deposition Technologies
3.2.1. Overview

To form a coating on the surface of a solid substrate, particles of the coating material must travel through a carrier medium in intimate contact with the surface. Upon striking the surface a substantial fraction of the coating particles must adhere to it or by chemical reaction at the surface form a new compound that adheres. The particles may be atoms, molecules, atomic ions, ionized fragments of molecules, or grains of material, both charged and uncharged. The carrier medium may be a solid, liquid, gas, or vacuum.

In the case of a solid carrier intimate contact between the substrate and carrier surfaces is often difficult to form, and only particles that can diffuse through the solid at a reasonable rate are useful. This restricts its usage to only a few specific combinations. A common example is the diffusion of oxygen through silicon dioxide, which is used in growing silicon dioxide layers on silicon (see Section 1.3.2). The oxygen reaching the surface of the silicon reacts, forming silicon dioxide, thereby increasing the thickness of the oxide layer from beneath. The diffusion rate decreases as the oxide layer grows thicker and essentially ceases when the oxide layer is about 1.3 μm thick.

A liquid carrier medium is much more versatile, since many compounds can be dissolved and intimate contact with the substrate surface is more easily obtained. Microgranules of compounds that cannot be dissolved can be put in suspension. In the case of dissolved ions or charged grains the transport rate to the substrate surface may be substantially increased by the application of an electric field.

We distinguish between a gaseous carrier medium and a vacuum by the mean free path of the transported particles. At high carrier gas pressures the particle has many collisions with the carrier gas molecules before arriving at the surface to be coated, but in a vacuum the particle moves from its source to the substrate surface with a very low probability of collision with residual gas molecules. The coating particles may be premixed, suspended, or dissolved in the carrier medium, which is depleted of coating particles as deposition proceeds. Often the coating material is continuously replaced by introduction into the carrier medium from some source such as an electrode in electrolytic deposition.

Thus, in order to characterize any given deposition process, we must specify the following:

- The carrier medium (solid, liquid, gas, vacuum)
- The nature of the coating particles (atom, molecule, ion, grain)
- The method of introducing coating material into the carrier medium (premixed or dissolved material, precipitation of premixed material in carrier, evaporation, electrochemical reaction at a supply electrode surface, bombardment of a supply surface by energetic particles)
- The surface reaction involved (simple condensation, chemical reaction between transported components at the surface, evaporation of carrier liquid, electrochemical reaction at surface, implantation)

- The mechanism by which the coating particles are transported from source to substrate (free flight, gaseous diffusion, liquid diffusion)

Table 1 characterizes a number of common coating methods. The choice of the deposition process will be dictated by the composition, purity, stoichiometry, and morphology required of the coating and physical properties sought.

If a crystal surface is exposed to air at a partial pressure on the order of 10^{-6} Torr, each atom composing the surface will on average be struck by one gas molecule per second. All technical surfaces are covered with impurity molecules on the substrate surface, which should be removed prior to any deposition process. Hence the partial pressure of unwanted impurities in the carrier fluid must be held substantially below 10^{-6} Torr during the deposition process if an atomically clean interface is to be preserved. Problems due to impurities in the carrier fluid are most easily solved by using UHV methods of deposition (less than 10^{-8} Torr), since once cleaned in the vacuum the time for the surface to become contaminated by residual gases can be made quite long.

We emphasize that there are no really general methods for thin-film deposition that can be used in all or even a majority of situations. Each device requires specific physical/chemical properties for the film and must be treated individually. Research to develop the optimum technique for a given film–substrate combination often takes years. A number of excellent books and papers that review the current status are available.[2–6] However, certain techniques have been of prime importance to the evolution of microscience, and it is the purpose of the following sections to introduce the key physical ideas behind the use of several of them, namely, UHV evaporation, sputtering, and chemical vapor deposition. Organic films are becoming of increasing importance and are also discussed.

The preparation of surfaces for thin-film deposition is often of key importance for obtaining good adhesion and other required properties of the interface. The removal of films in such a way as to create specific patterns or shapes in the film material uses etching technologies that are essentially the inverse of the deposition technologies. Discussions of these concepts are also included in this section.

3.2.2. Evaporation in Ultrahigh Vacuums

Evaporation in ultrahigh vacuum is particularly useful when it is necessary for the film material to remain uncontaminated by residual gases and where the interface between the substrate and the film must have specific prop-

TABLE 1. *Characterization of Common Coating Methods*

Process	Example	Carrier Medium	Coating Particles	Method of Injecting Coating Particles into Medium	Surface Reaction	Transport Mechanism
Oxidation	Growth of SiO_2 on silicon	Solid	Atoms	Diffusion from substrate	Chemical reaction with gas at surface	Diffusion
Settling	Cathodoluminescent screens (for CRT)	Liquid	Grains	Premixed	—	Settling under gravity
Electrophoresis	Coating of insulation on heater wires (tungsten, alumina)	Liquid	Charged grains	Premixed	Particle neutralization	Electromigration
Electroplating	Copper on steel	Liquid	Ions	From copper electrode	Ion neutralization	Electromigration
Liquid pyrolysis	CdS on metal	Liquid	Molecules	Predissolved	Chemical reaction at surface	Diffusion and stirring
Electrostatic toning	Xerography	Liquid or air	Charged grains	Premixed	Charge neutralization	Electromigration
Spray pyrolysis	CdS on metal	Air	Droplet of liquid containing dissolved reactants	Premixed	Chemical reaction at surface	Carried by air flow
Chemical vapor deposition	Tungsten on metal	Gas	Molecules	Premixed	Chemical reaction at surface	Diffusion
Evaporation	Aluminum on glass	Vacuum	Atoms	Heating source material	Condensation	Diffusion vacuum
Sputtering	Gold on silicon	Gas	Atoms	Bombardment by positive ions of carrier gas	Condensation	Diffusion
Painting	Spin coating of resists	Liquid	Organic molecules	Dissolved	Evaporation of solvents	—
Liquid-phase epitaxy	Gallium arsenide	Liquid	Atoms/molecules	Acts as its own carrier medium	Crystal growth	Diffusion
Molecular beam epitaxy	Gallium arsenide	Vacuum	Atoms/molecules	Evaporation from heated source	Condensation	Free flight in vacuum
Flame spraying	Crankshaft buildup (steel)	Gas	Molten droplets	Injection from heated source	Condensation	Transfer from carrier gas
Ion implantation	Boron into silicon	Vacuum	Ions	Plasma or ion gun	Implantation	Free flight in vacuum

erties that are influenced by contaminants. For this latter case the substrate surface once cleaned must remain uncontaminated until deposition is started.

Table 2 gives the relevant data for air at various pressures, illustrating that pressures of less than 10^{-8} Torr are necessary to maintain the substrate uncontaminated for a reasonable time after cleaning. In vacuum evaporation the following physical factors must be considered:

- The "cleanliness" of the vacuum
- The method of heating the evaporant
- The method for outgassing the evaporant
- The method for cleaning the substrate surface
- The method for ensuring film uniformity

While oil diffusion pumps have high pumping speeds and produce adequate pressure for many applications, they are also recognized as being quite "dirty," since even with careful trapping there is a good chance that the high-boiling-point silicone oils used in these pumps will eventually find their way into the evaporation chamber and contaminate the evaporant and substrate surface. Hence for coatings with stringent purity specifications oil-containing systems are avoided. In pumping down a system the major portion of the air to be removed is from atmospheric pressure to, say, 10^{-3} Torr. This is preferentially done by sorption pumping in which a material with a large surface-to-volume ratio (Xeolite) is cooled to liquid-nitrogen temperatures. The air is adsorbed on the Xeolite surfaces, and with this method very fast pumping speeds and pressures as low as 10^{-4} Torr may be obtained without contaminating the system. The pressure may be further reduced cleanly by the use of sputter ion pumps, titanium getter pumps, or cryopumps (operating at liquid-helium temperatures). Turbomolecular pumps are also increasing in popularity despite the minor possibility of contamination by bearing lubricants and back-

TABLE 2. Kinetic Data for Air

Pressure (Torr)	Mean Free Path (cm)	Number Impingement Rate $(s^{-1} \cdot cm^{-2})$	Monolayer Impingement Rate (s^{-1})
10^{-2}	0.5	3.8×10^{18}	4400
10^{-4}	51	3.8×10^{16}	44
10^{-5}	510	3.8×10^{15}	4.4
10^{-7}	5.1×10^4	3.8×10^{13}	4.4×10^{-2}
10^{-9}	5.1×10^6	3.8×10^{11}	4.4×10^{-4}

ing pumps. The walls of the vacuum chamber and all of the components within it contain gases adsorbed on their surfaces and dissolved inside. Much of this material can be removed by baking the whole system under high vacuum. A 150°C bake for 24 hr is extremely effective for stainless steel systems, but bakes up to 600°C or greater have been employed to obtain the lowest possible residual pressures.

If a solid or liquid is heated to a temperature T (°K) in vacuum, the number of molecules leaving a unit area per second is given from thermodynamic considerations by the relation

$$N = N_0 \exp - \left(\frac{\phi_e}{kT} \right) \tag{3.1}$$

where N_0 may be a slowly varying function of T and ϕ_e is the energy required to remove one molecule of the material from the bound state in the surface to the vapor state above the surface; ϕ_e is called the activation energy for evaporation and is usually given in electron volts. It is related to the latent heat of evaporation of a given molecule, Q by $Q = \phi_e e \times$ Avogadro's number J/mol. Data are usually available in the form of plots of the equilibrium vapor pressure p as a function of temperature. Figure 3.1 shows the data for a few elements. Data on more elements are available in ref. 7. The vapor pressure may be converted to evaporation rates from kinetic theory considerations by the relation

$$N = \frac{p}{(2\pi MkT)^{1/2}} = \frac{3.513 \times 10^{22} p}{(MT)^{1/2}} \text{ molecule·cm}^2/\text{s} \tag{3.2}$$

where M is the molecular weight of the vapor molecule in atomic mass units and p is in Torr.

For the deposition of thin films it is often useful to think of the deposition process in terms of the number of atomic layers of the film arriving per second at the substrate surface. The number of atoms per unit area corresponding to a monolayer can be calculated from the lattice constants and is about 10^{15} atom/cm^2.

Typically, vapor pressures at the source on the order of 10^{-2} Torr or greater are required for useful rates of film buildup. Heating the evaporant can present difficulties if high temperatures are required to obtain a reasonably fast evaporation rate or if the evaporant is highly chemically reactive at the operating temperature. Methods of heating that are commonly employed include resistive, eddy-current, electron bombardment, and laser heating.

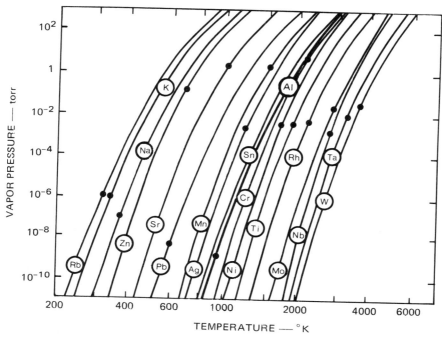

FIGURE 3.1. Temperature dependence of vapor pressure for several metal elements. Source: ref. 7. Note: ●—melting point.

Boats or holders are made of a highly refractory metal that can be heated by passing a current through it. Tungsten, tantalum, platinum, and graphite are preferred boat materials. Of course, if the evaporant is a conductor and remains solid at temperatures at which the vapor pressure is high, it can be made in the form of a wire or other resistive element and evaporated directly. Electron-beam heating offers the greatest flexibility (see Figure 3.2). In the preferred technology the material to be evaporated is placed in a recess in a water-cooled copper hearth. An electron current of about 100–500 mA is generated by a tungsten filament, hidden from direct viewing of the evaporant, and accelerated to a high voltage of 3–20 kV. The electron beam is magnetically directed onto a small spot on the evaporant material, which melts locally. Thus the material forms its own crucible and the contact with the hearth is too cool for chemical reaction. The main disadvantage of this type of source is that the substrate and coating film will be bombarded by X rays and possibly energetic ions as well as the neutral thermal evaporant. This may be avoided by using a focused high-power laser beam instead of the electron beam.

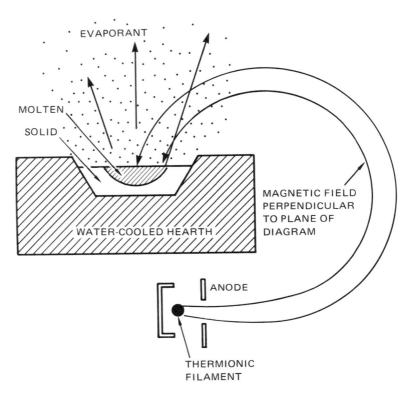

EVAPORANT

MOLTEN

SOLID

WATER-COOLED HEARTH

MAGNETIC FIELD
PERPENDICULAR
TO PLANE OF
DIAGRAM

ANODE

THERMIONIC
FILAMENT

FIGURE 3.2. Diagram of a 270° magnetized deflective electron-beam evaporation system.

Due to its high resistivity and ease of fabrication, graphite crucibles are particularly useful for eddy-current heating, since large amounts of rf power may be coupled into the crucible for heating evaporants that do not react with graphite.

Some compounds dissociate before vaporization and hence the higher vapor-pressure component is preferentially distilled from the evaporation source. To overcome this the various components are evaporated from separate sources at a rate appropriate to the molecular composition required in the condensate. This requires very precise temperature control, since evaporation rates depend exponentially on the temperature. In such circumstances it can be advantageous to elevate the temperature of the substrate during deposition to a temperature at which the compound condenses but the components do not.

Many materials contain absorbed gases and volatile impurities. For this reason a shutter is usually placed between the evaporation source and the substrate to protect the substrate surface from contamination during preheating of the source. The shutter is swung open when the experimenter is satisfied that the impurities in the vapor stream have been reduced to a tolerable level.

Having the substrate surface atomically clean before deposition can be of vital importance for adhesion and determining other physical properties that are dependent on the interface between the substrate and the film, such as electrical contact. Thermal removal of the contaminant materials can be used only with highly refractory materials, such as tungsten or graphite. Atomically clean single-crystal faces can be exposed by cleavage in a high vacuum. Bombardment by energetic inert gas ions (sputtering) is a good general way of removing surface layers but can result in rough surfaces due to preferential erosion of the different crystal faces of polycrystalline substrates and in unwanted ion implantation.

The geometrical relationship between the source and the substrate determines the uniformity of deposition and the ratio of the source evaporation rate to the substrate arrival rate. Most evaporation sources can be treated as point or line sources, so that the uniformity of deposition on the substrate can be estimated by simply using the Knudsen relation

$$\text{Arrival rate} \backsim \frac{\cos \theta}{r^2} \tag{3.3}$$

where r is the distance between the source and the substrate and θ is the angle between the normal to the substrate and the radial vector joining the source to the point being considered. A compromise must be effected between obtaining a film of uniform thickness over the substrate while maintaining a reasonable condensation rate without excessive waste of the evaporant. This topic is treated comprehensively in the literature.[2]

On the other hand, film uniformity can be maintained even for nonuniform sources by the simple expedient of moving the substrate so that each part of the sample integrates material from different source directions. Various combinations of rotary motions are typically used.

In summary, an evaporation source in which the desired material is transported in an ultrahigh vacuum to condense on a clean substrate is a preferred technique in cases where freedom from contamination both in the film and at the interface between film and substrate is of prime importance.

3.2.3. Sputtering

In Section 2.7 we discussed in some detail the physical processes involved when energetic ions strike the surface of a target material. One important effect is sputtering, namely, the ejection of atoms from the target surface. If the surface of a target material is bombarded by a uniform flux of energetic ions, the target may be utilized as a source of the material of which it is composed. In this section we discuss the properties of a sputter source for the deposition of thin films on a substrate and how these properties depend on the method used for producing the bombarding ions. We are concerned with the rate at which sputtered material leaves the target surface, the angular distribution of the ejected atoms, and the energy distribution of the ejected atoms. In addition, the effects of secondary particles produced at the target and the effects of the pressure of transport medium on the purity and morphology of the film are of interest.

3.2.3.1. Sputter Rate, Angular, and Energy Distributions

The number of atoms N per unit area per second leaving the target is given by

$$N = \frac{J_+}{ge} S(V, A, B) \tag{3.4}$$

where J_+ is the current density of the bombarding ions, g is the number of electronic charges per ion, and S is the sputter yield in atoms per incident ion, which is a function of the ion energy V, the ion species A, and the target material B.

The features of a typical sputter yield characteristic as a function of ion energy are shown in Figure 3.3. In the low-energy region the yield increases rapidly from a threshold energy, which is essentially independent of the ion species used, with a value about four times the activation energy for evaporation, and usually in the range 10–130 eV. The threshold energy decreases with increasing atomic number within each group of the periodic system. Above a few hundred volts there is a changeover region whose value depends on the ion species and the target used, after which the sputter yield increases much more slowly with increasing ion energy up to tens of kilovolts. In this region the sputter yields are typically in the range of 0.1 to 20 atoms per ion. At ion bombardment energies in the range 10 keV–1 MeV the sputter yield reaches a maximum, after which it gradually declines in value as a result of deep implantation of the incident ion.

When a target is bombarded by an ion stream, most of the energy is dissipated in the target. The rear face of the target is usually put in good thermal contact with a water-cooled copper holder; nevertheless, an important limitation to the sputtering rate is set by the maximum temperature to which the

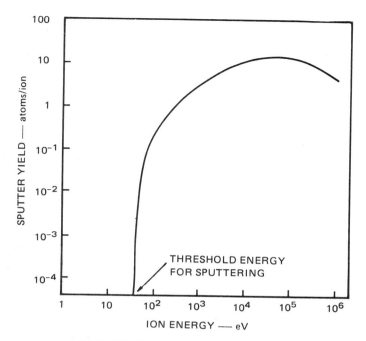

FIGURE 3.3. Typical sputter yield characteristic.

target surface may be heated. To be most efficient, ion energies in the range in which the sputter yield changes more slowly are utilized, namely, 0.5–3 kV, so as to minimize the energy required to eject an atom from the target. Even so, the energy dissipated at the target surface to liberate one atom can be 100–1000 times the activation energy for evaporation. Considerations such as this appear to limit useful sputtering rates to about one atomic layer per second. This is considerably less than the rates of about 1000 atomic layers per second available from a typical evaporation source operating at a vapor pressure of 10^{-1} Torr.

If a target is composed of several atomic species, the ratio of the species in the sputtered material may initially differ greatly from their ratio in the target because of their different probabilities of being ejected by the impinging ion. However, this condition will not persist for long, since the surface will be more rapidly denuded of the species with the higher yields until an equilibrium condition is reached in which the atoms are removed in the same ratio as they are present in the bulk. This is true whether the target is in the form of a compound, alloy, or mixture.

The process of sputtering appears to have two underlying mechanisms that contribute to the ejection of the sputtered atoms, namely, thermal spikes and collision sequences.[8] Thermal spikes arise from hot spots, where the energy of an incident ion heats up a small volume as it is slowed down in the lattice; evaporation occurs from this region for a short time until the heat has diffused away. The contribution to the sputter yield from thermal spikes is negligible at ion energies below 10 keV, but correlation between the energy distribution of the sputtered atoms at low energies with the heat of vaporization of volatile target materials (such as potassium) has demonstrated that the effect exists. Collision sequences, which are the major source of ejected atoms, are essentially cascades of atomic collisions resulting from the primary recoil of the lattice with the incident ion (see Section 2.7.2). The cascade ends with transfer of momentum to an atom at another part of the surface, resulting in its ejection.

In the case of an amorphous target material, the cascades take a random path, but for crystalline materials it has been shown that there is a focusing action related to the crystal planes.[9] Thus, polycrystalline or amorphous target materials should give a cosine angular distribution to the ejected atoms. However, since many polycrystalline materials have some preferential orientation to the grains of which they are composed, measurements of the angular distribution often show deviations from the cosine law, being "under cosine" or "over cosine."[10] For the purposes of calculating the uniformity of a thin film deposited by sputtering the cosine distribution gives sufficiently accurate results so the results generated for evaporation may be used directly.[2] Single-crystal targets should not be used for sputter sources because of their preferential directions of ejected atoms.

The theories predict and experiment confirms that the energy spectrum of the ejected atoms between the surface-binding energy of the target material and the primary recoil energy of the incident ion takes an inverse-square form.[8] Average ejection energies, however, are much higher than thermal energies, typically being between 10 and 100 eV (see Figure 3.4). This results in better adhesion of the sputtered atoms when condensing on the substrate and greater disorder in the condensed film and interface.

3.2.3.2. Implementation of Sputtering for Thin-Film Deposition

The major advantage of sputtering is its generality. In principle, any vacuum-compatible material can be coated by this method. Conceptually, the use of an ion gun focusing a beam of ions of controlled voltage, current density, and area on a target electrode might appear to be the simplest way to produce a source of sputtered atoms. However, this approach has practical difficulties,

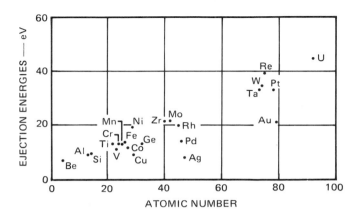

FIGURE 3.4. Average ejection energies under 1200-eV Kr-ion bombardment. Source: ref. 5.

and the utilization of electrical discharges in an inert gas with the target and substrate as electrodes has proved more useful. Both dc and rf gas discharges in a variety of configurations have been used. Power is required to generate the plasma from which the useful ions are to be extracted and accelerated to the target. In the case where the target material is insulating, means must be found to neutralize the surface charge built up on it by the impinging ions. The substrate is bombarded by inert gas molecules, electrons, photons, and negative ions from the plasma as well as the desired atoms from the target. Means must be found for avoiding deleterious effects from these extraneous sources.

The physics of electrical discharges in gases is highly complex,[11] and the following simplistic discussions are introduced only to enable the reader to grasp some basic principles. In its simplest form the electrical discharge is initiated between two parallel-plate electrodes, one composed of the target material and the other of the substrate that is to be coated. In the case of dc discharge, by applying a potential of several kilovolts between planar electrodes in an inert gas atmosphere (argon or xenon) at pressures between 10^{-1} and 10^{-2} Torr, a glowing plasma is formed, separated from each electrode by a dark space. The plasma is essentially at a constant positive potential with respect to the cathode. Electrons are drawn out of the boundary to the plasma close to the anode and accelerated across the anode dark space. These electrons then strike the anode, where the energy they carry is dissipated. Ions are drawn out of the other boundary of the plasma, accelerated across the cathode dark space, and strike the target cathode, causing sputtering. The plasma is replenished by electron–ion pairs formed by the collision of neutral molecules with the secondary electrons generated at the cathode surface, accelerated across the cathode dark space, and injected into the plasma region. The cathode dark space has a thickness of about that for an electron free path for an electron–ion collision. The anode dark space has a thickness determined by space-charge-limited flow of electrons from the plasma to the anode (see Section 2.4.6).

Direct current discharges have difficulties with initiating the discharge and with the sputtering of insulating materials. These disadvantages are eliminated by using high-frequency ac discharges (5 to 30 MHz). If one of the electrodes is capacitatively coupled to a rf generator, it develops a negative dc bias with respect to the other electrode. This is because electrons are much more mobile than ions, which hardly respond to the applied rf field at all. When the rf discharge is initiated, the capacitatively coupled electrode becomes negatively charged as electrons are deposited on it. However, this charge is only partially neutralized by positive ions arriving during the negative half-cycle. Since no charge can be transferred through the capacitor, the electrode surface retains a negative bias such that only a few electrons arrive at the surface during the next positive half-cycle, and the net current averaged over the next full cycle is zero.[12]

Ions diffusing from the plasma behave as if only the negative dc bias voltage existed and sputter the negatively charged electrode. The conducting anode, being directly connected to the power supply, receives pulses of energetic electrons accelerated essentially to the full amplitude of the rf field. If the target is an insulator, the target itself provides the capacitative coupling.

In the case of gas-discharge sputtering, where the substrate is an electrode of the system, the energy dissipated in the sample by electrons from the plasma can be substantial.[13] A small amount of heating can be advantageous, since the coating is continuously annealed, resulting in a stress-free film. However, excess heating can cause distortion of the film by differential expansion with the substrate, recrystallization of the film material, and eventually evaporation of the film material. Heating of the deposited film from electron bombardment limits the rate of film growth that can be utilized in sputtering, even before the target heating effect. To reduce electron bombardment of the film magnetic fields are used to deflect the electrons from striking the substrate. The presence of the magnetic fields has the additional benefit of increasing the efficiency of generating ions. The evolution of the various magnetically confined discharge configurations is given in ref. 6 (pp. 76–170).

Originally the coaxial magnetron configuration was used (Figure 3.5). It consists of electrodes made in the form of concentric cylinders, the target cathode being the innermost electrode and the substrates placed on the inside surface of the outer cylinder. The magnetic field is axial. Electrons from the cathode surface are trapped by the crossed electric and magnetic fields in the

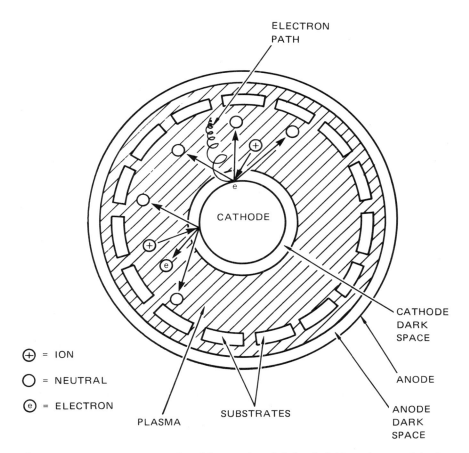

FIGURE 3.5. Coaxial magnetron. In axial-magnetic radial-electric fields an electron originating at the cathode and accelerated across the cathode dark space makes spiral orbits of decreasing radii as it loses energy to the plasma and drifts toward the anode.

cathode dark space and can escape only by losing energy in ionizing collisions. The loss of energy results in the electrons moving in a tighter spiral path as they drift towards the anode. As this process continues the energy carried by the electrons becomes smaller and the spiral path becomes tighter. This process results in a greater efficiency of ionization and a substantial lowering of the power dissipated by the electrons at the anode. The configuration can be used for dc or rf discharges.

The planar magnetron (Figure 3.6) is a simple extension of the parallel-plate discharge in which magnets (permanent or electromagnets) are placed behind the cathode. The magnetic field lines enter and leave normal to the

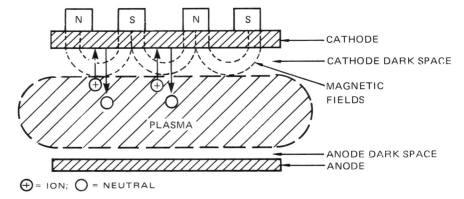

FIGURE 3.6. Planar magnetron. Electrons originating at the cathode surface are confined by the fields of permanent magnets placed behind the cathode.

cathode surface. The transverse component of the magnetic field in front of the target is arranged to be in the range 200–500 G. Electrons are trapped in the crossed electric and magnetic fields, both enhancing the efficiency of the discharge and reducing the energy dissipated by electrons arriving at the anode.

Circular magnetrons (sputter guns or "S" guns) in which the cathode is in the form of a ring that surrounds a planar disk-shaped anode are also popular (Figure 3.7). Magnets behind the cathode enhance the discharge and the electrons are collected by the anode. A large fraction of the sputtered material from the target cathode is ejected in a forward direction and deposited on a substrate that no longer needs be an electrode of the system. In this way very high power densities (50 W/cm²) in the erosion zone of the target can be dissipated.

The main disadvantage of planar and circular magnetrons is the fact that the deposition rate is not uniform across the region where the substrates are placed. This, as in evaporation, is overcome by an appropriate mechanical motion of the substrate table so that each substrate is exposed to the same average flux.

If an active gas such as oxygen, water, or hydrogen sulfide is intentionally added to the inert working gas, it can react with the target material either at the target surface or in the gas phase, resulting in the deposition of a new compound. By varying the pressure of the active gas, control can be gained on the stoichiometry of the film. This technique is called reactive sputtering.

Another variant of the sputtering process is ion plating. This essentially involves purposeful bombardment of the substrate with positive ions from a glow discharge so that the film is being sputter eroded at a slow rate at the same time as it is receiving a film material from an appropriate source, usually a simple evaporation source. This technique supposedly improves adhesion and film uniformity.

3.2.4. Chemical Vapor Deposition

In chemical vapor deposition (CVD) the constituents of a vapor phase, often diluted with an inert carrier gas, react at a hot surface to deposit a solid film.[14] Its importance lies in its versatility for depositing a large variety of elements and compounds at relatively low temperatures and at atmospheric pressure. Amorphous, polycrystalline, epitaxial, and uniaxially oriented polycrystalline layers can be deposited with a high degree of purity and control.

Aspects of CVD include the chemical reactions involved, the thermodynamics and kinetics of the reactors, and the transport of material and energy to and from the reaction site. The following is a list with examples of some of the common types of chemical reactions used in CVD.

Pyrolysis The simplest CVD process is pyrolysis, in which a gaseous compound decomposes on a hot surface to deposit a stable residue. Examples are the following: deposition of pyrolytic graphite from methane (CH_4), which

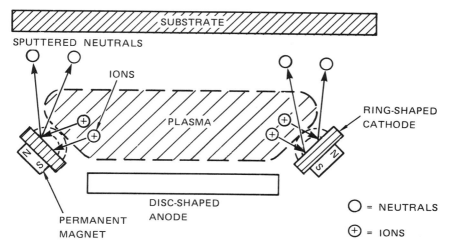

FIGURE 3.7. Circular magnetron. Electrons originating at the cathode are confined by fields from permanent magnets placed behind the cathode. They lose energy in spiral paths in the plasma and are collected by the disk anode. The substrate, not being an electrode of the system, collects only sputtered neutrals.

takes place at a substrate temperature of 2200°C; deposition of silicon from silane (SiH_4), which takes place in the range 800–1350°C; and deposition of nickel from the carbonyl form [$Ni(CO)_4$], which takes place at about 100°C.

Reduction   Hydrogen is the most commonly used reducing agent. Examples are deposition of silicon by the hydrogen reduction of silicon tetrachloride, which takes place at about 1000°C, and deposition of tungsten by the hydrogen reduction of tungsten hexafluoride, which takes place at about 800°C. Hydrogen reduction is also used to accelerate the pyrolytic process by removal of the unwanted byproducts as gaseous hydrogen compounds, for which less energy is required.

Oxidation   Silicon dioxide films can be deposited by the reaction of silane with oxygen.

Nitridation   Silicon nitride films can be deposited by reaction of silane with ammonia.

Carbidization   Titanium carbide films can be deposited by reaction to titanium tetrachloride with methane at a substrate temperature of 1850°C.

Chemical-transport reaction   For these processes the transport of the desired material from the source to the substrate on which it is to form a film depends on the difference in equilibrium constants between the reactant source and a carrier phase, and the substrate and the carrier phase, when each are held at different temperatures. For example, the deposition of gallium arsenide by the chloride process depends on the reversible reaction

$$6GaAs_{(g)} + 6HCl_{(g)} \underset{T_2}{\overset{T_1}{\rightleftharpoons}} As_{4(g)} + As_{2(g)} + GaCl_{(g)} + 3H_{2(g)},$$

where T_1 is the temperature of the solid GaAs source, T_2 is the temperature of the solid GaAs substrate, and $T_1 > T_2$. This allows, in effect, indirect distillation of the gallium arsenide from the hot source at temperature T_1 to the cooler substrate at temperature T_2 through an intermediate gas phase of different chemical composition.

Spray pyrolysis   In this process the reagents are dissolved in a carrier liquid, which is sprayed onto a hot surface in the form of tiny droplets. On reaching the hot surface the solvent evaporates and the remaining components react, forming the desired material. An example is the formation of cadmium sulfide films by spray pyrolysis of cadmium chloride and thiourea dissolved in water with the substrate at about 300°C.[15]

The choice of chemical reactions and the design of the reaction chamber to deposit specific films by CVD is a highly specialized field. The reader who wishes to explore this field more thoroughly is recommended to read the excellent review article in ref. 6, p. 258, before proceeding to the open literature to find out what processes may have been used previously for the specific film he has in mind.

3.2.5. Organic Thin Films

Organic compounds are used extensively in microfabrication, principally as photo- or electron-beam resists. These are materials whose structure and solubility change when exposed to radiation. Materials that are rendered insoluble in a developer solution by the action of radiation are called negative resists, while those that are rendered soluble are called positive resists. The material left behind after pattern generation must be chemically inert, or treated to become inert, so that removal of the unprotected substrate may proceed without affecting the resist or the substrate under the resist.

The principal constituents of a resist solution are a polymer, a sensitizer, and a solvent. For negative resists the polymers are characterized by unsaturated carbon bonds capable of forming longer or cross-linked molecules under the action of the radiation and in the presence of the sensitizer. For positive resists saturated carbon bonds are broken by the same actions rendering the resist film soluble. Resists are generally applied by uniformly coating the surface with the resist solution and allowing the solvent to evaporate. Wafers of silicon are usually rotated on a spinning wheel at high speed so that centrifugal forces throw excess solution to the edge of the wafer and the residue on the wafer is held by surface tension. In this way thin, uniform organic films down to 0.1 μm thick may be formed. Because of its importance to pattern generation, resist technology is discussed in more detail in Chapter 5.

Pinhole-free polymer dielectric films down to 100 Å thick have been made by radiating monomer layers with ultraviolet light or energetic electrons.[16] Materials that have been polymerized by electron bombardment include dimethyl polysilosane, butadiene, styrene, and methacrylate. Direct deposition by electrical discharges in gases containing organic vapors can also be utilized to make polymeric films. A large number of substances have been explored and the mechanism has been identified as the formation of cross-linked, high-molecular-weight polymers by activation of the adsorbed monomer by ionic bombardment of the surface. The organic vapor simply dispenses the monomer to the surface.[17]

Organic insulator films from 1 to 10 monolayers thick have also been formed by the classical Langmuir–Blodgett process.[17] This process consists of spreading a monolayer of film-forming molecules (such as stearic acid) on an aqueous surface, compressing the monolayer into a compact floating film, and

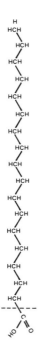

FIGURE 3.8. Structure of stearic acid $[CH_3(CH_2)_{16}COOH]$.

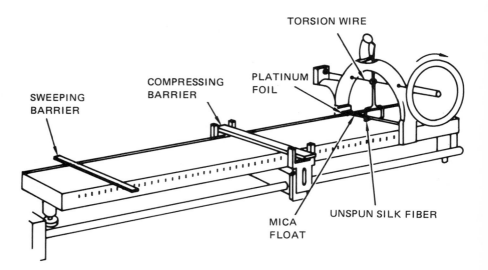

FIGURE 3.9. Film balance apparatus. Source: ref. 18.

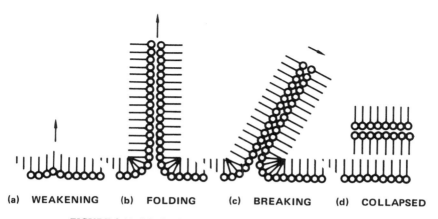

(a) WEAKENING   (b) FOLDING   (c) BREAKING   (d) COLLAPSED

FIGURE 3.10. Mechanism of monolayer collapse. Source: ref. 18.

transferring it to a solid substrate by passing the substrate through the water surface.[18] Film-forming molecules have a polar- or water-soluble group attached to a long hydrocarbon structure that is sufficiently large to make the complete molecule insoluble in water. The archetypical molecule used is stearic acid, whose molecular structure is shown in Figure 3.8. Monolayer films are studied using a film balance (Figure 3.9). This consists of a long, shallow trough filled with pure water. A mica strip floats on the water and is attached to a stirrup by an unspun silk fiber and to the sides of the trough by thin, flexible platinum foils. The stirrup is fixed to a calibrated torsion wire that indicates the surface pressure, usually measured in dynes per centimeter. A compressing barrier of brass can also be moved along the trough as shown. A small amount of the film-forming material in a volatile solvent is spread between the float and the main barrier. The barrier is then moved toward the mica float, thus compressing the film. Compression results in the molecules forming a carpetlike structure, with the soluble group attached to the water surface and the long chains perpendicular to the surface, as shown in Figure 3.10(a). The balance is used to measure the pressure–area isotherms that characterize the film material. The pressure against the mica float increases with decreasing area until it reaches a constant value or even begins to fall. This is called the collapse point and is usually in the rage 10–100 dyn/cm. Electron microscopy has suggested that collapse occurs as the monolayer distorts and forms more layers. Figure 3.10 shows the probable mechanism of collapse of a monolayer of molecules consisting of a polar group (small circle) attached to the water surface and a long chain hydrocarbon (line).

If a cylinder is rotated in a monolayer held at constant surface pressure, the monolayer is transferred quantitatively from the water surface to the cylinder and films of precisely known monolayer thicknesses can be built up. Langmuir films have been utilized for rust prevention and as very thin insulating films for MOS devices,[18] and it seems likely that other interesting uses for such films should be forthcoming as more microdevices requiring them are conceived.

3.2.6. Surface Preparation

The condition of a surface prior to deposition of a film upon it is of prime importance for adhesion of the film and other interface properties. The major culprits are oil or greasy residues left from protective coatings or fingers. These are best removed by warm ionic detergents that are ultrasonically agitated to render them water-soluble, followed by rinsing in pure water, which is usually in plentiful supply through modern filtering and deionizing techniques. Other sources of surface contamination are corrosion or the reaction products formed between the material of the surface and the ambient to which it has been exposed. These can be removed by using chemical or electrochemical processes, which react at the surface with the corrosion or the underlying material-forming compounds that can be dissolved in a carrier fluid (if possible, water). Removal of the reactants and reaction products by rinsing with pure water is again important.

Even after such careful treatment most surfaces remain contaminated at levels from one to several molecular layers with adsorbed oxygen, water, or other air contaminants. At worst, adherent layers of oxide many molecular layers thick may be found, such as form on aluminum or silicon surfaces. These layers can only be removed in high vacuum to prevent them from reforming. Outgassing and the removal of weakly adsorbed layers can be accomplished by baking at moderate temperatures of up to 600°C, However, heating to remove the last monolayer of oxygen is practical in only a few cases of highly refractory materials, such as tungsten, that can be heated to the very high temperature (3000°K) required to desorb the last monolayer of oxygen, or in materials that have low adsorption energies for oxygen, such as SiO_2. Sputtering away the surface with an inert gas such as argon is the most popular way of removing this last atomic layer of contamination, but low bombardment energies (< 500 V) should be used to prevent surface damage and low current densities (100 $\mu A/cm^2$) should be used to prevent heating of the surface, which may give rise to thermal etching.[19] Electron bombardment can also be effective in removing certain residual monolayers.

The cleavage of single crystals in a high vacuum produces perhaps the most perfect surfaces from the point of view of cleanliness. However, many cleaved crystal surfaces exhibit cleavage steps. If atomically smooth surfaces are required, these can be obtained only by the cleavage of strongly layered crystals such as mica. Atomically smooth surfaces can also be obtained on glasses and metals by long annealing close to the melting point (see ref. 4, p. 63).

3.2.7. Etching

Removing a film from a substrate, usually in a pattern defined by a protective resist, is called etching. Etching can be achieved by chemical methods, whereby an easily removable compound is formed with the film material, physical methods, whereby the film is physically removed, as by abrasion or sputtering, or by a combination of both. Certain highly chemically active molecules or ions can only be formed in a glow discharge (or plasma), and the usage of such species is called reactive plasma etching.

3.2.7.1. Chemical Etching

Wet chemical etching is widely used. A solution is found that dissolves the film material but not the resist or the substrate. The reaction products must dissolve in the carrier fluid or otherwise be carried away. In etching very small channels or cavities the solution may quickly become saturated and must be removed by ultrasonic or other means of agitation so that fresh solvent may continuously be brought to the surface to be etched. If one of the reaction products is gaseous, small bubbles will form at the interaction surface. Ultrasonic removal of such bubbles also helps to keep the solution stirred. Making the film an electrode of an electrolytic cell allows the chemical reaction to be electrically driven, resulting in the deplating of the film layer and the removal of the reaction products by ionic flow. The choice of reagents for particular film removal properties is extremely vast. For application to specific insulators, dielectrics, semiconductors, and metals the reader is referred to the excellent tables compiled in ref. 6, p. 433.

Some etchants attack a given crystal face of a material much faster than others, giving rise to anisotropic etching.[20] Anisotropic etchants for silicon have been studied extensively. For example, in the case of the water–ethylene diamine pyrocatechol (EDA) etchant the etch rates of $\langle 100 \rangle$, $\langle 110 \rangle$, and $\langle 111 \rangle$ oriented faces of silicon are in the ratio of approximately $50 : 30 : 3$ μm/

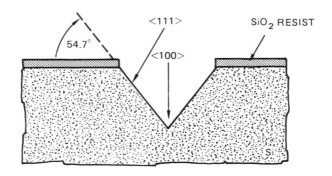

(a) **THROUGH A PINHOLE <100> ORIENTED Si**

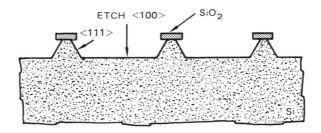

(b) **THROUGH A SQUARE WINDOW FRAME PATTERN <100> ORIENTED Si**

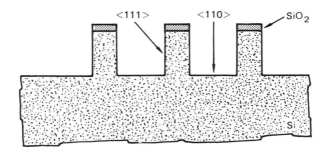

(c) **THROUGH A SQUARE WINDOW FRAME PATTERN <111> ORIENTED Si**

FIGURE 3.11. Anisotropic etching of silicon. Source: ref. 22.

hr, respectively, at 100°C. Anisotropic etching of ⟨100⟩ Si through a patterned SiO$_2$ mask creates precise V-shaped grooves, the edges being ⟨111⟩ planes at an angle of 54.7° from the ⟨100⟩ surface, as shown in Figure 3.11(a).[21]

If there is a pinhole in the oxide, the etch front through the hole will propagate into ⟨100⟩ silicon to form a perfect square-topped pit in the form of an inverted pyramid. On the other hand, if a square window frame pattern is opened in the oxide mask, a square-based pyramid structure can be etched, as shown in Figure 3.11(b). If ⟨110⟩ oriented silicon is used, essentially straight-walled grooves with sides of ⟨111⟩ orientation can be formed, as shown in Figure 3.11(c). In this way 0.6-μm-wide openings on 1.2-μm centers and 600 μm deep have been made.[22] Scanning electron micrographs of the intersection of anisotropically etched planes show them to be sharp to the resolution of such instruments (about 50 Å) and likely to intersect in a single row of atoms.

An important disadvantage of chemical etching is the undercutting of the resist film that takes place, resulting in a loss of resolution in the etched pattern (see Figure 3.12). In practice, for isotropic etching the film thickness needs to be about one-third or less of the resolution required. If patterns are required with resolutions much smaller than the film thickness, anisotropic or other special etching techniques must be sought.

Gas-phase etching methods using vapors that react with the substrate and forming gaseous reaction products have been used. For example, ahydrous hydrogen fluoride between 150–190°C at a pressure between 0.1 and 3.0 Torr reacts specifically with silicon dioxide, but has no effect on silicon, silicon nitride, aluminum, and other common materials.[23] At high pressures gas-phase reactions tend to be isotropic, but directing the gas flow in jets could conceivably help to reduce undercutting.

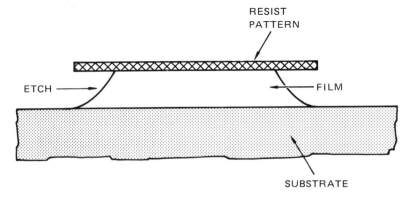

FIGURE 3.12. Undercutting of the resist film through chemical etching.

3.2.7.2. *Sputter Etching*

Directed ion bombardment of the surface (or sputter etching) is used to obtain good aspect ratios. Sources of ions for sputter etching are generally based on the Kaufman ion thruster developed by NASA (Chapter 1, ref. 47) (Figure 3.13). A magnetically confined gas discharge is obtained between a thermionic cathode and a concentric-anode cylinder. To extract the ion beam from the discharge region a potential difference is applied between a pair of grids with aligned holes. The ions are injected into the working space as a well-collimated energetic beam. The beam is made neutral by extracting low-energy electrons from an auxiliary thermionic cathode so that it can be used for sputtering insulators as well as conductors.

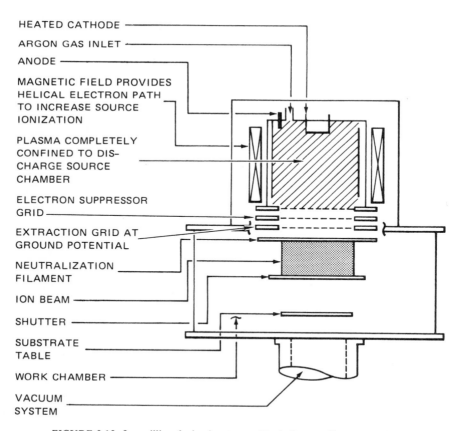

HEATED CATHODE

ARGON GAS INLET

ANODE

MAGNETIC FIELD PROVIDES HELICAL ELECTRON PATH TO INCREASE SOURCE IONIZATION

PLASMA COMPLETELY CONFINED TO DIS- CHARGE SOURCE CHAMBER

ELECTRON SUPPRESSOR GRID

EXTRACTION GRID AT GROUND POTENTIAL

NEUTRALIZATION FILAMENT

ION BEAM

SHUTTER

SUBSTRATE TABLE

WORK CHAMBER

VACUUM SYSTEM

FIGURE 3.13. Ion milling device (sputter etching). Source: Chapter 1, ref. 47.

Ion optical systems utilizing liquid-ion sources are now being built that are capable of forming intense energetic beams focused into very small diameter spots (see Section 2.4.5). The spot can be positioned on a sample surface by deflector electrodes to directly generate and sputter-etch a pattern on the surface. This technology, while still in its infancy, holds promise for the controlled fabrication of submicron structures without the use of resists (see Section 4.9.2). The selection of substrates and resists for sputter etching is based simply on relative sputter coefficients. However, this offers limited opportunity for control, since the sputter yields of most materials do not differ by large factors.

3.2.7.3. Reactive Plasma Etching

Interest in reactive plasma etching is currently very great because, in contrast to sputtering, large differences in etch rates for different materials can be obtained. In this technique the substrate is exposed to a plasma of a reactive gas, but there is no high-voltage acceleration of ions toward the substrate. Since the ions are not highly energetic, sputtering is not in itself an important surface-removal mechanism. Rather, the reactive gas component appears to be first adsorbed on the surface, where it is dissociated by electron or ion bombardment from the plasma, then it reacts with the film material, forming a gaseous product molecule. The reaction products are desorbed as gases and pumped away. Anisotropic etching can occur because the impingement rate on the horizontal surfaces is much greater than at the sidewalls at the low pressures used, where the mean free path of the molecules is typically much larger than the depth to be etched. Because of its chemical nature, a high degree of control over the relative etch rates for the film, resist, and substrate can be obtained by the choice of suitable materials.

The first important application of the technique was the stripping of photoresists in a low-pressure ($\sim 10^{-2}$ Torr) oxygen rf glow discharge (13.56 MHz).[24] Hardened photoresists are required to be extremely inert polymers so as to resist the etchants that attack the substrate they cover. This same property makes them difficult to remove by wet chemical processes or even organic solvents. However, when oxidized in a glow discharge, the oxides of carbon, being gaseous, are easily pumped away, whereas the oxides of many substrate materials (silicon, aluminum) rapidly form a protective film. This technique is generally used to remove carbonaceous contaminants.

Glow discharges in low-pressure fluorocarbon (CF_4) vapors or mixtures of fluorocarbons with various additive gases (usually oxygen) are currently being used for patterning silicon, silicon dioxide, silicon nitride (Si_3N_4), and other materials used in IC fabrication.[25] The chemically active radicals formed by the glow discharge react with silicon and its compounds to form gaseous SiF_4. CF_4 is assumed to decompose into CF_2 and atomic fluoride.

$$CF_4 \rightarrow CF_2 + 2F \qquad \text{(in atomic form)}$$

The reactions are then assumed to proceed as follows:

$$Si + 4F \rightarrow SiF_4$$
$$SiO_2 + 4F \rightarrow SiF_4 + O_2$$
$$Si_3N_4 + 12F \rightarrow 3SiF_4 + 2N_2$$

However, the detailed mechanisms are apparently much more complex.[26] Additives to fluorocarbon such as O_2, H_2, N_2, H_2O, and CF_4 are used to enhance or suppress various chemical reactions. Table 3 indicates that photoresists can be used for patterning Si and Si_3N_4. However, for the patterning of SiO_2 the addition of hydrogen to the fluorocarbon instead of O_2 is required to increase the etch rate of SiO_2 over silicon.

Reactive plasma etching is currently being intensively studied, and it is anticipated that both applications and an understanding of the detailed mechanisms involved will be expanded.[26]

TABLE 3. *Relative Etch Rates in Fluorocarbon (96% CF_4 + 4% O_2), Glow Discharge for Common Microelectronic Materials*[a]

Material	Relative Etch Rate
Si_3N_4	100
Si (111)	690
Si (poly)	990
SiO_2 (thermal)	40
SiO_3 (CVD)	120
Al	0
W	100
Mo	100
Ti	100
Ta	100
Waycoat I (Photoresist)	20
A21350J (Photoresist)	40

[a]Source: ref. 24.

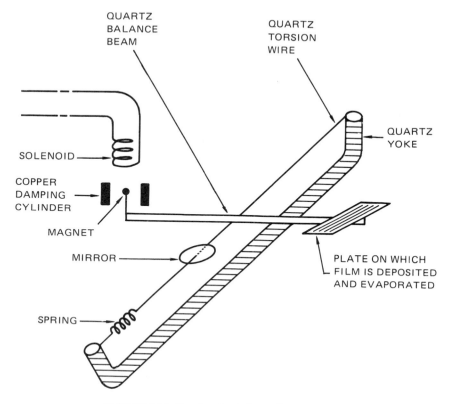

FIGURE 3.14. Simple microbalance. Source: ref. 5.

3.2.8. Thickness and Rate Measurement

While there are a very large number of physical phenomena that are functions of film thickness, there are only a few that are conventionally used for deposition control purposes because of their simplicity and convenience. These are usually based on classical measurements of mechanical, electrical, or optical properties. For ultrathin films less than a few tens of monolayers the classical concept of thickness may not be meaningful owing to the possible non-uniform structure of the films. The weight per unit area of a film, however, does provide an unambiguous measure of the quantity of material deposited, and an equivalent thickness may be inferred from an assumed density; it also has the advantage of complete generality; that is, it may be used with any film molecule. Weights on the order of 10^{-9} g/cm^2 or less than $\frac{1}{100}$th of a monolayer may be detected by gravimetric methods in vacuums.

The simple *microbalance* (Figure 3.14) consists essentially of a horizontal quartz torsion wire (120 mm long and 40 μm in diameter) fixed to a quartz frame and held taut by a quartz spring. The balance beam attached to the center of the torsion wire is also of quartz. At one end is a plate on which the material is evaporated, and on the other is a small magnet. Above the magnet is a small solenoid. Displacement, brought about by the deposition of the film on the plate, is magnetically compensated for by an appropriate current through the solenoid. A mirror on the torsion wire enables precise optical determination of the null position to be made. A copper cylinder surrounds the magnet to enable eddy-current damping of the balance beam oscillation to be effective. Such balances used in high-vacuum systems can be outgassed by baking to 500°C and are sensitive to 10^{-8} g. They are, however, very delicate and have largely been replaced by *quartz crystal oscillator* monitors. These devices operate on the principle that the resonant frequency of a piezoelectric crystal is changed by added weight. A quartz crystal plate is accurately ground to a thickness t, and metal electrodes evaporated on each side. When the electrodes are attached to an oscillating circuit, resonance occurs at a frequency f, given by

$$f = \frac{v_T}{2t} = \frac{A}{t} \text{ Hz} \tag{3.5}$$

where v_T is the velocity of sound across the quartz wafer and A is the frequency constant. For the usual orientation along which quartz crystals are cut (called the AT cut) $A = 1670$ kHz/mm. If an amount of material Δm is deposited on the crystal area S, the effective increase in film thickness is given by

$$\Delta t = \frac{\Delta m}{\rho S} \tag{3.6}$$

where ρ is the quartz density. Using Eqs. (3.5) and (3.6) we get

$$\Delta f = \frac{-f^2 \Delta m}{A \rho S} \quad \text{or} \quad \frac{\Delta m}{S} = \frac{-\Delta f}{f^2} A \rho \text{ g/cm}^2 \tag{3.7}$$

Since the resonant frequency can be monitored continuously, the instrument can be used to measure the rate of deposition. Instruments based on this principle are commercially available and ruggedly built, yet they can measure routinely to 10^{-9} g/cm^2 and, in circumstances in which errors due to extraneous effects are reduced as much as possible, to 10^{-12} g/cm^2.

Electrical and optical methods are useful for films above the thickness at which the films may be considered coherent. The *resistance* of a metallic film on an insulating substrate may be measured by bridge methods and the thickness inferred, provided that a thickness–resistance calibration has been previously performed. The thickness of an insulating film on a conducting substrate may be inferred from its *capacitance* per unit area. This may be obtained by depositing a metallic electrode of known area on its surface and again using ac bridge methods.

A portion of the vapor from the deposition stream may be allowed to pass through a slit and be ionized by energetic electrons in an ionization gauge arrangement. The *positive ion current* measured in the gauge will be proportional to the deposition rate. Such devices have to be calibrated but are useful when repetitive work has to be done.

Electron-beam evaporation sources (see Section 3.2.2) generate their own ion current, since the beam of bombardment-heating electrons also passes through the vapor. By arranging to collect a portion of the ion current, the source can be made self-monitoring. If the substrate being coated has a portion of its surface appropriately shadowed, a sharp step may be produced whose height is of the thickness of the film. The step height can be measured with a *diamond stylus* arrangement. In commercially available instruments the vertical movement of a tip of radius about 1 μm and weight 0.1 g is converted by a transducer to an electrical signal, which is amplified and recorded. Film thicknesses and surface irregularities as small as 25 Å can be detected by this method.

Another method of measuring the step height uses *optical interference* methods. The film and step are first coated with aluminum to make them highly reflecting. Then an optical flat is placed up against the surface at a small angle, forming a wedge-shaped air gap (Figure 3.15). As is well known, illumination with parallel monochromatic light wavelengths produces evenly spaced interference fringes as intervals where the air-gap thickness differs by $\lambda/2$. The displacement of the fringes at the step (Figure 3.15) gives a direct measure of the height of the step in terms of λ. If L is the fringe spacing and ΔL is the displacement of the fringes at the step, then the thickness t is given by $t = \Delta L / L(\lambda/2)$. Film thicknesses of less than 100 Å can be estimated by such methods.

3.3. Nucleation, Composition, Morphology, and Structure

3.3.1. Overview

Thin films usually have physical properties that differ substantially from bulk samples of the same material. This is due to both the influence of the surface and the interface and to the structure formed by the initial condensa-

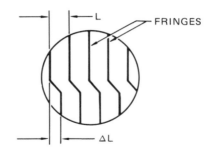

VIEW THROUGH THE MICROSCOPE

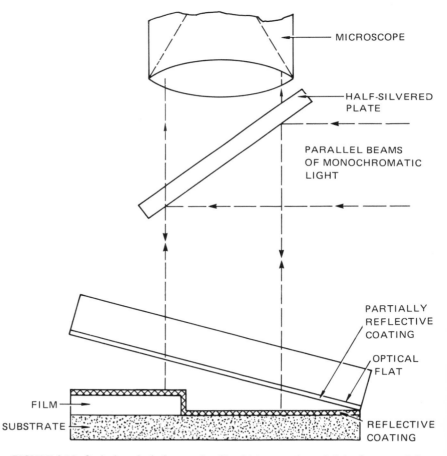

FIGURE 3.15. Optical method of measuring film thickness and step height. Source: ref. 5.

tion process. In condensation from a vapor source film growth has been observed by electron microscopy (see Chapter 6) to proceed according to the following stages:

- Nucleation: Small nuclei appear (5 Å diameter) statistically distributed over the surface.
- Growth of nuclei: Nuclei grow into larger, three-dimensional islands that are often in the form of small crystals.
- Coalescence of islands: The interconnected islands form a network containing empty channels.
- Filling of the channels.

Until the last stage is complete we expect novel differences in such physical properties as the electrical conductivity or film density. Completion of this stage takes place in the range of a hundred to a few thousand angstroms apparent thickness, depending on the substrate, its temperature, and the evaporant. Beyond the last stage the film may become amorphous (short-range order), randomly oriented polycrystalline, uniaxially oriented (in direction of growth) polycrystalline, or indentically oriented (epitaxial), depending upon the conditions of growth. When attempting to produce very thin films with specific properties, the initial growth phase is obviously of crucial importance.

3.3.2. Nucleation and Initial Growth

When a particle impinges upon a surface, a number of interactions between them may take place, depending upon the chemical binding energy between the surface atoms and the impinging particles, the surface electric fields due to the asymmetry of the lattice at the surface, the polarizability of the impinging particle, the lattice constants of the substrate and film materials, the temperature of the substrate, and the energy of the impinging particle.[27]

In general, we may suppose that the impinging particle after striking the surface will be held to it by polar forces resulting from the surface electric fields. Its excess energy will enable it to move a certain distance along the surface until it loses energy to the lattice and gets chemically bound to a permanent site, or gains sufficient energy from the lattice to escape the surface electric fields. On its travels along the surface the particle may be temporarily bound to sites with weak bonds. The time τ' it spends on the weakly bonded sites depends upon the activation energy for migration, ϕ_m and the surface temperature T, and typically follows the relationship

$$\tau' = \tau'_0 \exp\left(\frac{\phi_m}{kT}\right) \tag{3.8}$$

TABLE 4. *Comparison of Activation Energies for Migration (ϕ_M) and Desorption (ϕ_D)*

System	ϕ_m (eV)	ϕ_D (eV)
Barium on tungsten	0.65	3.8
Cesium on tungsten	0.61	2.8
Tungsten on tungsten	1.21	5.83

The total time τ the particle spends on the surface is characterized by the activation energy for desorption, ϕ_D, that is,

$$\tau = \tau_0 \exp \left(\frac{\phi_D}{kT} \right) \tag{3.9}$$

Table 4 gives some typical values. At high substrate temperatures τ' is so short that the adsorbed particles essentially behave as a two-dimensional gas; at intermediate temperatures, where τ' is finite, the adsorbed particles hop from site to site in a random walk; whereas at low substrate temperatures, where τ' is very long, the particles soon lose their excess energy to be essentially permanently bonded. These processes have been well illustrated in the case of barium adsorbed on tungsten, where the low work function of barium on tungsten enables the movement of the barium to be observed in thermionic-emission microscopes (see Section 6.5).

Impinging particles are usually trapped on a cold surface, with kinetic energies corresponding to a higher temperature than that of the substrate, depending on their source, and move over the surface as a two-dimensional gas until it is possible for them to give up this excess energy and condense into a solid. Condensation can only occur if sites for nucleation exist where the adsorbed particles can give up their excess energy. If the supply of the impinging particles exceeds their desorption rate, the film becomes supersaturated and condensation becomes more likely.

The conditions under which condensation starts depend on the ratio of the desorption energy ϕ_D to the heat of sublimation of the condensed film ϕ_S. If $\phi_D \ll \phi_S$, condensation occurs without supersaturation. If $\phi_D \cong \phi_S$, condensation requires a moderate level of supersaturation. If $\phi_D > \phi_S$, a high degree of supersaturation is required to start condensation. Obviously, after the film has started to form we soon reach the condition $\phi_D = \phi_S$.

Theoretical models for the condensation process usually assume that when the adsorbed atoms collide, there is a certain probability of their forming a bonded pair. Further collisions of atoms with the pair results in the formation of greater aggregates. The reverse process of aggregate disintegration proceeds simultaneously. The capillary theory (ref. 5, p. 77), valid for aggregates containing a large number of atoms (>100), uses thermodynamic methods to show that an aggregate will be stable if its radius exceeds a value r^* given by

$$r^* = \frac{2\sigma_{cv}V}{kT \ln p/p_e} \qquad (3.10)$$

where σ_{cv} is the condensate–vapor interfacial free energy for a molecular volume V condensed from the unsaturated vapor at pressure p to its adsorbed state at equilibrium pressure p_e at a surface temperature T.

The theory has been modified to take into account the nonspherical shape taken by the aggregate as a result of surface tension forces at the substrate. When the critical nucleus contains only a few atoms (1–10), it is necessary to consider the effects of individual bonding of the atoms to themselves and the substrate lattice. One approach has been to modify the capillary theory to take these into account[28] whereas another has been to attempt to consider in detail the statistics of aggregate formation.[29] In the range appropriate to the theories they yield results qualitatively in accord with observations.[30] The possibility for future quantitative comparison of theory and experiment exists, as the assumptions made in the theories can be made to correspond better to reality.

The process of nucleation and growth has been confirmed by electron microscopy. As evaporation proceeds, the number of resolvable islands (> 5 Å in diameter) increases with a random position distribution until a saturation island density in the range 10^{10}–10^{12} atoms/cm^2 (for metal vapors condensing on insulating substrates at room temperature) is obtained, that is, the islands are 100–1000 Å apart. As evaporation proceeds further, the islands grow and the island density decreases by mutual coalescence. The density of nucleation sites can be increased by external agents, such as by electron bombardment of the substrate. Nucleation also occurs preferentially in crystalline surfaces at dislocations, impurity sites, and other irregularities in the lattice. This effect is utilized for the "decoration" of dislocations in single crystals to render them visible.

Impinging particles originating from sputtered sources have much higher effective temperatures than from evaporation sources and hence can be used to obtain epitaxial films.[6] The nucleation site density is also higher at lower substrate temperatures as a result of the electron and ion bombardment of the substrate surface that takes place in most sputtering systems.

Nucleation processes for films deposited by reactive and electrochemical methods appear to have been studied less than the physical deposition methods and are obviously much more complex. We conclude that although at this time the phenomenon of nucleation is understood in general terms, the kind of detailed understanding necessary for quantitative prediction requires more experimental knowledge and more sophisticated theories.

3.3.3. Postnucleation Growth Processes

When the nucleation sites on the substrate are filled, three-dimensional islands begin to grow from these sites. Depending on the substrate temperature, the islands may be liquid droplets or single crystals of the condensing materials. The melting point of the islands is lower than that of the bulk material, T_m, and has been observed to be $2T_m/3$. Coalescence is assumed to take place only when the islands touch. Heat is liberated by coalescence, so that solid islands of many materials melt on contact. After coalescence they cool down, forming a new solid crystal, usually with the orientation of the larger of the two original island crystallites.

The occurrence of epitaxy, or the oriented growth of single-crystal films on monocrystalline substrates, suggests that in this case the orientations of most nucleation islands are similar and recrystallization with a preferred orientation takes place on coalescence. This is confirmed by the observation that the orientation of an epitaxial film is strongly dependent on the crystal structure of the substrate (see Sections 1.4.3 and 5.3.5). On the other hand, it is possible to grow single-crystal oriented films on amorphous or even liquid substrates as well as single-crystal surfaces.

For every pair of condensate and substrate there exists for a given deposition rate a critical substrate temperature above which a single-crystal oriented film grows, regardless of the degree of lattice mismatch. Condensation at temperatures below the critical temperature produces increasing disorder in the film, until at sufficiently low temperatures ($T_m/3$) the incident atoms condense close to the point of impingement, producing an amorphous (highly disordered) film. A laterally applied dc field on the order of 100 V/cm can lower the average film thickness at which coalescence occurs for certain combinations, such as silver on sodium chloride.[31] The applied field appears to flatten the islands, thus increasing their diameter and forcing them to touch earlier. Irradiating the surface with electrons or ions also promotes coalescence, although it is not immediately clear why this occurs. Films deposited from a sputter source are generally accompanied by electron and ion bombardment of the substrate, depending on the configuration being used. The earlier coalescence induced may produce thin films with properties different from those of evaporated films.

3.3.4. Structure and Composition

In this section we shall simply define the terms used to describe structure and composition and indicate what measurement techniques are used to obtain these descriptions.

3.3.4.1. Morphology

Here we wish to describe whether the deposit consists of three-dimensional islands, a channeled layer film, or a uniformly dense film. This is best observed by high-resolution electron microscopy, but optical diffraction, electrical conductivity, and density measurements can offer clues.

3.3.4.2. Crystallinity

The film may be amorphous, randomly oriented polycrystals, uniaxially oriented polycrystals, or single crystals (epitaxial). This condition is measured by electron or X-ray diffraction. Amorphous films give no well-defined diffraction rings; randomly oriented polycrystals give well-defined diffraction rings. From the width of the diffraction lines it is possible to infer the average size a of the crystallite from the relationship

$$a = \frac{\lambda}{D \cos \theta} \tag{3.11}$$

where λ is the X-ray wavelength, D is the angular width of the diffraction line, and θ is the Bragg angle.

Single-crystal films or identically oriented polycrystals give rise to round diffraction spots (Laue patterns). As the orientation of the polycrystal deviates from perfection, the spots become distorted to take elliptical shapes, and new spots appear. In the extreme case of complete randomization the elliptical spots join to form the diffraction rings. If one axis of the crystal remains the same but the orientation of the other axes are random, this gives rise to missing or low-intensity rings, compared with the completely random case. If the electron beam is made small enough, the crystallinity of individual islands in the post-nucleation stage can be ascertained by electron diffraction.

3.3.4.3. Lattice Constants

The lattice constants for the film material can, of course, be deduced from the angular positions of the diffraction patterns. The lattice structure of very thin films is affected by the presence of the substrate and surface.

3.3.4.4. Composition

Analysis of films is carried out by several methods:

- Conventional chemical analytical techniques whereby the film is dissolved off the substrate and subjected to analysis. In this category we include volumetric, gravimetric, and optical spectrographic methods.
- X-ray fluorescence, a procedure in which the film is bombarded by electrons sufficiently energetic to excite inner-shell radiation, which is detected by solid-state counters and where an energy spectrum of the emitted X-rays is displayed. By using a focused electron beam, very small areas can be analyzed.
- X-ray and (high-energy) electron-beam diffraction allow a comparison of the diffraction patterns obtained with those of known materials.
- Auger spectroscopy is the analysis of the energy spectrum of the secondary electrons emitted from the surface upon bombardment by primary electrons and enables identification of the surface atomic layer to be made. By sputtering the film away in steps, followed by Auger spectroscopy, a profile of the composition from the surface to the substrate may be obtained (see Chapter 6).
- Mass spectroscopy may be used if the surface is sputtered away by a beam of energetic ions. The sputtered material may be subsequently ionized and injected into a mass spectrometer for analysis.

X-ray fluorescence, Auger spectroscopy, and mass spectroscopy have been adapted for scanning systems so that variations in the composition of a film over an area can be obtained. With modern instrumentation submicron resolutions may be obtained. However, impurity levels below a few percent are difficult to detect by other than the conventional techniques.

3.3.5. Surfaces and Interfaces

The topography of surfaces can be visualized by scanning electron microscopy, with resolutions down to 50 Å (Section 6.3). A significant advantage of this device is its extraordinary depth of focus, which enables structures with large aspect ratios to be viewed clearly. Higher resolutions can be obtained with transmission electron microscopy, but the substrate must first be dissolved from the film. Surface structures can be seen by reflection electron microscopy, using shadowing and replication techniques.

The arrangements of the atoms in the surface can be detected by low-energy electron diffraction (LEED), and the atomic constituents of the surface atomic layer can be unambiguously identified from the energy spectrum of the secondary electrons (Auger spectroscopy). Small movements of the peaks in Auger spectroscopy can indicate how the atoms are bound together. These techniques, which view only the surface atomic layer due to the small penetration of low-energy electrons (less than a few keV), have revolutionized our understanding of surfaces over the last 15 to 20 years since the instruments have become commercially available. This availability itself turned on the development of UHV technology, which enables experiments to be carried out in vacuums of less than 10^{-9} Torr so that the surface is not seriously contaminated by its environment during observation.

The substrate surface can be studied before deposition. After deposition the interface can be reached by sputtering away a portion of the film, step by step, with inert gas ions until the substrate is penetrated. However, this sputter "scalpel" is not precise enough to indicate exactly when the interfacial atomic layers are reached.

Pattern Generation

4

4.1. Introduction

While thin-film deposition enables one dimension of a device to be made with remarkable precision, down to, say, 100 Å, the other two dimensions are more difficult to form with the same degree of accuracy. Two-dimensional patterns are usually made on the surface by the process of lithography, which derives from printing-plate technology. The trend in microfabrication is towards increased circuit complexity and reduced pattern dimensions. Linewidths of 0.5 μm and tolerances of 0.1 μm are currently being considered for microcircuits. In order to meet these requirements the need has arisen for pattern generation and lithography systems with superior performance specifications.

Patterns are formed by exposing a wafer coated with a thin film of resist material that is light, X-ray, or electron-beam sensitive to a corresponding pattern of the appropriate radiation and developing the resultant latent image. Depending on the radiation source and the type of resist used, we characterize these techniques[1,2] as optical, X-ray, electron beam, or ion beam. The ways in which these lithographies are utilized for mask making, image transfer, and direct writing are summarized in Figure 4.1.

The optical lithography process usually starts with a computer-aided pattern design, followed by photoreduction of the physical pattern layout. The mask or "reticle" pattern (demagnified 2–10 times) made by photoreduction is used directly in the printing process. In the "step-and-repeat" reduction process the sample (mask or wafer) is mechanically stepped between exposure sites to cover the entire sample so that a limited-field-diameter lens system can be used.

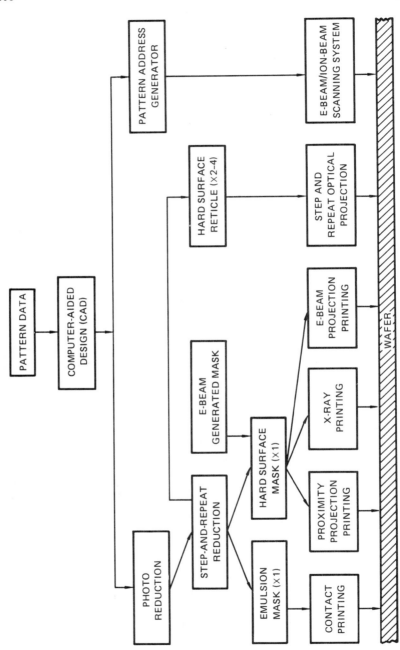

FIGURE 4.1. Use of lithography techniques for mask making, image transfer, and direct writing.

In scanning systems (electron, ion beam) the computer-stored pattern is directly converted to address and switch (blank) the beam, enabling the pattern to be exposed sequentially point by point over the wafer area.

In this chapter reference will be made to the resolution, linewidth, edge acuity, accuracy, distortion, and precision of the pattern. The physical definition of resolution as the distance at which two points can be clearly recognized as distinct has little utility in lithography, but the engineering resolution measurement in terms of resolvable high-contrast line pairs per unit distance is directly related. In lithography the term *resolution* is interchangeable with *linewidth* and indicates the smallest width of a line, or the smallest distance between two lines that is reproduced. *Accuracy* is an indicator for spatial dimensions in the standard unit. *Distortion* is a measure of the relative dimensions of a pattern, and *precision* specifies how closely the patterns made by the same process match one another.

4.2. Optical Lithography
4.2.1. Contact Printing

An early method used for producing patterns on semiconductor wafers was contact printing. A mask transparency containing the circuit patterns is positioned on the top of the photoresist-covered wafer and exposed to light, creating exposed and unexposed areas. Development (selective etching) removes the resist according to its exposure state. Resolution is satisfactory for features as small as 2 μm, the limit being set by diffraction effects between adjacent lines (see Figure 4.2). If partially coherent light is used for illumination[3] we obtain increased contrast, increased ringing in the image due to coherent diffraction effects, and sharper edge gradients (intensity versus position). Uniformity of both the exposure and the photosensitive material can be maintained so that linewidth control is within acceptable tolerances.

Mask defects accumulated during successive mask uses are the main limitation of contact printing. The pressure during contact printing damages both the mask and substrate. The accrued damage and particles of resist adhering to the mask are printed on the following exposures, causing a rapid buildup of defects. Depending on the scale of integration and the durability of the mask surface, only a given number of uses can be tolerated.

Emulsion masks are commonly used only for 10 or less exposures for large silicon ICs. Chromium, iron oxide, or other hard-surface masks have become practical alternatives because they can be cleaned periodically and used for

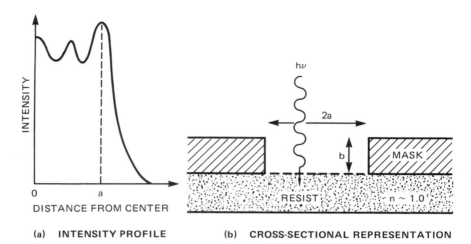

FIGURE 4.2. Contact printing geometry.

more exposures during their lifetimes. In practice, up to 100 uses may be possible with these hard-wearing materials.

Resolution in contact printing can be improved by reducing the wavelength of the exposing radiation,[4] thus decreasing the diffraction effects. Deep-uv radiation (200–260 nm) has been used to produce 0.5-μm linewidths, roughly twice the wavelength of the radiation.

As chip areas grow larger and features grow smaller, the need for lower defect densities and higher dimensional integrity to maintain yields has encouraged the development of alternate exposure methods, particularly proximity and projection printing.

4.2.2. Proximity Printing

Spacing the mask away from the substrate minimizes contact and eliminates most of the defects that result from contact. However, diffraction of the transmitted light causes a reduction in resolution and increases the distortion of individual photoresist features. The degree to which this occurs depends on the actual mask-to-wafer spacing, which can be variable across the wafer.

In large wafers flatness variations, which are especially troublesome at small nominal spacings, and diffraction effects, which are troublesome at large spacings, make approximately 7-μm features the smallest ones practical for proximity exposure (see Figure 4.3).

4.2.3. Projection Printing

Projection is a method[4] by which an image of the photomask is projected directly onto the photoresist-covered wafer by means of a high-resolution lens. In this system the mask life is potentially unlimited except for handling damage. As a consequence, the use of the highest-quality masks can be justified. The depth of focus must cover the ± 10.0-μm flatness errors common in wafers after high-temperature processing. This limits the aperture of the lens, and therefore the resolution, to marginal values. It is also difficult to fabricate a lens systems with both a uniform (diffraction-limited) image quality and a uniform light intensity over the full area of currently used wafers (10 cm diameter). The scattering of light associated with the use of glass optics has generally limited the choice of light-sensitive material to positive resists. Conventional projection systems have been used with success for features greater than 5 μm. However, as wafer diameters have increased, stationary projection systems have become less attractive.

In a scanning (or step-and-repeat) projection system a portion of the mask is imaged onto the corresponding part of the wafer. A much smaller area (on the order of 1 cm^2) is exposed and the exposure is repeated by either scanning or stepping the image over the wafer.[6,7]

$$\theta_D = \text{DIFFRACTION HALF POWER} \approx \frac{50°}{W_M/\lambda}$$

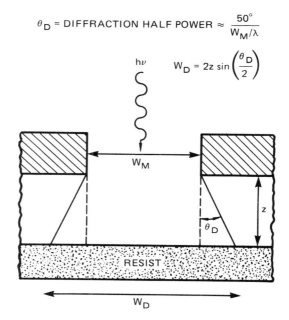

$$h\nu \qquad W_D = 2z \sin\left(\frac{\theta_D}{2}\right)$$

FIGURE 4.3. Proximity printing geometry.

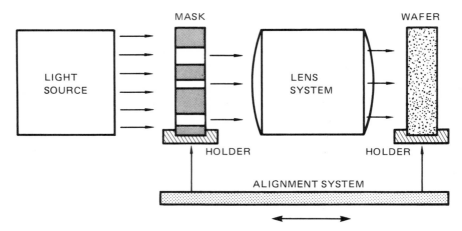

FIGURE 4.4. Basic components of a projection system.

Figure 4.4 shows the basic components of a projection–alignment system. The major components are as follows:

- A light source
- A lens system
- A mask holder
- A wafer holder
- An alignment system

In conventional photolithography light of the spectral region 3300–4000 Å is commonly adopted. Shorter wavelengths are not used because almost all the known combinations of photoresists, intensive light sources, and illuminating optics cease to be effective at wavelengths shorter than 3300 Å. Useful light sources for projection systems are Hg-arc lamps in the 3300–4000 Å spectral region and Xe–Hg-arc lamps and deuterium spectral lamps for deep-uv lithography.

Modern projection printers employ optics that are essentially diffraction limited. This implies that the design and fabrication of the optical elements are such that their imaging characteristics are dominated by diffraction effects associated with the finite apertures in the condenser and projection optics, rather than by aberrations. Figure 4.5 shows the basic printing lens performance parameters.

It is useful to recall the definitions[7] of several parameters used to specify a multielement imaging system. The numerical aperture (NA) is given by NA $= n \sin \alpha$, where n is the refractive index (usually unity) in image space and

2α is the maximum cone angle of rays reaching an image point on the optical axis of the projection system. The effective F number of a projection system is given by $F = 1/(2NA)$. It can be shown that $F = (1 + M)f$, where f is the F number of the system with the object at infinity and M is the operating magnification.

When light from a very small point is imaged by a diffraction-limited lens, the image consists of alternating rings of light surrounding a central bright spot, called the Airy disk. The width w at the half-intensity points of the Airy disk is given by $w = 0.5\lambda/NA = 1.0\,\lambda F$ and is a rough measure of the smallest dimension that can be printed in a resist by a diffraction-limited projection system.

The defocusing aberration (or error) of

$$\Delta a = \pm n\lambda/2\,(NA)^2 = \pm 2n\,\lambda\ F^2$$

causes an optical path difference (OPD) of $\pm\lambda/4$ in the image plane. This OPD of $\pm\lambda/4$ does not seriously affect the image quality, since it causes a loss of only 20% of the light from the Airy disk, with hardly any change in its diameter. Δa is known as the depth of focus. For a $F/3$ optical system $w = 1.2\ \mu m$ and $\Delta a = \pm 7.3\ \mu m$. Although Δa is a useful guide, it is more satisfactory in photolithography to define the depth of focus of a projection system

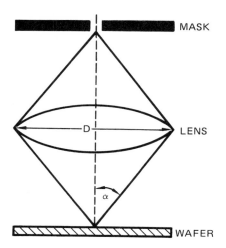

PARAMETER	SPECIFICATION
FOCAL LENGTH	f
APERTURE DIAMETER	D
F NUMBER	$f/D = F$
NUMERICAL APERTURE	$N.A. = n \sin \alpha = \dfrac{D}{2f}$
RESOLUTION	$1.22\ \lambda F = \Delta x$
DEPTH OF FIELD	$\pm 2\lambda F^2 = \pm \Delta a$

FIGURE 4.5. Basic printing lens performance parameters.

as the range over which the image plane can vary for a specified variation in image size.

The optical imaging quality of a projection printer can also be characterized in terms of the modulation transfer function (MTF) curve,[8] in which the modulation of the output image is plotted against the spatial frequency of a square-wave many-bar pattern having 100% modulation. In practice, the MTF

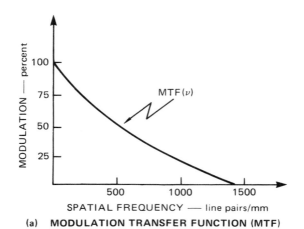

(a)   MODULATION TRANSFER FUNCTION (MTF)

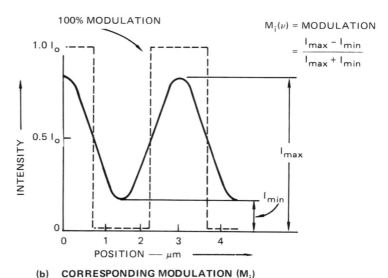

(b)   CORRESPONDING MODULATION (M$_i$)

FIGURE 4.6. MTF(ν) and the $M_i(\nu)$ for output light intensity for 333 line pairs/mm corresponding to 1.5-μm lines and 1.5-μm spaces between lines.

curve is obtained by imaging sinusoidal gratings placed in the object plane (see Figure 4.6). Each grating is characterized by a frequency (measured in line pairs per millimeter) and a modulation $M_0 = I_{max} - I_{min}/I_{max} + I_{min}$, where I_{max} and I_{min} are the local maximum and minimum light intensities emerging from the spaces and lines. The ratio I_{max}/I_{min} is called the contrast C. The corresponding modulation $M_i(\nu)$ in the image plane can be measured by scanning a very small photodetector across an image of a grating. The MTF at frequency ν is given by $MTF(\nu) = M_i(\nu)/M_0$. For a diffraction-limited optical system the MTF also can be calculated from[9]

$$MTF(\nu) = \frac{2}{\pi} \left\{ \cos^{-1} \frac{\nu}{\nu_0} - \frac{\nu}{\nu_0} \left[1 - \left(\frac{\nu}{\nu_0} \right)^2 \right]^{1/2} \right\} \qquad (4.1)$$

where ν is the spatial frequency variable and ν_0 is the optical cutoff frequency of the system, determined by its numerical aperture NA and the wavelength λ, and is given by $\nu_0 = 2NA/\lambda$.

Optical imaging systems for projection printers are classified as either coherent or incoherent, depending on the type of illumination employed. If a mask (or grating) is illuminated by a narrow-angle light beam originating from a point source, the imaging of the projection printer is coherent or partially coherent[8,10] [see Figure 4.7(a)], since the light diffracted by the grating is coherent in amplitude at the image (wafer) plane. Information about the spatial frequency of the grating is contained only in the diffracted light beam. The direction of the first diffraction peak is given by the grating formula

$$n(a + b) \sin \theta = N\lambda$$

and the spatial frequency ν is given by

$$\nu = (a + b)^{-1} = \frac{n \sin \theta}{\lambda}$$

for $N = 1$. For imaging it is required that $\theta = \alpha$, where α is defined by the numerical aperture of the projection optics $(NA = n \sin \alpha)$. Therefore, the highest grating frequency that can be imaged by a coherent illumination system is given by

$$\nu_{max} = \frac{n \sin \alpha}{\lambda} = \frac{NA}{\lambda} = \frac{1}{2\lambda F} = \nu_c$$

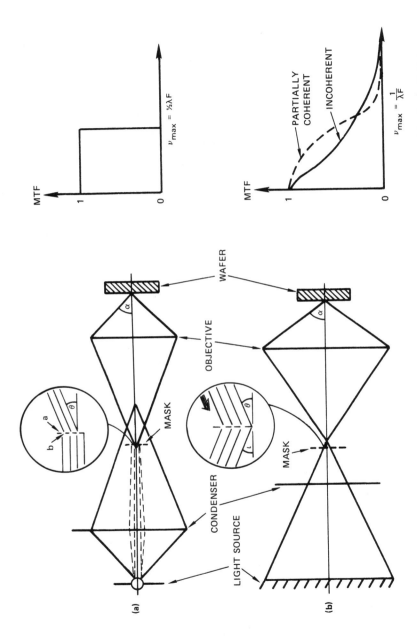

FIGURE 4.7. Coherent and incoherent projection printers. The inserts show the diffraction of coherent and incoherent light by a grating pattern. Modulation transfer functions also shown for coherent and incoherent printers.

where ν_c is the cutoff frequency. Figure 4.7(b) shows a projection system where the grating is illuminated by rays from all portions of an extended incoherent light source. In this case each ray is diffracted by the grating and forms its own image in the wafer plane. Since there is no fixed phase relationship between the different rays from the extended source, the various images at the wafer plane add in an incoherent manner.

To illuminate rays incident at an angle i the direction of the first diffraction maxima is given by

$$n(a + b)(\sin i + \sin \theta) = \lambda$$

For image formation it is required that $i, \theta \leq \alpha$. Therefore

$$\nu_{max} = (a + b)^{-1} = \frac{2NA}{\lambda} = \frac{1}{\lambda F} = \nu_c$$

Hence ν_{max} (incoherent) $= 2\nu_{max}$ (coherent).

MTF curves for coherent, partially coherent, and incoherent illuminations are shown also in Figure 4.7.

The MTF for an incoherent source is at $\nu = 0$ and decreases monotonically to zero at the cutoff frequency $\nu_c = 1/\lambda F$. In contrast, the MTF for a coherent source is unity out to its cutoff frequency $\nu_c = \frac{1}{2}\lambda F$, where it drops to zero. Mathematically the incoherent MTF is the weighting function for the intensity components of the image spectrum. If the object spectrum and MTF are known, an incoherent image can be reconstructed. Coherent systems are more difficult to interpret, being sensitive to the phase of the image spectrum components as well as to their amplitude. Since the MTF does not contain phase information, an image cannot be reconstructed from its MTF.

It is generally accepted that a 60% (MTF) is required for a minimum working feature in a positive resist.

If the incoherent MTF(ν) is known from measurements or from calculation, the intensity distribution of the image (line pattern) can be calculated from the intensity distribution of the object (mask) by the use of the following transformation sequence.

The object spatial frequency distribution $I_r(\nu)$ is given by the Fourier transform F_T of the intensity distribution of the object, $I_0(x)$, as in

$$I_0(\nu) = F_T I_0(x)$$

In practice, a discrete Fourier transform is used with an implied periodicity. The image spatial frequency distribution $I_i(\nu)$ is given by

$$I_i(\nu) = \text{MTF}(\nu) \, [I_0(\nu)]$$

and the intensity description, $I_i(x)$, is given by the inverse Fourier transform of the image spatial frequency distribution,

$$I_i(x) = F_T^{-1} \, [I_i(\nu)]$$

Figure 4.8 shows the calculated intensity distribution for the image of a nominal 1-μm line object produced by a NA 0.45 lens with a 435.7-nm wavelength.[8]

The incident intensity distribution (in W/cm^2) multiplied by the exposure time (s) gives the incident energy density distribution (J/cm^2) or dose across the surface of the resist film.

Most projection printers are designed so that the source only partially fills the objective. The imaging behavior of the system is then described as partially coherent.

Contact and near-contact printers make use of highly collimated coherent illumination. If incoherent illumination were employed, the image would be blurred by its penumbra in the wafer plane. In practice, the illumination system is decollimated by a few degrees in order to blur out the otherwise intense diffraction effects.

4.3. The Physics of Photolithography

4.3.1. Positive Photoresist Exposure

Optical lithography, as applied in microelectronics, uses photoresists that can be applied as thin-film coatings on surfaces (wafers) where a pattern is to be delineated. The photoresist film is then exposed to an optical pattern, using blue or ultraviolet light, thus creating exposed and unexposed areas. Development selectively removes the resist according to its exposure state. The remaining pattern of photoresist on the surface is used to delineate etching, plating, sputtering, evaporation, or other processes commonly used in microelectronics.

Optical lithography is similar to conventional photography[11] in that a light-sensitive thin film is chemically altered by light so that an image can be produced by a subsequent development process. However, both the chemical

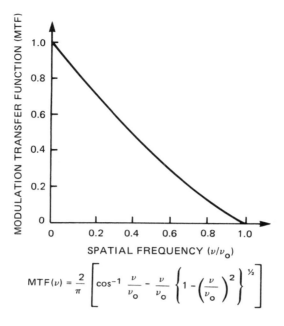

$$MTF(\nu) = \frac{2}{\pi} \left[\cos^{-1} \frac{\nu}{\nu_o} - \frac{\nu}{\nu_o} \left\{ 1 - \left(\frac{\nu}{\nu_o} \right)^2 \right\}^{1/2} \right]$$

(a)   MODULATION TRANSFER FUNCTION (MTF)

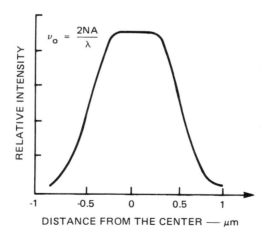

(b)   INTENSITY DISTRIBUTION

FIGURE 4.8. Calculated intensity distribution for the image of a nominal 1-μm line object produced by a NA 0.45 lens with a 435.8-nm wavelength. Source: ref. 8.

TABLE 1. *Differences Between Photography and Optical Lithography*

Process or Characteristic	Photography	Optical Lithography
Exposure	Multiphoton absorption	Single-photon absorption (breaking bonds for positive resists; polymerizing for negative resists)
Image	Optical density	Surface profile
Development	Plating silver crystals on latent image	Etching process

processes and the resulting images are different. Photography involves multi-photon absorption by silver compounds, whereas optical lithography for positive photoresists involves the destruction of an organic compound by single-photon adsorption. The photographic image is one of optical density; the lithographic one is one of surface profile. Development in photography involves converting light-activated silver halide crystals to silver; the development of positive photoresists is an etching process. The differences, as shown in Table 1, are sufficient so that a different mathematical basis must be built for understanding optical lithography.

Photoresists are available for both positive and negative processes, that is, development causes removal where the photoresist was exposed or unexposed. First we will discuss a quantitative basis for the lithographic process as applied to positive photoresists. Significant differences between the behavior of positive and negative resists make the mathematical treatment of exposure and development for one inapplicable to the other.

Positive photoresists are typically three-component materials consisting of a base resin, which gives the resist its film-making properties, a photoactive compound, and volatile solvents to make the material liquid for application. In a dried film (typically 0.3–2 μm thick), the photoactive compound serves to inhibit dissolution of the photoresist in an alkaline aqueous developer solution. We refer to this compound as an inhibitor to emphasize this role. Destruction of the inhibitor by light creates byproduct compounds that allow dissolution of the resist, resulting in an increased rate of removal of the resist by the developer[12]

Lithography can be better understood if we consider the exposure and development processes separately. Exposure modifies the photoresist chemically by destruction of the inhibitor by light in a suitable wavelength range. The effect of chemical modification is highly localized around the point where the photon is absorbed, and the exposure level can vary widely within the resist film. Development is an etching process that selectively removes the photoresist at a rate related to the amount of inhibitor destroyed. The link between these two processes is the inhibitor distribution after exposure.

4.3.2. Resist Characterization

The photoresist film, after application to the substrate and drying, must be a uniform isotropic medium. It must have a uniform thickness (on both micro and macro scales) over the region used for characterization. It should be chemically isotropic so that its response to exposure and development is uniform throughout the film. Particularly, radial (starburst) thickness variations often associated with the application of photoresist by spinning must be avoided.

Since optical absorbtion measurements are an important part of resist characterization, several qualifications must be met. Optical scattering must be small so that the material can be described optically by its index of refraction and an absorption coefficient α.

The complex index of refraction for the photoresist can be written[9] $\bar{n} = n - ik$, where n is the real part of the index ($n = 1.68$ for AZ1350J photoresist at a 404.7-nm wavelength) and k is the extinction coefficient:

$$k = \frac{\alpha\lambda}{4\pi} \qquad k = 0.02 \text{ for AZ1350J at 404.7nm}$$

so that initially the incoming light is attenuated in the resist by a factor

$$\exp\left(-\frac{4\pi kx}{\lambda}\right) = \exp(-0.58x) = 0.56$$

per micron of path length.

After exposure the inhibitor concentration is much reduced and the attenuation factor is typically $\sim 0.10/\mu$m. For much of the exposure, therefore, the resist will be transparent.

The inhibitor is the principle contributor to the absorption and it is usually assumed that the absorption spectra and spectral sensitivity curves for positive resists are equivalent.

4.3.3. Exposure

The inhibitor compound typically represents about 30% of a dried photoresist film. The strong absorption associated with its photosensitivity contributes significantly to the optical absorption of the material at exposing wavelengths. As the inhibitor is destroyed, this absorption is removed as well. This can be described in terms of a relative inhibitor term $M(z, t)$, which is the fraction of the inhibitor remaining (at any position z and exposure time t) as compared to the inhibitor concentration before exposure. There is little scattering in most photoresist films, so that the absorption constant α is given by

$$\alpha = AM(z, t) + B$$

where A and B are measurable material parameters that describe the exposure-dependent and exposure-independent absorptions.

The rate of destruction of the inhibitor is dependent on the local optical intensity $I(z, t)$, the local inhibitor concentration, and a measurable optical sensitivity term C, as given by

$$\frac{\partial M}{\partial t} = -I(z, t) \, M(z, t) \, C \qquad (4.2)$$

A, B, and C depend upon the photoresist material and exposure wavelength. For Shipley Chemical Company's AZ1350J photoresist, one of the commonly used high-resolution resist materials, $A = 0.86 \ \mu m^{-1}$, $B = 0.07 \ \mu m^{-1}$, and $C = 0.018 \ cm^2/mJ$ for a 404.7-nm exposure wavelength. The light intensity in an infinitely thick resist, or in a film on a matched substrate, is described by

$$\frac{\partial I(z, t)}{\partial z} = -I(z, t) \, [AM(z, t) + B] \qquad (4.3)$$

Figure 4.9 shows the normalized inhibitor concentration as a function of the distance from the resist–air interface for different total exposure energies.[13] In the matched environment the inhibitor distribution does not depend upon resist thickness, making these calculations useful for films up to 2 μm. Note that even for this relatively simple exposure environment, inhibitor concentration is never uniform once exposure has started.

Generally, Eqs. (4.2) and (4.3) can be solved by straightforward numerical integration techniques for $M(z, t)$ and $I(z, t)$ once A, B, C, and I_0 are specified. The experimental techniques for measuring the parameters (A, B, C) used in this image-forming model are described in the literature.[13]

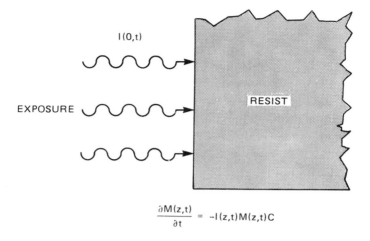

$$\frac{\partial M(z,t)}{\partial t} = -I(z,t)M(z,t)C$$

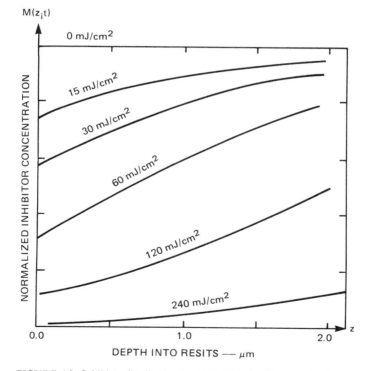

FIGURE 4.9. Inhibitor distribution in a "thick" resist. Source: ref. 13.

4.3.4. Interference Effects

Optical interference can cause close-spaced intensity variations in the photoresist material when the index of refraction of the resist is not optically matched to the index of the substrate. The optical intensity of the exposing light on bare silicon has an intensity profile as shown in Figure 4.10(a).[9,11] The absorption causes a decrease in the average intensity from the surface to the substrate. However, this intensity pattern is modulated by the interference of the standing waves reflected from the resist–substrate interface. The inhibitor concentration also shows a modulation effect due to the intensity changes in the resist film [Figure 4.10(b)]. These interference effects can be analyzed with some simplifying assumptions to illustrate the main points.[10]

The incident monochromatic light is imaged normally into a weakly absorbing film of thickness d, coating a perfect reflector (substrate, $r_s = 1$) located in the plane $z = d$, as shown in Figure 4.11(a). If incident light wave 1 has unit amplitude, then light wave 2 in the film can be represented by

$$E_2(z) = E_2 \sin(\omega t - kz + \phi) \qquad (4.4)$$

where $E_2 = (1 - r^2)^{1/2}$ and absorption effects are neglected. Here $r = (n - 1)/(n + 1)$ is the reflection coefficient at the air–film interface, $k = 2\pi n/\lambda$ and n is the real part of the film dielectric constant, $\bar{n} = n - ik$. The amplitude of reflected wave 3 is

$$E_3(z) = E_2 \sin[\omega t - k(2d - z) + \phi + \pi] \qquad (4.5)$$

A phase change of π is assumed during reflection. Adding E_2 and E_3 gives a standing wave $E_{23}(z)$,

$$E_{23}(z) = 2E_2 \sin[k(d - z)] \cos(\omega t - kd + \phi) \qquad (4.6)$$

The locations of the maxima (antinodes) and minima (nodes) are clearly independent of the arbitrary phase $\omega t + \phi$ of the incoming wave. The envelope function $(I \sim E^2)$ for the intensity of the standing wave is

$$I_{23} = 4I_2 \sin^2[k(d - z)] \qquad (4.7)$$

Measured from the substrate, the locations of the intensity extrema are given by the following conditions, where $N_1 = 0, 1, 2, \ldots$:

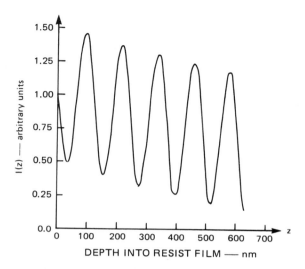

(a)   PLOT OF INTENSITY OF EXPOSING LIGHT WITHIN A 630-nm AZ1350J
      PHOTORESIST FILM ON BARE SILICON AT THE BEGINNING OF EXPOSURE
      AT A 404.7-nm WAVELENGTH

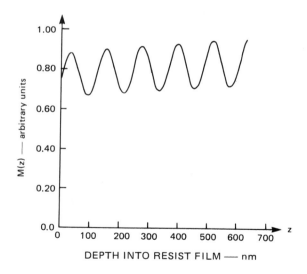

(b)   INHIBITOR CONCENTRATION AS A FUNCTION OF DEPTH INTO AN
      AZ1350J PHOTORESIST FILM ON BARE SILICON EXPOSED WITH
      15.7 mJ/cm^2 AT A WAVELENGTH OF 404.7 nm

FIGURE 4.10. Interference effects of the standing waves reflected from the resist–substrate interface. Source: ref. 9.

$$n(d - z) = \lambda/4, 3\lambda/4, \ldots, (2N_1 + 1)\lambda/4 \qquad (4.8)$$

for antinodes (maxima) and

$$n(d - z) = \lambda/2, \lambda, \ldots, N_1 \lambda/2 \qquad (4.9)$$

for nodes (minima). Taking into account additional waves (4, 5, . . .) does not affect the location of the maxima and minima but changes the amplitude of the standing wave.

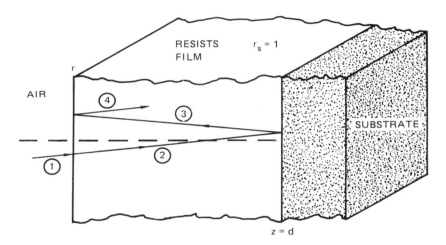

(a)   MULTIPLE REFLECTION AT THE RESIST–SUBSTRATE INTERFACE

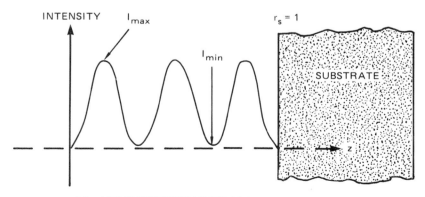

(b)   LIGHT INTENSITY DISTRIBUTION IN THE RESISTS

FIGURE 4.11. Standing-wave effects of incident monochromatic light ($r_s = 1$).

Using "nonideal" substrates (Si, Al, Pt, and Au), we cannot assume total reflection ($r_s = 1$) at the film–substrate interface. Also, the phase change is less than π. These cause a node shift and the appearance of finite light intensity at the minima. It can be shown[10] that the locations of the intensity nodes relative to the substrate and the maxima/minima intensity conditions are unaffected by absorption in the resist. However, the modulation of the standing waves is, in general, dependent on resist absorption.

To calculate the absolute intensity distribution in the resist material the source (lamp) and the reflective characteristics of the optical elements have to be taken into account. In optical lithography the light source is usually a high-pressure mercury-arc lamp, which emits in the 200–600 nm range. For printers using reflective optics the light spectrum at the wafer is essentially the same as the source spectrum, $S(\lambda)$ is the transmission characteristic of the printer, and the light spectrum reaching the photoresisted wafer is $S(\lambda) F(\lambda)$. If $R(\lambda)$ is the spectral sensitivity of the resist, the effective light intensity at the resist surface in the wavelength range $\lambda \pm \delta\lambda/2$ is proportioned to $S(\lambda) F(\lambda) R(\lambda)\delta\lambda$. For a given wavelength, standing waves in the resist films cause light intensity variations given by Eq. (4.7). Therefore the resulting light distribution in the resist due to several wavelengths (λ_i) is proportional to

$$I(z) \backsim \Sigma_i \, S(\lambda_i) F(\lambda_i) R(\lambda_i) \, \delta\lambda_i \sin^2[\, k_i(d - z)] \qquad (4.10)$$

Since the refractive indices of common resists ($n_r \backsim 1.6$) are very close to those of SiO_2 and Al_2O_3, this simplified model is also useful for resist–SiO_2/Si systems.

An important parameter used to describe the interaction of radiation $I(z)$ with the resist is the depth-dose function. This relates the rate of energy dissipation (energy loss dE/dz in the electron-beam, ion-beam resist) to the penetration distance z into the resist film.

The dose at penetration z, $D(z)$, is the photon or electron intensity at z, $I(z)$, multiplied by the exposure time τ, that is,

$$D(z) = I(Z)\tau$$

The intensity at z for the photon case is $I(z) = I_0 F(z)$, where I_0 is the incident intensity and $F(z)$ is the depth-dose function. Hence

$$D(z) = I_0 F(z)\tau = D_0 F(z)$$

where D_0 is the incident dose at $z = 0$. $F(z)$ varies as a function of z owing to light interference in the resist film.

The absorbed energy per unit volume is responsible for the chemical reactions (polymerization, bond breaking) in the resist. If dI is the light intensity in photons $\cdot$ cm$^{-2} \cdot$ s^{-1} absorbed by a slab of resist of thickness dz at a depth z, then the rate of energy absorption is

$$\left(\frac{dE}{dt_z} \right) = h\nu \left(\frac{dI}{dz} \right)_z = \alpha h\nu I(z) = \alpha h\nu I_0 F(z) \; \text{erg} \cdot \text{cm}^{-3} \cdot \text{s}^{-1}$$

where $h\nu$ is the photon energy and α the linear absorption coefficient. The energy absorbed in time τ is obtained by integration. For chemical reactions at penetration z the energy that must be absorbed per unit volume is

$$E_a(z) = \alpha h\nu D_0 F(z) \; \text{erg} \cdot \text{cm}^{-3}$$

requiring an incident dose D_0. The depth-dose function in photolithography is the result of the interference in the resist film. In electron-beam lithography the depth-dose function $\Lambda(z)$ is determined by scattering and energy loss. The linear absorption coefficient (corresponding to α for light) is R_G^{-1} for the case of electron-beam lithography, where R_G is the Grün range. The photon energy $h\nu$ is replaced by kinetic energy of the electron, $E(z)$, at penetration z.

4.3.5. Resist Development

The development of a positive photoresist is a surface rate-limited etching reaction. The parameters that control this rate are the resist and developer chemistry and the inhibitor concentration of the resist at the surface exposed to the developer. An experimentally determined curve relating the development rate R to the inhibitor concentration M provides the link between exposure and development. From experimental data[13] the $R(M)$ relationship can be written in the form

$$R(M) = \exp(E_1 + E_2 M + E_3 M^2) \tag{4.11}$$

The values of E_1, E_2, and E_3 are shown[13] with the development profile in Figure 4.12.

4.3.6. Negative Photoresists

A negative photoresist can be modeled as a photosensitive material with an effective threshold energy E_T.[14] If the energy E incident on the resist is

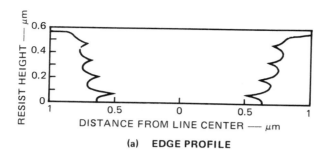

(a)   **EDGE PROFILE**

PARAMETER	SPECIFICATION
SUBSTRATE	Si
SiO$_2$	60 nm
PHOTORESIST	AZ1350J
	Lot 30000J
THICKNESS	583.6 nm
EXPOSURE CONSTANTS	A = 0.54/μm
	B = 0.03/μm
	C = 0.014 cm^2/mJ
	n = 1.68
EXPOSURE WAVELENGTH	435.8 nm
EXPOSURE ENERGY	57 mJ/cm^2
DEVELOPMENT	1:1 Az DEVELOPER: H$_2$O
	22°C
DEVELOPMENT RATE	E$_1$ = 5.27
	E$_2$ = 8.19
	E$_3$ = - 12.5

(b)   **EXPOSURE PARAMETERS**

FIGURE 4.12. Positive photoresist development profile. Example is for an edge profile for a nominal 1-μm line in AZ 1350 photoresist developed for 85 s 1:1 AS developer–water. Source: ref. 13.

less than E_T, the resist will be washed away in the developing cycle, but if $E > E_T$, the resist will be insoluble in the developer and the resulting image will be an effective etching or plating mask. This definition of E_T embraces a host of resist processing variables. For example, E_T will increase as the resist gets thicker and will decrease if the substrate has a greater than "normal" reflectance. It obviously also depends on the resist formulation. E_T can be found approximately from the characteristic curve of the resist (developed thickness as a function of exposure energy).

The size of the image defined in photoresist is determined by combining the effective threshold exposure energy of the photoresist with the energy distribution in the diffraction pattern from a mask edge, as shown in Figure 4.13. In the vicinity of the mask edge the intensity incident on the photoresist can be approximated by

$$I = kI_0 \exp(-mV)$$

or, in terms of energy,

$$E = kE_i \exp(-mV) \tag{4.12}$$

where E_i is the energy incident on the mask, E is the energy incident on the resist, V is the dimensionless parameter defined in Figure 4.13, k is the value of I/I_0 at $V = 0$, and m is the slope of I/I_0 near the mask edge. The constants k and m can depend on the optical system in the exposure tool and/or on the pattern that is printed. The effective edge of the image in photoresist will occur

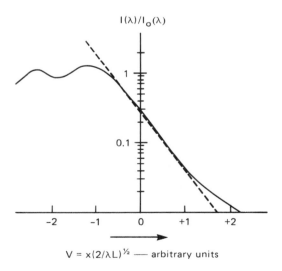

$I(\lambda)/I_0(\lambda)$

$V = x(2/\lambda L)^{1/2}$ — arbitrary units

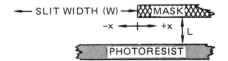

FIGURE 4.13. Fresnel diffraction pattern at a straight edge. Source: ref. 14.

at a value of V for which the energy incident on the resist is equal to the resist's effective exposure threshold, $E = E_T$. This value of V is found to be

$$V = \frac{1}{m} \ln \left(\frac{E_i}{E_T} \right) + \ln(k) \qquad (4.13)$$

In Figure 4.13 V is defined by

$$V = x \left(\frac{2}{\lambda L} \right)^{1/2}$$

where x is the distance from the mask edge, λ is the wavelength of the incident light, and L is the mask-to-wafer separation. The effective resist edge occurs a distance x from the mask edge and δ, the difference between the size of a clear mask feature and the size of its image in the photoresist, is $2x$ or

$$\delta = 2x = \frac{2}{m} \left(\frac{\lambda}{2} \right)^{1/2} L^{1/2} [\ln E_i - (\ln E_T - \ln k)] \qquad (4.14)$$

For $\delta > 0$ the resist image of a mask slit is larger than the mask slit, and smaller if $\delta < 0$. Note that $\delta = 0$ when $E_i = E_T/k = E_0$.

4.4. Projection Systems

The most widely used optical projection system[6] is shown in Figure 4.14(a). This system operates at unity magnification and utilizes a mirror–lens system. Only a crescent of the mask is illuminated at one instant, and this portion of the pattern is transferred to the sample with a parity that allows the mask and wafer to be scanned in the same direction and therefore to be mounted on a single holder. The entire wafer is exposed in a single scan. The magnification of the mirror optical system is $1\times$, so the mask has the same dimensions as the pattern. The resolution is adequate to reproduce 3-μm linewidths or comparable to conventional uv proximity printing. Alignment accuracy is ± 1 μm and the depth of focus is ± 5 μm for 2-μm lines and ± 12.5 μm for 3-μm lines.

A broad spectrum of radiation from the mercury arc is used to expose the resist. This produces an exposure time of under 1 min, with a minimum of 6 s. The time for manual alignment must also be considered when estimating throughput. A full description of the system is given in ref. 6.

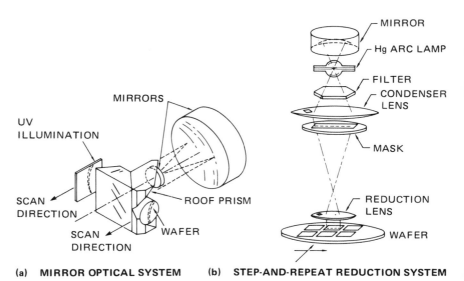

(a) MIRROR OPTICAL SYSTEM   (b) STEP-AND-REPEAT REDUCTION SYSTEM

PARAMETER	SPECIFICATION
PROJECTION OPTICAL SYSTEM	ALL-REFLECTING 1:1, f/1.5
IMAGE FIELD	3-INCH DIAMETER
RESOLUTION	2.0 μm WIDE LINES AND SPACES IN THE PHOTORESIST
DEPTH OF FOCUS	± 5.5 μm FOR 2.0-μm RESOLUTION ± 12.5 μm FOR 3.0-μm RESOLUTION
UNIFORMITY OF ILLUMINATION	± 10.0 percent
DISTORTION PLUS 1:1 MAGNIFICATION TOLERANCE	± 1.0 μm
PHOTORESIST CAPABILITY	NEGATIVE AND POSITIVE
EXPOSURE TIME	FASTEST EXPOSURE: 6 SECONDS
MASK CAPABILITY	CHROME, EMULSION, AND IRON OXIDE
ALIGNMENT ACCURACY	± 1.0 μm
VIEWING	SPLIT-FIELD BINOCULAR MICROSCOPE; MULTICOLORED DIRECT WAFER VIEWING

FIGURE 4.14. Common projection systems. Source: ref. 6.

Step-and-repeat reduction–projection systems similar in principle to that shown in Figure 4.14(b) are designed to produce master masks for contact printing and to write directly on the wafer.[7] Demagnifications between 4× and 10× are used between the mask and the wafer. The sample is mechanically stepped between exposure sites to cover the entire sample. The depth of focus for the case of 1-μm linewidth is very small (about ± 1 μm), which makes refocusing for every chip essential. Accurate sample leveling is also required. Current step-and-repeat systems have used two forms of alignment. In the first case an initial mask-to-wafer alignment is made and an accurate laser interferometric system is used to keep track of the sample during the stepping process. Systems of this type have relatively high throughput because only one alignment is required per wafer.

The new step-and-repeat system[7,15] exposes fifty 10-cm-diameter wafers per hour (0.3 s exposure, 0.3 s table stepping), with a minimum linewidth below 2 μm and an alignment accuracy better than ± 0.5 μm. In these printers the mask-to-wafer alignment is performed at each chip site. An automated system that aligns at every chip can also compensate for any wafer distortion error that might occur due to hot processing and/or wafer chucking. An overall accuracy of 0.25 μm may be achievable and, in principle, the resolution of a reduction-projection system should be adequate to fabricate linewidths approaching 1 μm.

4.5. Holographic Lithography

In holographic lithography[1,16] a substrate is exposed by placing it in a region where two beams from a laser interfere to produce a standing wave. This technique is useful primarily for exposing periodic and quasiperiodic patterns. Grating periods approaching 50% of the laser wavelength can be exposed. Refractive index-matching techniques can be employed to shorten even more the effective wavelength.[17] To calculate the spacing of the interference fringes produced in the photoresist film denote the complex amplitudes of the waves (see Figure 4.15) as

$$E_1 = A \exp[ik(x \sin \theta - z \cos \theta)]$$
$$E_z = A \exp[ik(-x \sin \theta - z \cos \theta) - i\phi]$$

respectively; the intensity on the surface is

$$|E_1 + E_2|^2_{z=0} = |A|^2 |2 + 2 \cos[2(kx \sin \theta - \phi)]|$$

Hence the intensity is modulated in the x direction with a period

$$\Lambda = \frac{\lambda}{2n \sin \theta} \qquad (4.15)$$

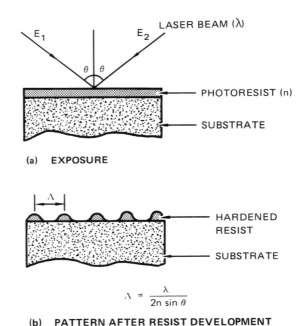

(a)   **EXPOSURE**

$$\Lambda = \frac{\lambda}{2n \sin \theta}$$

(b)   **PATTERN AFTER RESIST DEVELOPMENT**

FIGURE 4.15. Holographic lithography setup for grating fabrication. Source: ref. 17.

where n is the index of refraction of the resist ($n = 1.6$) and λ is the vacuum wavelength of the laser beams.

Using visible light, gratings with up to 10^3 line/mm have been produced.[18]

If two branches of a sufficiently coherent segment of a synchrotron-generated x-ray beam are used to produce the interference pattern, gratings with up to about 1.5×10^5 line/mm could be manufactured.[19]

Such a grating could serve as a diffractive or focusing element in x-ray imaging. In addition, it may be utilized as a circuit element in nanoelectronic devices in the 10–100 Å range or as building blocks to produce variable two-dimensional model "crystals."

4.6. X-Ray Lithography

4.6.1. The X-Ray Lithography System

X-ray lithography[20] uses contact/proximity printing with low-energy (1–10 keV) X rays rather than optical radiation. By using X rays, the diffraction problems common to photolithography are avoided, as well as is the backscattering problem encountered in electron-beam lithography. Figure 4.16 shows the basic principle of the X-ray technique.

Because of the X ray's short wavelength ($\lambda = 4$–50 Å), there are no convenient mirrors or lenses that can be used to collimate the rays as in optical, uv, or electron-beam systems. Thus an X-ray source of finite size has to be far enough away from the mask and resist for the X ray to appear to arrive with small divergence. The source size and beam divergence cause penumbral and geometric distortions, respectively. These distortions are illustrated in Figure

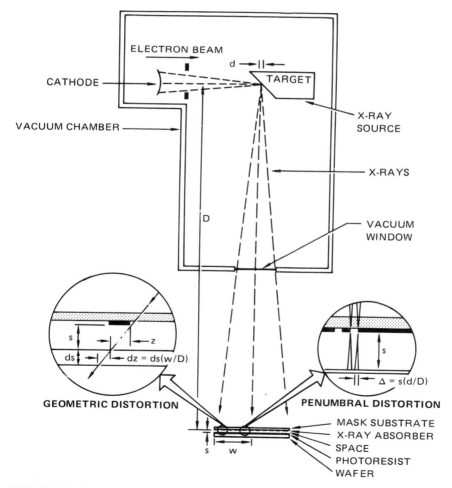

FIGURE 4.16. Arrangement for exposing resists, illustrating penumbral and geometric distortion in projection printing.

4.16. Penumbral distortion blurs the definition of lines in the resist and determines the minimum attainable lithographic feature resolution Δ:

$$\Delta = s(d/D) \qquad (4.16)$$

where s is the spacing between the mask and the wafer, d is the source diameter, and D is the distance from the source to the mask.

In a high-resolution system Δ should be controlled to within 0.1 μm. The spacing s should be large enough so that large-diameter masks can be used without much risk of contacting the resist, which would greatly increase the occurrence of defects.

A noncontact projection system is desirable to maintain a low-defect density. Geometric distortion arises from the fact that the image is projected from the mask to the wafer by a diverging beam. The degree of distortion z depends upon the distance of the image from the central axis of the beam.

$$z = s(w/V) \qquad (4.17)$$

where w is the distance on the wafer from the image to the central axis of the beam. For whole wafer exposure w is equal to one-half the wafer's diameter. The distortion itself is not as important as the registration problem when multiple projections must be made on a single wafer. The variations in the gap, ds, lead to variations in the geometric distortion dz, since

$$dz = ds(w/D) \qquad (4.18)$$

Since the variation in geometric distortion also affects the minimum attainable resolution, dz must be held to 0.1 μm.

Step-and-repeat lithography, of course, would keep the value of w small. However, in this analysis we consider a whole 3-in. diameter wafer where w = 38 mm. The mask-to-wafer spacing and the degree of control of this parameter, ds, are quantities less readily defined. For purposes of estimation here we chose a spacing $s = 10$ μm that is controlled to an accuracy of $ds \cong 1$ μm. Putting these values into Eqs. 4.16 and 4.17, we get

$$D = 380 \text{ mm} \quad \text{and} \quad d = 3.8 \text{ mm}$$

Three additional parameters are required to define the needed source strength. These are the fraction of X rays transmitted through the window and mask to the resist, the desired wafer throughput rate, and the resist sensitivity.

4.6.2. X-Ray Sources

X-rays are produced when electrons incident on a target material are suddenly slowed down. The maximum X-ray energy is the energy E of the incident
electrons. If E is greater than the excitation energy (E_c) of characteristic line
radiation of the atoms of the material, then the X-ray spectrum will contain
these lines.

X-ray generation by electron bombardment is a very inefficient process,
most of the input power being converted into heat in the target. The X-ray
intensity that can be produced is generally limited by the heat dissipation possible in the target. With electrons focused to a spot 1 mm in diameter onto an
aluminum target on a water-cooled stem, 400–500 W is a typical upper limit
for the input power. The X-ray power produced is still only on the order of 10
mW, and this is distributed over a hemisphere. The X-ray power is proportional to the electron current. The power in the line radiation is also proportional to $E - E_c$ for a target thin enough so that the absorption of X-rays by
the target itself can be neglected.[21] However, as E increases and the electrons
penetrate a thick target more deeply, the characteristic X rays produced must
on the average pass through more material on the way out and are thus
absorbed more strongly.

A reasonable thermal model for the soft X-ray source is one where the
electron-beam energy is dissipated in a circle of diameter d (cm) on a semi-
infinite solid of thermal conductivity K (W/$°$C$\cdot$cm). If the electron beam has
a Gaussian profile with Gaussian diameter d, then the temperature rise ΔT at
the center of the circle is $\Delta T = W/1.78 dK$. The maximum input power in
watts is

$$W_{max} = 1.78 \,\Delta TdK$$

Water-cooled rotating anodes provide the most intense X-ray sources for
the most demanding situations.[22] One chooses the anode material on the basis
of its fatigue strength, thermal capacity, and thermal conductivity, as well as
on the characteristic and bremsstrahlung X-radiation it produces. A 20-cm-
diameter aluminum anode operating at 8000 rpm can dissipate up to about 20
kW in a 6-mm spot. The maximum beam power scales as the $\frac{3}{2}$ power of the
spot diameter, so the brightness of the source actually decreases with increasing beam power. To reduce penumbra blurring, short turnaround time exposure systems must be built with small source diameters and short working distances between the source and the mask.

Figure 4.17 shows a rotating-target X-ray lithography system developed

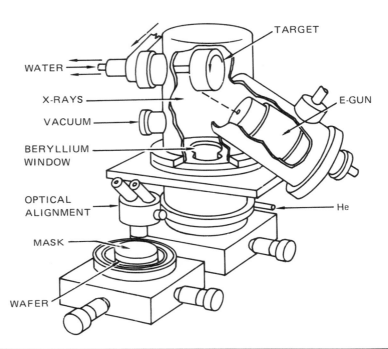

PARAMETER	SPECIFICATION
ANODE MATERIAL	Pd, ℓ = 4.37 Å
INPUT POWER	4.4 kW IN 3 mm SPOT AT 25 kV
VACUUM WINDOW	50 μm Be
MASK SUBSTRATE	25 μm KAPTON (POLYIMID)
EFFECTIVE TRANSMISSION OF WINDOW AND SUBSTRATE	0.7
SOURCE WAFER DISTANCE	50 cm
X-RAY FLUX AT WAFER	4 mJ cm^{-2} min^{-1}
X-RAY FLUX ABSORBED IN PMMA	0.5 J cm^{-3} min^{-1}
EXPOSURE TIME FOR PMMA	1000 min FOR VERTICAL WALLS
POWER CONVERSION EFFICIENCY TO X-RAY (E)	4.76 x 10^{-4}

FIGURE 4.17. X-ray lithography system developed at Bell Telephone Laboratories. Source: ref. 23.

at Bell Laboratories with a built-in optical alignment. In this system the mask–wafer combination is aligned under an optical microscope and then shifted under the X-ray source for exposure. It is capable of 1-μm resolution printing for bubble memory production.

In a new version of the Bell X-ray printer a 4-kW beam is focused onto a stationary water-cooled palladium cone target, which in turn emits X rays at a characteristic wavelength of 4.36 Å. With this printer a resolution as small as 0.3 μm in 0.3 = μm-thick resist containing chlorine has been achieved, using a 1.5-min exposure to $Pd_{L\alpha}$ X rays.[24]

4.6.3. Plasma Sources

Hot plasmas generated by high-power lasers or electric discharges produce pulses of soft X rays.[25] These sources, although as yet untapped, may play an important role in microfabrication. Because conventional X-ray tubes radiate very inefficiently at low photon energies (1 keV), much attention is currently centered on pulsed devices. In these machines the efficiency of conversion of electrical energy to soft X rays with a laser is found to be an order of magnitude higher than for electron-beam bombardment sources. The process of plasma formation from a solid surface starts with vaporization of the surface layer.[26] The energy necessary for this step is negligible compared to that required for ionization and heating the plasma to keV temperatures. The laser energy absorbed by the plasma goes almost entirely into ionization and heating of the resulting electrons, with the thermal energy of the ions being negligible for high-Z plasmas.

Three processes contribute to the radiation from the plasma: bremsstrahlung continuum from free–free transitions, recombination continuum from free–bound transitions, and line radiation from bound–bound transitions. Each of these rates of radiation depends on the level of ionization attained within the short times available.

Following the absorption of laser energy by the target, the plasma comprises heavy ions and electrons with Maxwellian energy distributions. Figure 4.18 shows the required energy conversion processes for plasma radiation.

The limiting factor for plasma heating by laser beam is the electron density in the plasma, which has to be less, but near, a critical value n_p for which the plasma frequency ν_p equals the laser frequency ν_L, namely,

$$n_p = 1.24 \times 10^{-8} \, \nu_L^2 \, \text{cm}^{-3}$$

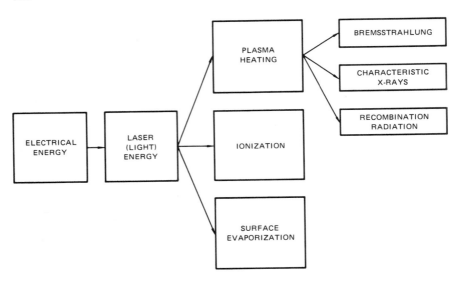

FIGURE 4.18. Energy conversion processes in plasma X-ray generation.

At present the most powerful lasers employ Nd glass emitting at a wavelength of 1.06 μm, corresponding to $n_p = 1.0 \times 10^{21}$ cm^{-3}. The absorption proceeds through the process of inverse bremsstrahlung, in which electrons accelerated by the electric fields of the focused laser light undergo momentum-transfer collisions with the ions via their Coulomb interaction.

In another approach[27] a plasma is generated in a coaxial discharge tube, with the ions being heated primarily through the collapse of the quasicylindrical current sheet in a manner reminiscent of the linear pinch effect.

In addition to the electron bombardment and plasma X-ray sources, the synchrotron radiation source is currently being explored for high-intensity exposures (see Section 4.7).

4.6.4. X-Ray Masks

Many different approaches have been taken in fabricating X-ray lithography masks. The primary problem is to produce a thin but strong substrate that is transparent to X rays. Two types of thin films have been tried: organic and inorganic membranes. Organic films include Mylar, Kapton, Pyolene, and Polymide and are held stretched to form a planar pellicle. Inorganic types include silicon, silicon oxide, silicon metals, aluminum oxide, and silicon carbide. The X-ray absorber pattern can be ion-beam etched, sputter etched, or electroplated through a resist pattern. The absorber film typically consists of two metal layers: first a thin layer of chromium for adhesion to the substrate, and then a layer of gold. The X-ray attenuation of the mask material is a function of the X-ray wavelength, as shown in Figure 4.19.[28] Figure 4.20 shows the energy loss as function of the photon energy for light elements usually used in the mask support material.[29]

The selection criteria for mask materials can be summarized in the following simple optimization rules.
Minimize the transmission of the mask:

$$T_M = \exp(-\mu_M z)$$

Minimize the absorption of the support materials (Si, Be, polymers):

$$A = 1 - \exp(-\mu_s z) = 1 - T_s$$

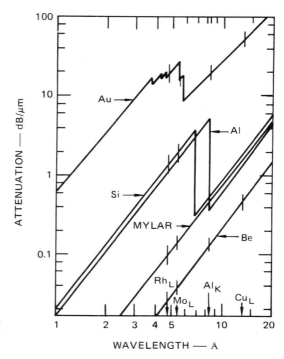

FIGURE 4.19. X-ray attenuation as a function of the X-ray wavelength.

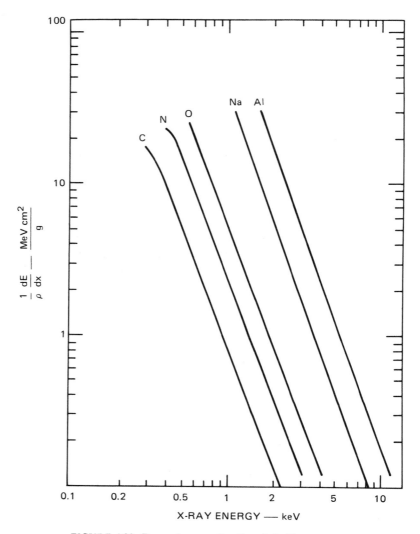

FIGURE 4.20. Energy loss as a function of the X-ray energy.

Maximize the contrast:

$$c = \frac{T_M}{T_s}$$

Figure 4.21 shows the fabrication steps for a silicon membrane with a gold absorber pattern[28] and the processing sequence of a gold mask using an intermediate titanium mask.[23]

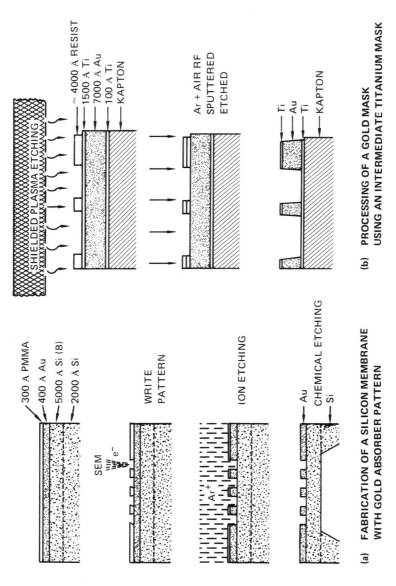

FIGURE 4.21. Processing sequence for the fabrication of a gold mask on silicon and Kapton substrates. Source: refs. 23 and 28.

(a) **FABRICATION OF A SILICON MEMBRANE WITH GOLD ABSORBER PATTERN**

(b) **PROCESSING OF A GOLD MASK USING AN INTERMEDIATE TITANIUM MASK**

4.6.5. X-Ray Resists

X-ray resists are basically electron resists that are sensitive to the photo-electrons created when X rays are absorbed in the resist. The long exposures (hours) that are commonly used are necessary because of the low energy conversion of the X-ray source, the low absorption of currently available X-ray resists, and the flux loss associated with the small solid angle that the substrate must subtend at the X-ray source for collimation reasons.

All the many resists reported absorb less than 10% of the incoming flux and are therefore relatively insensitive to X rays. Some resists have been formulated to increase absorption by resonance with certain characteristic wavelengths. For example, the palladium characteristic line at 4.37 Å is strongly absorbed by chlorine-containing resists. Although there is considerable room for improvement, current resist technology suffices for the fabrication of ICs.

Conventional electron bombardment sources generate X-ray fluxes incident on the wafer in the range of 1 to 10 $mJ/cm^2 \cdot min^{-1}$. In comparison, conventional photoresists have sensitivities of 100 mJ/cm^2; electron resists COP and PBS have 175 and 94 mJ/cm^2, respectively, at the 4.37 Å $Pd_L\alpha$ wavelength.

More sensitive X-ray resists can be designed in several ways.[30] The most important is to increase the absorption coefficient for the resist by the incorporation of high weight percentages of absorbing elements for a given X-ray source. At short wavelengths resonance absorption can result in the target. Examples are Cl at 4.36 Å, S at 5.41 Å, and F at 8.34 Å. Heavy-metal atoms (bromine) can increase the absorption in the entire energy range. However, high sensitivity is not the only requirement; the X-ray resist must have the following qualities:

- High sensitivity to X rays
- High resolution
- Resistance to chemical, ion, and/or plasma etching

No present resist satisfies all three requirements. Sensitivity ranges from 1 mJ/cm^2 for experimental resists to 2 J/cm^2 for poly(methylmethacrylate) (PMMA) at a wavelength of 8.34 Å.

A PMMA resist is a positive resist in which the X rays split large polymer molecules. The smaller molecules in the exposed areas have a faster dissolving rate during development. The mainstay of electron and X-ray lithography, the PMMA resist has a demonstrated resolution of 500 Å.[28]

The copolymer resist—poly(glycidal methacrylate-coethyl acrylate) or COP—represents a negative resist in which cross-linking occurs in areas exposed to X rays and prevents dissolution during developing. Although the COP resist is at least 20 times as sensitive to 8.34-Å X rays as PMMA, the control of fine geometries with COP is difficult.

Line-edge profiles can be calculated using computer simulation[31] of the exposure, development, and surface etching of the resist. Unlike in the optical case, no visible interference patterns are observable in the X-ray-exposed line-edge profiles.

4.6.6. Alignment

The attainment of the extremely high positional accuracy that X rays require for reregistration has also proved difficult. Alignment techniques using X rays themselves require the use of an active fabricated alignment structure (on the wafer or a second material) that absorbs X ray, fluroesces, or photo-emits electrons.

Figure 4.22 shows a scheme for the registration of multiple masks in X-ray lithography.[28] This system utilizes alignment marks on the mask and

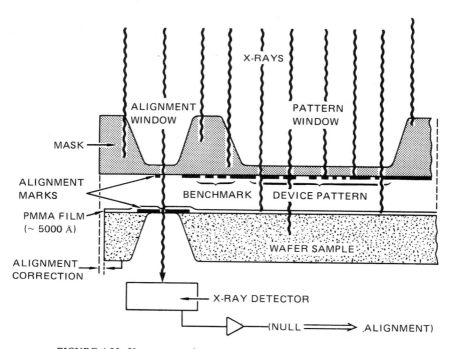

FIGURE 4.22. X-ray system for multiple mask alignment. Source: ref. 28.

wafer and a proportional counter for X-ray detection. With this registration system 0.1-μm registration precision has been achieved. Other mask-alignment systems using fluorescent X rays or backscattered electrons have been built in laboratory X-ray printers.[32]

4.6.7. Linewidth Control

The trend toward higher packing densities and smaller features in microelectronic devices has demanded tighter linewidth tolerances at critical patterning steps. As features are packed closer together, the margin for error on feature size is obviously reduced. Furthermore, the successful performance of most devices depends upon the control of the size of critical structures, such as gate electrodes in MOS devices. Since the allowed size variation on such structures is generally a fixed fraction of the nominal feature size, reduced dimensions also imply tighter tolerances. Linewidth control of better than ± 0.5 μm from the design value is not an uncommon requirement for critical mask levels.

One of the more difficult problems with resist pattern generation is to achieve good linewidth control, high resolution, and good step coverage simultaneously. Often the requirements appear to be mutually exclusive; good step coverage requires a thick resist; high resolution, however, is more easily obtained in a thin resist. This is true for all resists, both positive and negative.

With any resist the necessary conditions to obtain high resolution and good linewidth control are a flat surface and a thin resist (3000–4000 Å). The flat surface ensures that the resist has very little variation in thickness, and as a result there will be little variation in resist linewidth. Even so, resist linewidth variations still occur when lines traverse a step. As device wafers have steps due to previously patterned layers, thick resists (7000–15000 Å) must be applied to achieve a flat surface when covering the stepped features on the substrate.

Bell Laboratories developed a technique[33] for generating high resolutions and steep profile patterns in a few-μm-thick organic layer that covers the steps on a wafer surface and is planar on top. A thick layer (2–3 μm) of photoresist serves as the thick organic layer. It is patterned using an intermediate masking layer (0.12 μm SiO_2) covered by 4000-Å-thick layer of chlorine-based negative X-ray photoresist.

After X-ray exposure and development of the top X-ray resist layer the intermediate layer of SiO_2 is etched by CHF_3 reactive plasma etching. The thick organic layer is then etched by O_2 reactive plasma etching, with the SiO_2 acting as a mask. The various steps required to define a steep resist profile are shown in Figure 4.23.

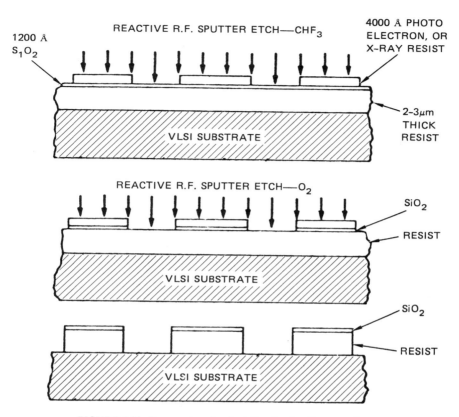

FIGURE 4.23. Process steps for the trilevel resist. Source. ref. 33.

The X-ray resist is a mixture of a host polymer and a volatile monomer that is locked into the host during exposure.[34] Pattern development is accomplished by heating and exposure to an oxygen plasma. This X-ray resist allows exposure times as short as 1.5 min when exposed to a 2.9 mJ/cm² · min flux of palladium X ray (4.37-Å wavelength).

This new resist with the multilevel structure ensures submicron resolutions (0.3 μm) combined with good step coverage and linewidth control. This three-layer technique has various advantages for other lithographic systems as well.[24,33]

For optical lithography the thick underlying layer of resist and the layer of SiO₂ are sufficient to reduce the reflection from the wafer surface and, as a result, reduce standing-wave problems. The flat surface of the thick organic layer keeps scattering down and the top resist can be made thin for high resolution.

For electron-beam lithography backscattering from a substrate covered by 0.1 μm of SiO_2 on top of 2–3 μm of polymer should be less than that from a Si or SiO_2–Si substrate. Thus proximity effects existing in that technology might be reduced and better linewidth control can be expected.

Although a three-layer technique [i.e., top radiation sensitive resist, SiO_2 mask, bottom organic (resist) layer] has been discussed here, a two-level scheme without SiO_2 also works if the top patterned resist is very resistant to the reactive rf sputter etch. In addition, there are many choices for the intermediate and thick organic layers. The intermediate layer could be silicon nitride, boron nitride, or some other suitable low-defect material. For the bottom layer many positive and negative resists familiar to photo and electron lithography may be used, as well as other polymers.

4.7. Synchrotron Radiation for X-Ray Lithography

4.7.1. Introduction

Synchrotron radiation sources are by far the brightest sources of soft X rays.[35] Electron storage rings and synchrotrons emit a much higher flux of usable collimated X rays than does any other source, thereby allowing shorter exposure times (on the order of seconds), large throughputs, less critical resist exposure conditions, and simplified geometrical conditions for applications requiring registration. Because of the high collimation of synchrotron radiation, the spatial resolution of X-ray lithography with synchrotron radiation is not limited by penumbral blurring, and rather large distances between the mask and the wafer can be tolerated (about 1 mm for 1-μm linewidth patterns).

The basic characteristics of synchrotron radiation are[36] the following:

- High intensity over a broad spectal range from the infrared through the visible and ultraviolet and into the X-ray part of the spectrum
- High polarization
- Extreme collimation
- Pulsed time structure
- Small source size
- High vacuum environment

4.7.2. Properties of Synchrotron Radiation

Synchrotron radiation is emitted by high-energy relativistic electrons in a synchrotron or storage ring, which are accelerated normal to the direction of

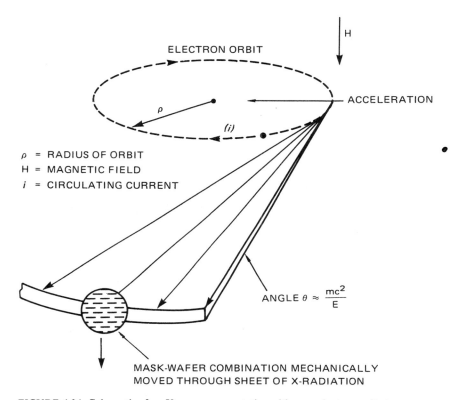

FIGURE 4.24. Schematic of an X-ray exposure station with a synchrotron radiation source.

motion by a magnetic field. Figure 4.24 shows a schematic of a X-ray exposure station with a synchrotron radiation source.

The flux available from a large synchrotron source is 10^4 times as powerful as from a rotating-anode-type X-ray source (10 W of X rays).

The energy loss per turn for the highly relativistic electrons ($\beta = v/c \approx 1$) is[36]

$$\Delta E \text{ (keV)} = \frac{88.47 E^4}{\rho} \qquad (4.19)$$

where E (in GeV) $= 10^9$ eV, ρ is the radius of the orbit (m), and i is the circulating electron current in the ring (A). Multiplying Eq. (4.19) by the circulating current i gives the total radiated power

$$P \text{ (kW)} = \frac{88.47 E^4 i}{\rho} \qquad (4.20)$$

or in terms of the magnetic field B (in kGauss)

$$P \text{ (kW)} = 2.654 \, B E^3 \qquad (4.21)$$

For machines such as SPEAR (Stanford Positron–Electron Annihilation Ring) at Stanford University the total radiated power is 105 kW, where $E = 3.5$ GeV, $\rho = 12.7$ m, $B = 9.2$ kG, and $i = 0.1$ A.

Synchrotron radiation is emitted in a cone with an opening angle θ, given by

$$\theta \approx \frac{mc^2}{E} = \frac{0.5}{E \text{ (GeV)}} \text{ (mrad)} \qquad (4.22)$$

The spectral distribution of the synchrotron radiation extends from the microwave region through the infrared, visible, ultraviolet, and into the X-ray region, decreasing in intensity below a critical energy E_c or, alternatively, the critical wavelength λ_c, where

$$E_c \text{ (keV)} = \frac{2.218 \, E^3 \text{ (GeV)}}{\rho \text{ (m)}} \qquad (4.23)$$

or since

$$\lambda \text{ (Å)} = \frac{12.4}{E \text{ (keV)}}$$

$$\lambda_c \text{ (Å)} = \frac{5.59 \rho \text{ (m)}}{E^3 \text{ (GeV)}} \qquad (4.24)$$

For SPEAR operating at 3.5 Gev, $E_c = 7.4$ keV and $\lambda_c = 1.7$ Å. The spectrum of the synchrotron radiation from SPEAR is shown in Figure 4.25. The high polarization and the pulsed time structure available from synchrotron radiation, however, are not relevant for X-ray lithography.

Using a synchrotron X-ray source for lithography, penumbral blurring can be completely eliminated and one can tolerate large distances between the mask and the wafer without loss in resolution. However, at very large distances diffraction effects determine the resolution limit.[35]

Synchrotron radiation can form high-quality submicron images, but its main advantage is the short exposure time (a few seconds) compared to a few

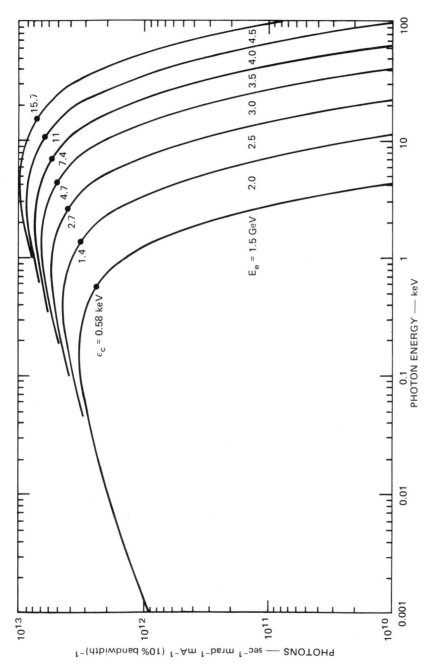

FIGURE 4.25. Spectral distribution of synchrotron radiation from SPEAR (ρ = 12.7 m). Source: ref. 36.

hours using conventional X-ray sources. Synchrotron radiation sources that provide powerful, well-collimated beams of soft X rays for research are in operation or being built at Stanford University, Brookhaven National Laboratory, the University of Wisconsin, and Cornell University. Research facilities for X-ray work have also been set up in Italy (Frascati), Germany (Hamburg), Japan (Tokyo), France (Orsay), and in the U.S.S.R. (Novosibirsk).

4.8. Electron-Beam Lithography[37]

4.8.1. Introduction

There are two ways in which electron beams can be used to irradiate a surface and create a pattern.[38,39] There are the parallel exposure of all pattern elements at the same time and the sequential (scanning) exposure of one pattern element (pixel) at a time.

Projection systems generally have a high throughput and are less complex than scanning systems. The pattern information is stored in masks. The scanning systems are under computer control, where a finely focused beam (or beams) of electrons is used to generate the pattern, correct the distortions and proximity effects, and register the wafer position. The pattern information is stored in a digital memory.

Although an electron beam can be used in a variety of ways, direct control by a computer offers the ability to generate patterns without the need of a mask. This enables the scanning-electron-beam systems to be used for both mask making and direct wafer writing. These machines combine high spatial resolution (<0.1 μm) with accurate registration (<0.1 μm). The small electron beam (or beams) are either rastered or manipulated in a vector mode. The vector system is more efficient, since scanning is only performed over areas that are to be exposed on the resist. In the vector mode, however, the deflection system accuracy is limited, and to apply corrections a computer system is needed.

Shaped-electron-beam machines[40] that combine vector scanning and projection (hybrid machines) have been designed to further improve the efficiency. In this approach the fundamental picture elements from which a device can be built up are projected onto an appropriate wafer location by the use of a variable electron-beam shape. In character projection systems a selected character from a mask is imaged onto the wafer, and this process may be repeated many times (bubble memory chips, random access memories (RAMS), read only memories (ROMS), or for any repetitive pattern). Here the scanning-type system is combined with a projection system in which special apertures (beam shaping) or masks form the picture elements.

Multibeam scanning systems use an optical element (screen lens,[41] single crystal[42]) to obtain multiple beams for simultaneously generating patterns on the wafer. A classification of the currently used electron-beam machines is shown in Figure 4.26.

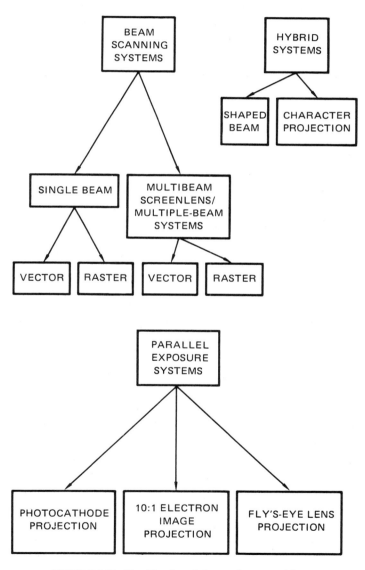

FIGURE 4.26. Classification of electron-beam machines.

4.8.2. *Major System Components*

4.8.2.1. *Projection Systems*

Electron beam projection lithography provides a means of replicating, in a single large-area exposure, submicron linewidth patterns originally made by scanning-electron-beam lithography. The technique is complementary to optical and X-ray projection lithographies. Two types of beam projection systems have been developed specifically for semiconductor device fabrication: the 1 : 1 and reduction projection systems.

Work on 1 : 1 projection systems has been under way for many years. It began at Westinghouse[43] and has been more recently pursued by Mullard (Philips) in England.[44] A schematic of the Mullard system is shown in Figure 4.27. This projection system employs a photocathode masked with a thin metal pattern. Photoelectrons from the cathode are accelerated onto the sample by a potential of about 20 kV applied between the cathode and the sample. A uniform magnetic field focuses these photoelectrons onto the wafer (anode) with unity magnification. The mask and wafer can, in principle, be as large as desired.

In this system image position is detected by collecting characteristic X rays from marks on the wafer with the X-ray detectors shown. (During this process the photocathode is masked so that only the alignment marks are illu-

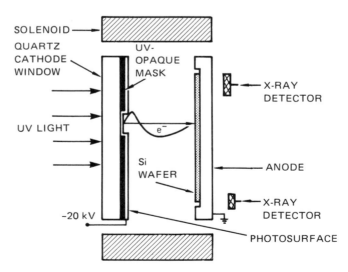

FIGURE 4.27. Schematic of the Mullard 1 : 1 electron-beam projection system. Source: ref. 43.

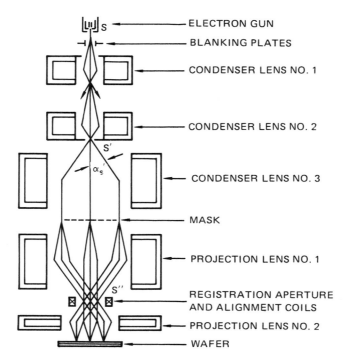

FIGURE 4.28. Reduction-type electron-beam projection system developed by IBM. Source: ref. 45.

minated.) Magnetic deflection is then used to position the pattern with an accuracy of 0.1 μm. The image current density is about 10^{-5} A/cm^2 (1-s exposure for 10^{-5} C/cm^2 resist sensitivity) for cesium iodide photocathodes, which have the best lifetime and resistance to poisoning. The dominant aberration limiting resolution is chromatic aberration, and theoretical estimates of minimum linewidths vary from 0.5 to 1 μm. So far the technique has been used to make operating semiconductor devices with lines 1 μm wide.

Figure 4.28 shows the reduction-type projection system developed at IBM Research.[45] The basic concept for this system, which is the electron-optical analogy of reduction–optical projection cameras, was first described by researchers at Tübingen University, Federal Republic of Germany.[46] The mask is a freely suspended metal foil. A special electron-optical system illuminates the mask and forms a sharp demagnified image of it on the wafer. The demagnification factor is 10$\times$, and a field 3 mm in diameter and linewidths of down to 0.25 μm can be produced.

The system uses its scanning mode of operation for alignment. In this mode the illuminating beam is focused onto the mask rather than flooding it, as in the case of image projection. The focused beam scans across the mask and an image of this focused beam scans across the sample. Scattered electrons are collected from the sample to detect sample position, and corrections are made by shifting the projected image, with deflection coils placed between the two projection lenses. This system offers very low distortion and very high resolution, factors that are problems in the 1 : 1 cathode projection system.

A parallel-image electron-beam projection system has been developed[47] at SRI International (see Figure 4.29). In this system a two-dimensional array of lenses is used for image multiplication and projection from an object mask. The screen lens, that is, holes in a planar electrode, constitutes a particularly simple form of a multiple-imaging element. The screen lens separates two regions of uniform electrostatic fields. This image projection system is used to fabricate thin-film field-emitter electron and ion sources for various applications.[48-51]

4.8.2.2. Beam-Scanning Systems

The basic operating principles of the scanning-electron-beam system will be discussed in the order electron-optical column, beam shaping, pattern generation, stage assembly, and registration.

4.8.2.2.1. The Electron-Optical Column Beam-forming systems use either the Gaussian round-beam approach or the shaped-beam (square or round) approaches[39] shown in Figure 4.30.

Gaussian systems use the conventional probe-forming concept of the scanning electron microscope, as in Figure 4.30(a). In general, two or more lenses focus the electron beam onto the surface of the wafer (or mask) by demagnifying the electron gun source. High flexibility can be achieved, since the size of the final beam can be readily varied by changing the focal lengths of the electron lenses. To ensure good line definition the beam size is generally adjusted to about a quarter of the minimum pattern linewidth.

In the square-beam approach, Figure 4.30(b), an electron source illuminates a square aperture at the center of a condenser lens placed immediately after the gun. The condenser lens images the gun crossover (1 : 1) into the entrance pupil of a second condenser lens. This lens, together with a third condenser lens, demagnifies the square aperture to form a square beam. A fourth lens images the square beam (1 : 1) onto the target plane. The size of the square beam is generally equal to the minimum pattern linewidth. To achieve an equivalent resolution the edge slope of the intensity distribution of the square beam is matched to the intensity distribution of a Gaussian round beam between the 10% and 90% points.

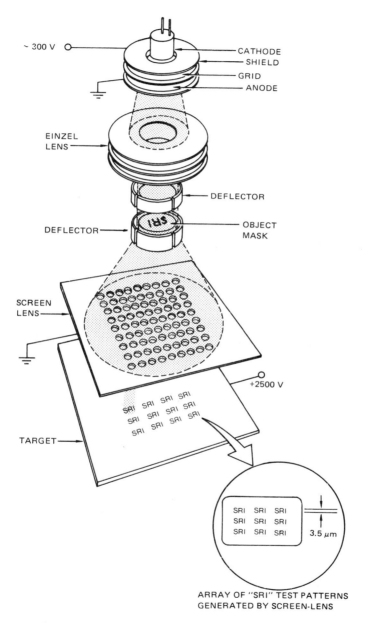

FIGURE 4.29. Screen lens parallel-image electron-beam projection system developed at SRI International.

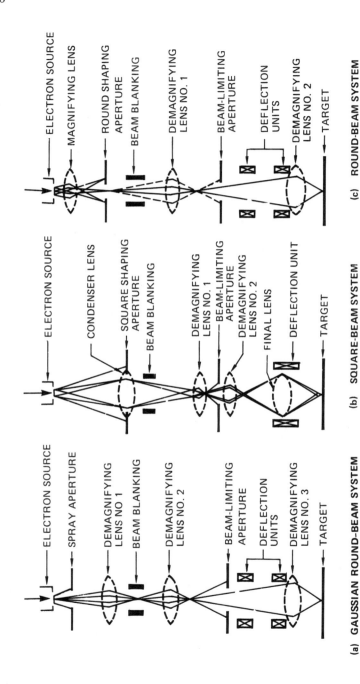

FIGURE 4.30. Beam-forming systems. Source: ref. 39.

(a) GAUSSIAN ROUND-BEAM SYSTEM

(b) SQUARE-BEAM SYSTEM

(c) ROUND-BEAM SYSTEM

In the case of the round-shaped beam shown in Figure 4.30(c) a lens focuses a magnified image of the gun crossover onto a round aperture and two condenser lenses demagnify the round aperture onto the plane of the target.

Round-beam systems (Gaussian or shaped) are generally simpler than shaped square-beam systems and have more flexibility. However, the square beam has more current in the spot (current is proportional to spot area for the same gun brightness) and therefore offers higher exposure speed in cases where speed is limited by the beam current and/or beam stepping rate. Difficulties with square-beam systems may arise when angle lines are required or when some lines have dimensions that are not integral multiples of the beam size. In the latter case overexposure in the overlapping regions will occur.

To calculate the beam spot size on the resist one has to know the crossover size ($d_0' = 2r_c$), the particular optical system for demagnification, and the lens errors introduced by the nonperfect optical elements. For a perfect system (no lens errors) the spot size for a three-lens system would be

$$d_0 = \frac{f_1 f_2 f_3}{L_1 L_2 L_3} d_0' \qquad (4.25)$$

where the optical system demagnifies the crossover $d_0' = 2r_c$ to d_0, f_i denotes the focal lengths of the lenses, and L_i is the distance between lenses. For a real system one has to take into account the enlargement of the demagnified spot resulting from the lens errors for each lens in the system (see Section 2.3.5).

Figure 4.31 shows the current limits in the electron-beam spot on target as a function of the beam spot size, using different cathodes.[52] Electron-beam machines operating in the 10^{-6}–10^{-7} A beam current mode are capable of writing 0.5–1.0 μm lines.

4.8.2.2. Beam Size and Shape Beam-shaping techniques[40] in electron-beam exposure systems have proved to be very effective in overcoming throughput limitations inherent to the serial exposure of scanning systems.

Figure 4.32 illustrates the spectrum of beam profiles used for pattern generation. One end of the spectrum is represented by the scanning electron microscope-type Gaussian round-beam exposure of one image point at a time. The size of the round beam, defined as the half-width of the Gaussian distribution, represents the spatial resolution and is typically four to five times smaller than the minimum pattern feature. For shaped-beam systems the spatial resolution given by the edge slope of the beam profile is decoupled from the size and shape of the beam. Consequently, a plurality of image points can be projected in parallel without loss of resolution.

In the case of the fixed-shape beam a square aperture is projected to fit

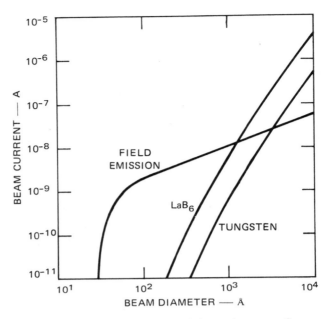

FIGURE 4.31. Diameter and current limits of electron-beam spot. Source: ref. 52.

the minimum pattern feature containing 25 or more image points. The variable-shaped beam is generated by projecting two superimposed square apertures. The compound image formed by both apertures can be varied in size and shape to produce the various pattern elements containing up to 100 (or even more) image points.

Character projection represents the most efficient beam-shaping technique. Complex pattern cells containing up to 2000 image points (pixels) are addressed and projected in parallel. This technique overcomes the restriction to rectilinear geometries and is very efficient in the printing of repetitive patterns.

Character projection is an extension of the dual-aperture technique in which the second aperture is replaced by a character plate with an array of complex aperture shapes. These can be selected electronically for each exposure in full or in part to compose the image. Nonrectilinear and curved patterns can easily be included.

4.8.2.2.3. Pattern Generation After the beam has been focused and shaped it must be deflected (scanned) over a wafer by a beam-writing technique. Deflection systems for electron beams are generally electromagnetic, though electrostatic systems and even partly electromagnetic, partly electrostatic systems exist.[52,40,41]

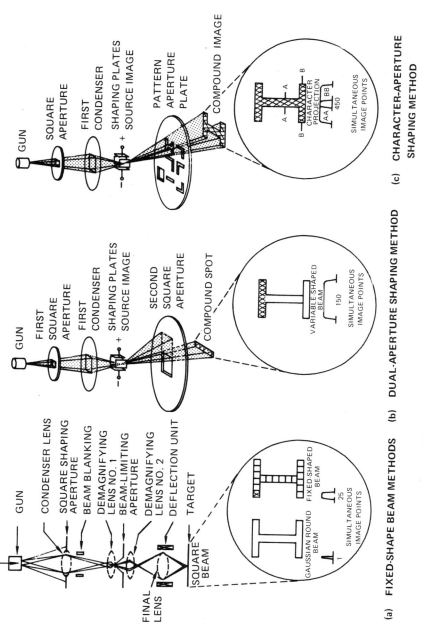

(a) FIXED-SHAPE BEAM METHODS   (b) DUAL-APERTURE SHAPING METHOD   (c) CHARACTER-APERTURE SHAPING METHOD

FIGURE 4.32. Beam-shaping techniques. Source: ref. 40.

The two basic beam-writing techniques are the raster and vector techniques. In the raster technique[53,54] the beam is scanned over the entire chip area and is turned on and off according to the desired pattern. In the vector technique[39] the beam addresses only the pattern areas requiring exposure, and the usual approach is to decompose the pattern into a series of simple shapes such as rectangles and parallelograms.

The raster scan places less stringent requirements on the deflection system because the scanning is repetitious and distortions caused by eddy currents and hysteresis can be readily compensated for. It can also handle both positive and negative images. The vector scan is more efficient but requires a higher-performance deflection system. In addition, it has several other advantages not readily available to the raster scan technique, for example, ease of correction for the proximity effects of electron scattering and a significant compaction of data that can lead to a much simpler control system.

Figure 4.33(a) shows the operating mode of a raster scan system [electron beam exposure system (EBES)]. In this machine a 0.5-μm-diameter writing spot exposes the area of a pixel (0.025 μm^2) in 100 ns, and thus it requires about 25 μs to expose a line consisting of 256 addresses. An interferometer-controlled table moves the substrate an address length in the x direction (2 cm/s) while the beam is sweeping out one scan line in the y direction. The alignment is performed for the full wafer.

Figure 4.33(b) depicts the operating mode for a vector-scan-type electron-beam system (IBM EL1). After the alignment of each chip (dye) in this step-and-repeat system the exposure is made with a two-dimensional electronic scan that covers the entire chip. Since the maximum area over which it is possible to electronically scan a submicron beam is smaller than a chip, it is necessary to move the sample mechanically. Two basic types of mechanical movement can be used: a step-and-repeat movement, where the substrate is stepped between individual exposure sites, or a continuous mechanical scanning, where the substrate moves continuously under the beam.

4.8.2.2.4. Stage Assembly   Two methods have been employed to control the relative position between the electron beam and the wafer, namely, field-to-field registration with bench marks and measuring the exact table location with a laser interferometer. The field-to-field registration with bench mark technique has proved to be acceptable in direct wafer processing, but when the electron beam is used for the direct generation of master masks, a laser interferometer is invariably used to obtain sufficient step-and-repeat accuracy. The laser interferometer, operating on appropriately located mirrors on the work table, can provide a monitoring accuracy down to several hundredths of a micron. Other important parameters such as the variation of the z position

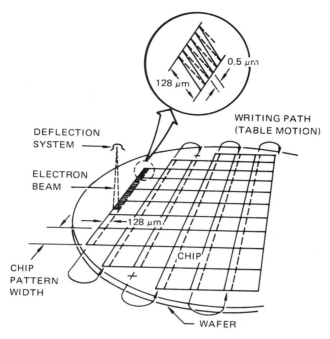

(a) **EBES RASTER SCAN FEATURING ONE-DIMENSIONAL LINE SCAN (RASTER), CONTINUOUSLY MOVING TABLE, FULL WAFER ALIGNMENT, AND LASER INTERFEROMETER FEED**

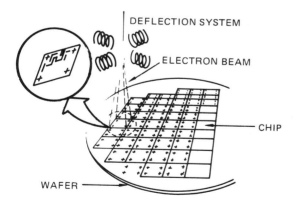

(b) **IBM EL1 VECTOR SCAN FEATURING TWO-DIMENSIONAL FULL-CHIP SCAN, DISTORTION-FREE DEFLECTION, STEPPED TABLE, AND ALIGNMENT AT EVERY CHIP**

FIGURE 4.33. Raster and vector scan electron-beam systems. Source: refs. 39 and 53.

while the table is being translated in the x and y directions and the translation speed vary according to the particular writing strategy. Sophisticated servo systems, complete with multiple-feedback loops, are generally used in the control of such work tables.

4.8.2.2.5. Registration In a high-resolution scanning lithography system registration can be made on a wafer global basis (a one-time registration) relying on extreme system stability, flat and undistorted substrates, and exact stage motion during the step-and-repeat process, or on a chip-by-chip basis, making fine corrections at each site.[55] The former system can have a special mark region and considerable time can be tolerated to achieve registration. The latter system requires marks that are fully compatible with all subsequent processing steps and usually completes a registration cycle in a time that is short compared to the sum of the pattern write time and stage chip-to-chip step time.

In direct microcircuit fabrication technology the registration mark will most likely be made of thin film strips of a high-atomic-number material or of special geometrical patterns in a material having an atomic number equal or similar to that of the substrate (see Figure 4.34). As the electron beam is scanned across the sample surface, it passes over these differing material or topographic registration marks. Two common types are a raised pedestal or an etched hole of Si or SiO_2. The signal detected by the registration system can be thought of as having three constituents. First is a steady (dc) part that represents the average number of electrons scattered and detected. The second part is a noise component that contains the noise of the primary beam (a small effect) and any other noise arising from the scattering and detection processes. The third component is the change in detected electrons resulting from a change in scattering as the beam crosses a mark edge.

Backscattered electrons offer more promise in detecting registration marks overcoated with resist than do the easily absorbed secondary emission electrons. The backscattered electrons emerge in many directions and possess energies varying from the primary-beam value down to about several hundred eV. This broad spectrum is caused by single and/or multiple inelastic scattering of the electrons in the substrate, registration mark, and resist layer(s). To be detected an electron emerging from the sample must be traveling in a given direction and have an energy above the detector threshold. Several different types of detectors (e.g., silicon diode, scintillation) have been described in the literature for detecting the backscattered electrons.[56,57] Assuming a step-function edge as the primary beam crosses over the mark edge, the backscattered electrons must travel a much reduced distance within the base material before emerging. Hence they have an energy nearer to the primary-beam value, and a rapid increase in backscattered electron signal is observed. The detectors would usually be used in pairs.

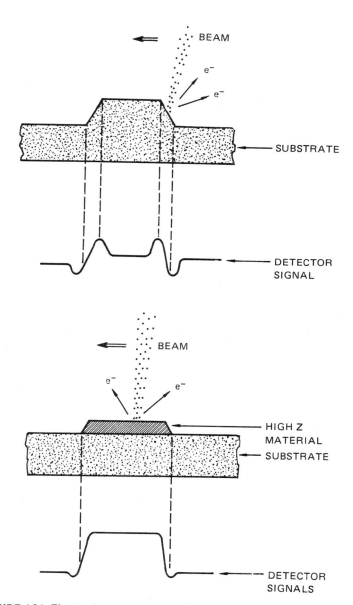

FIGURE 4.34. Electron-beam alignment marks and detector signals for registration.

There is no standard set of signals or detector systems used in electron-beam lithography; each developer tends to prefer his own systems. The most commonly used are the following:

- Signals
    Backscattered electrons
    Secondary electrons
    Luminescent radiation
- Detectors
    Scintillation detector
    Solid-state detector
    Photomultiplier

The types of alignment (registration) marks also vary from system to system. There is no standard set of marks in the industry.

4.8.3. Parallel-Scanning Systems

Scanning-electron-beam lithography is used for making masks and experimental devices but has not found general usage for the general writing of patterns on wafers, primarily because of the long time it takes to scan a production-size wafer with a single beam.

Figure 4.35 shows a scheme for multiple-beam lithography developed by SRI International.[41] In this approach a flood beam emerging from an aperture is incident on an array of holes, each of which forms an immersion lens with the substrate.[47] This type of lens array is called a screen lens. Each lens of the array addresses a die of the wafer and focuses on it a demagnified image of the aperture. The flood beam is deflected by a double electrostatic system, enabling the beamlets to scan the dies. The amplitude of the deflection, d, is given by

$$d = \frac{l' l_D E_D}{2\sqrt{V_A}\sqrt{V_{sc}}} \tag{4.26}$$

where l' is the distance between the screen lens and the Si wafer, l_D is the effective length of the deflector electrodes, E_D is the deflecting field V_D/cm, V_A is the accelerating voltage (nominally 1 kV), and V_{sc} is the screen lens voltage (nominally 10 kV). All the beams are scanned and blanked in parallel so that identical patterns are written simultaneously over each die. A laser interferometer stage under computer control enables dies of large areas to be completely covered, despite limitations on the deflection amplitude of the beamlets.

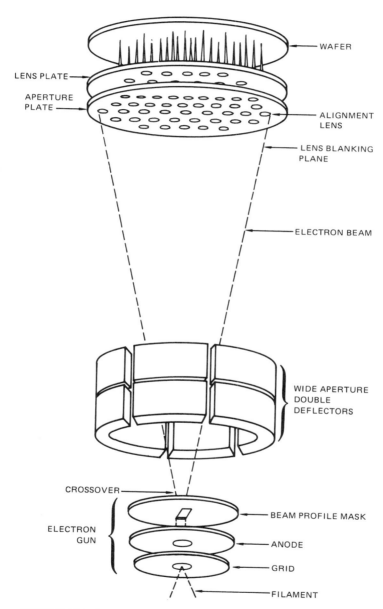

FIGURE 4.35. Optical column for a parallel-scan system. Source: ref. 41.

This machine is capable of submicron lithography and fast printing (tens of seconds) with PBS resist (see Table 4.2).

Another scheme using beam multiplexing and the diffraction effect in a thin single-crystal film has recently been proposed for parallel writing.[42] In this system several patterns are addressed simultaneously, using diffracted beams (see Figure 4.36). The basic geometry for printing in the transmission

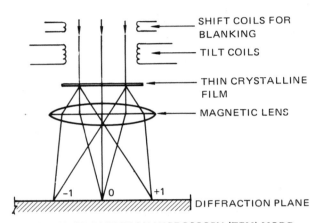

(a) **TRANSMISSION-ELECTRON MICROSCOPY (TEM) MODE**

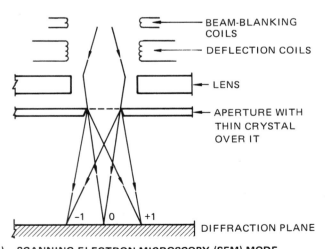

(b) **SCANNING-ELECTRON MICROSCOPY (SEM) MODE**

FIGURE 4.36. Beam-multiplexing method using the diffraction effect of a thin film. Source: ref. 42.

electron microscope (TEM) mode is shown in Figure 4.36(a). For the scanning electron microscope (SEM) mode the gold crystal was suspended across the final lens-limiting aperture, as shown in Figure 4.36(b). The applicability of the diffraction effect for electron-beam multiplexing has been demonstrated for both projection and scanning systems. The number of patterns that can be obtained by pattern multiplexing is an increasing function of the electron energy. At 100 kV at least 36 patterns can be obtained.

4.8.4. Electron-Beam Resists

4.8.4.1. Resist Characterization

Electron resists must be capable of forming uniform pinhole-free films on a substrate by simple processes such as spinning, dip coating, or spraying, and usually only polymeric materials fulfill these requirements. When energetic electrons interact with organic materials, the molecules may either be broken down to smaller fragments or link together to form larger molecules. Both types of change occur simultaneously, but when any given material is irradiated, one of these processes predominates and determines the net effect upon the material. For polymers it is known that if every carbon atom in the main chain is directly linked to at least one hydrogen atom, then larger molecules are formed upon irradiation by the process of cross-linking. If a polymer is cross-linked on irradiation, it forms a three-dimensional network and becomes insoluble and infusible. This type of polymer, which is the most common, forms the basis of negative-working electron resists, as it is possible to dissolve and remove unirradiated material, while irradiated material cannot be dissolved away.

A relatively small number of polymers is known in which irradiation in moderate doses predominantly causes a break in the main polymer chain, resulting in irradiated material having a lower average molecular weight than unirradiated material. By judicious choice of a solvent or mixture of solvents, it is possible to dissolve the irradiated polymer selectively, which allows positive-working electron resists to be formulated. At higher electron exposures, however, cross-linking predominates again and the irradiated material becomes completely insoluble. Thus there is always a limited range of electron exposures in which a positive action may be observed.

The energy available for breaking and forming chemical bonds by electron beams with accelerating voltages of 10–20 kV is very large when compared with ultraviolet radiation. Therefore photoresists exposed by electron beams have completely different properties than do the same resists exposed to ultra-violet radiation and are normally not suitable for high-resolution delineation. Fortunately, almost all polymers and copolymers react in some manner to electron-beam irradiation, but until recently very few of these had been investigated for this application.

A positive electron resist is characterized by a reduction of molecular weight due to chain scission of the original molecules. Since polymers with different molecular weights have different solubility characteristics in a given solvent, the resist developer is required to distinguish between the fragmented molecular weight M_f and the original molecular weight M_n. The relationship between M_f and M_n and the exposure parameters is expressed as[58]

$$M_f = \frac{M_n}{1 + KQG(z)S(z)}$$

with

$$K = \frac{M_n}{100\,e\rho N_A}$$

where e is the electronic charge, ρ is the resist density, and N_A is Avogadro's number. Q is the incident charge per unit area, $G(z)$ is the radiation yield defined as the number of chemical events per 100 eV absorbed, and $S(z)$ is the depth-dose function for electrons expressed in terms of the normalized penetration $f = z/R_G$. Where the Grün range R_G (see Section 2.6.2.1) is a function of the electron energy E_0 in keV and is expressed as

$$R_G = \frac{0.046}{\rho} E_0^{1.75}\ \mu m$$

and ρ is the density of the resist in $g \cdot cm^{-3}$. Empirically, the rate of energy absorption in the resist as a function of the normalized penetration can be expressed as[59]

$$\frac{dE}{df} = -E_0 \Lambda(f)$$

where $\Lambda(f) = 0.74 + 4.7f - 8.9f^2 + 3.5f^3$.

The energy dissipation in terms of z is thus

$$\frac{dE}{dz} = \frac{1}{R_G} \frac{dE}{df} = -\frac{E_0}{R_G} \Lambda(f)$$

per incident electron. Backscattering is not included in this estimation. Since $(Q/e)(E/R_G)\Lambda(f)$ is the absorbed energy density (ϵ) in the resist, M_f can be written as

$$M_f = \frac{M_n}{1 + G_e M_n \epsilon / 100 \rho N_a}$$

According to this, the results of electron radiation on a positive resist is expressed in terms of molecular weight. But the solubility rate of PMMA ($M_m \sim 10^5$) is expressed by the following empirical formula[58]:

$$R = R_0 + \frac{\beta}{M_f^\alpha}$$

where R_0, β, and α depend on the temperature and on the developer solvent. Hence the development time at constant temperature is given by

$$\tau = \int_0^{T_0} \frac{dz}{|R_0 + \beta/M_f^\alpha|}$$

where T_0 is the original thickness and the dependence of the solubility rate on z is calculated from depth-dose functions. The density of PMMA is 1.2 g/cm³ and its initial molecular weight $M_n \approx 100,000$. For a typical film of 5000 Å the sheet density is 6×10^{-5} g/cm². For 20-keV electrons the $G(z)$ value is ~1.9. The most probable molecular weight after irradiation (5×10^{-5} C/cm²) is 7500. The molecular weight distribution curve typically extends over 40,000. However, in the case of PMMA resist, neither the average molecular weight nor its distribution is important so long as the molecular weight distribution curve is well separated from that of the scission fragments.

Polymethyl methacrylate (PMMA) resist has been routinely used in delineating microwave transistors, charge-coupled devices (CCDs), surface-wave devices (SWDs), bipolar memory cells, magnetic bubbles, and solar cells. For the economic fabrication of CMOS ICs by electron-beam writing it is necessary to use an electron resist much faster than PMMA. Several high-speed electron resists have been identified during the past few years, and some of those are listed in Table 2.[60] Some have shown that they are capable of both high speed and high resolution.

4.8.4.2. Negative Resists

The negative electron resist is characterized by sensitivity and contrast. The dose is defined as the number of electrons (coulombs) per square centimeter at a specific voltage required to achieve a desired chemical reaction:

$$D = it/A$$

where i is the electron current (A), t is the time (s), A is the area (cm^2), and D is the dose (C/cm^2).

Figure 4.37 illustrates the definition of sensitivity.[61] The correct feature size is achieved with a dose that results in approximately 50% of the thickness remaining. This dose is denoted by $D^{0.5}$. The contrast is expressed as

$$\gamma = |\log (D^0/D_m)|^{-1} \tag{4.27}$$

where D^0 is the projected dose required to produce 100% of the initial film thickness and is determined by extrapolating the linear portion of the thickness–dose plot to a value of 1.0 (normalized film remaining). D_m is the interface gel dose and the minimum dose required for any detectable gel formation.

The phenomenological depth-dose description (see Section 4.8.4.1) of electron energy loss can also be applied to the exposure of negative resists.[62] This model predicts a dependence of resist sensitivity on the electron energy and gives an expression for the minimum dose D_m.

The mechanism of primary-electron energy dissipation is the excitation and ionization of the resist atoms producing positive ions and electrons. The rate of energy absorption by the resist film, dE/dt, is related to the incident current density J_0 by

$$\frac{dE}{dt} = \frac{J_0}{e}\left(\frac{dE}{dz}\right) = \frac{E_0}{R_G}\frac{J_0}{e}\Lambda(f) \text{ eV} \cdot \text{cm}^{-3} \cdot \text{s}^{-1}$$

The rate of production of secondaries, dn_s/dt, with a mean excitation energy I_m, can be written as

$$\frac{dn_s}{dt} = \frac{1}{I_m}\frac{J_0}{e}\left(\frac{dE}{dz}\right) \text{ cm}^{-3} \cdot \text{s}^{-1}$$

TABLE 2. *Properties of Selected Electron-Beam Resist Materials*

Polymer[a]	Tone	Sensitivity (C/cm^2)	Resolution (μm)
PMMA	Positive	4×10^{-5}–8×10^{-5}	0.1
P(GMA-co-EQ)	Negative	3×10^{-7} (10 keV)	1.0
PBS	Positive	8×10^{-7} (10 keV)	0.5
COP	Negative	4×10^{-7}	1.0
P(GMA-co-EA)	Negative	3×10^{-7} (10 keV)	1.0
PGMA	Negative	5×10^{-7} (20 keV)	1.0
PCA	Positive	5×10^{-7} (20 keV)	0.5

[a]P(GMA-co-EA) = poly(glycidyl methacrylate-co-ethyl acrylate), PBS = poly(butene-1-sulfone), PCA = copolymer of α-cyano ethyl acrylate and α-amide ethyl acrylate, PGMA = poly(glycidyl methacrylate), PMMA = methylmethacrylate, COP = copolymer methylmethacrylate.

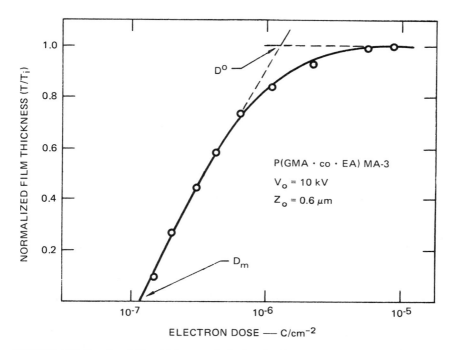

FIGURE 4.37. Fractional film thickness etched after development (normalized to final resist thickness) as a function of electron dose for P(GMA-co-EA) batch MA-3. The accelerating voltage was 10 kV and the initial resist thickness was 0.6 μm. Source: ref. 61.

or, using the G value, the number of ions or electrons produced per 100 eV lost by the primary:

$$\frac{dn_s}{dt} = \frac{G}{100} \frac{J_0}{e} \left(\frac{dE}{dz} \right)$$

and

$$G = \frac{100}{I_m}$$

If some type of chemical event (cross-linking) occurs concurrently with the production of internal secondaries, then the G value for cross-linking (G_c) is related to that for ionization as

$$G_c = \alpha G = \alpha \left(\frac{100}{I_m} \right)$$

With this the rate of crosslinking is found to be[62]

$$\frac{dn_c}{dt} = \left(\frac{G_c}{100} \right) \left(\frac{J_0}{e} \frac{dE}{dz} \right) \exp(-\eta t) \ \text{cm}^{-3} \cdot \text{s}^{-1}$$

where

$$\eta = \frac{G_c}{100 S_0} \frac{J_0}{e} \left(\frac{dE}{dz} \right)$$

and S_0 is the initial concentration of available sites. After integration the concentration of cross-links in the resist, n_c (which determines the degree of gelation, and hence the solubility of the irradiated material), is given by

$$n_c(t) = S_0(1 - \exp(-\eta t)) \ \text{cm}^{-3}$$

for small ηt

$$n_c \cong \frac{G_c D_0}{100 e} \left(\frac{dE}{dz} \right) \backsim \eta S_0 t \ \text{cm}^{-3}$$

Two parameters characteristic of a negative electron resist are the G value and the minimum number of cross-links per cubic centimeter, n_g, necessary to form an insoluble gel.

At the resist interface (i)

$$\frac{n_g}{G_c} = 10^{-2}\, \frac{D_g(i)}{e} \left(\frac{dE}{dz} \right)$$

or, using the energy absorption expression,

$$E_g(i) = 10^2\, \frac{n_g}{G_c} = \frac{D_g(i)}{e} \left(\frac{dE}{dz} \right) \text{eV} \cdot \text{cm}^{-3}$$

where $E_g(i)$ is the input energy density required to produce an insoluble gel at $z = i$ and $D_g(i)$ is the minimum incident dose necessary for gel formation. Typical gel energy densities are in the range 4×10^{18}–1×10^{22} eV·cm^{-3}.

4.8.5. Cross-Sectional Profiles of Single-Scan Lines

In high-resolution electron lithography it is important to know the cross section of the developed electron resist lines.[63] The shape of the cross section often has a direct influence on the results of other processing steps in the fabrication of electronic or optical devices. The cross-section shape depends, among other things, on the exposure pattern of the resist film. The simplest exposure pattern is a single line scanned by a focused electron beam. A more complex exposure pattern may be composed of many segments of such single-scan lines.

The profile of the single-scan resist line depends on the current density distribution of the incident beam as well as the electron scattering effects inside the resist and the substrate. Electron scattering broadens the beam radius in the resist films. The effect of electron scattering on the resulting resist line profile depends very much on whether the resist is positive or negative working (see Figure 4.38).

It has been shown[63] for negative resists that the line profile is parabolic when the incident radial current density and backscattered dose distributions in the resist–substrate interface plane are Gaussian. The base width W of the parabola after development is given as

$$W = 2r' \left[\ln\left(\frac{D^0}{D_m} \right) \right]^{1/2} \tag{4.28}$$

where r' is the Gaussian radius in the resist–substrate interface and is related to the Gaussian radius of the incident beam r by Eq. (2.180) as

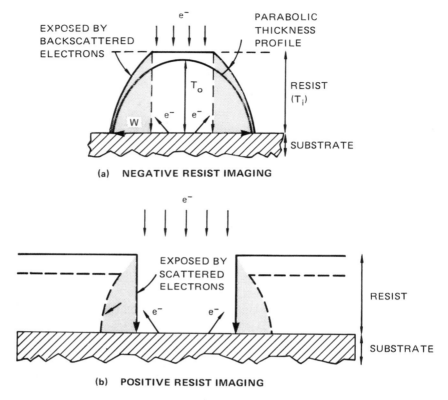

(a)   NEGATIVE RESIST IMAGING

(b)   POSITIVE RESIST IMAGING

FIGURE 4.38. Scattering effects in electron-beam imaging.

$$r'^2 = r^2 + \frac{4 z_0^3}{3\lambda}$$

where the z_0 is the resist thickness and λ is the scattering mean free path (see Section 2.6.1.4).

The height of the resist line (see Figure 4.38) is given by

$$T_0 = T_i \frac{\gamma}{\ln 10} \ln \left(\frac{D^0}{D_m} \right) \tag{4.29}$$

where T_i is the initial resist thickness and γ is the contrast of the resist. Equations (4.28) and (4.29) specify completely the cross-section profile of the resist lines if the parameters $I(r)$, r, r', γ, and D_m are given.

The resolution in electron-beam lithography is limited by the electron scattering in the resist and by the backscattering from the substrate, and not by the electron-optical system. Electron beams with diameters in the range of 20 to 200 Å can be routinely produced in electron microscopes. By the use of a thin support substrate (200–500 Å) the effect of backscattered electrons can be eliminated.

The electron scattering in the resist, which increases the dimensions of the exposed spot over the size of the beam, can be drastically reduced by the use of a thin (30–1000 Å) monomolecular resist prepared by the Langmuir method (see Section 3.2.5) and a reduced accelerating voltage (5 keV). Such a resist film built from $CH_2 = C—(C_2)_{20}—COOH$ molecules, superimposing monolayers, each 30 Å thick, gave patterns of 600-Å resolution after development.[64]

Lithography in the nanometer range (10–100 Å) can be achieved by the use of a thin salt resist deposited on a carbon support film. The salt evaporates under electron-beam bombardment and delineates the pattern on the substrate.[65]

Here the pattern writing and development can be carried out in one step. The selection of small molecules (alkali halides and aliphatic amino acids) for the resist material with molecular grain size yields resolutions in the 15–20 Å range.[66]

4.8.6. Development

When an organic polymer is subjected to electron-beam radiation, energy is absorbed and the polymer undergoes chemical changes involving two possible events. Polymer cross-linking results when adjacent polymer chains chemically cross-connect to form a complex, three-dimensional structure for which the average molecular weight increases. Chain scission (degradation) is the result of chemical bonds being broken by the electron radiation; this results in a lower average molecular weight. In general, both processes occur to different degrees in the same polymer. In PMMA chain-scission processes dominate cross-linking events under moderate (less than 2×10^3 C/cm^2) doses of kilovolt electron-beam radiation. In the case of chain scission the fragment molecular weight M_f is inversely proportional to the total absorbed energy loss density ϵ, as seen by[63] (Sec. 4.8.4.1)

$$M_f = \frac{1}{g\epsilon}$$

where g represents an efficiency factor that is a characteristic of a given polymer, indicating its susceptibility to electron-beam degradation.

Electron-beam-sensitive resists are developed by the method of fractional solution. A simple model states that for a given polymer organic solvents at a given temperature dissolve polymer chains of molecular weight below some critical value M_{fc}. For PMMA a solvent is used that removes the polymer chains in the irradiated volume while leaving the surrounding chains of higher molecular weight; it is therefore a positive resist.

4.8.7. Proximity Effects

As the incident electrons penetrate the resist, they undergo both elastic and inelastic scattering events. Consequently, a delta-function beam will spread, the amount of spreading depending upon the cumulative scattering history of the penetrating electrons.

Proximity effects[67] are created by scattered electrons in the resist and backscattered electrons from the substrate, which partially expose the resist up to several micrometers from the point of impact. As a result, serious variations of exposure over the pattern occur when pattern geometries fall in the micrometer and submicrometer ranges.

The proximity effect is dependent on the beam-accelerating voltage, the resist material and thickness, the substrate material, the resist contrast characteristics, and the chemical development process used. Where the beam diameter is appreciable, the current density profile of the beam must also be considered. To achieve accurate pattern delineation it is therefore necessary to apply some method of exposure adjustment to compensate for the proximity effect. One such method is to vary the charge reaching each pattern element by controlling and scanning speed or the beam current.

4.8.8. Specifications for Electron-Beam Exposure Systems

The most important quantities that must be specified for electron-beam lithography are the following:

$$\tau = \text{pixel exposure time (s)}$$
$$D = \text{dose for large area (C/cm}^2)$$
$$fD = \text{dose for one pixel } (f \le 1)$$
$$l_p^2 = \text{pixel area (cm}^2)$$

$$l_p^2/\tau = \text{area (cm}^2\text{) exposed per second}$$

$$j = \text{current density in the spot (A/cm}^2\text{)}$$

These quantities are not independent from each other. One can see, for example, that

$$\frac{1}{\tau} = \frac{j}{fD}$$

and

$$\frac{l_p^2}{\tau} = \frac{j l_p^2}{fD}$$

An important specification for these machines is the operating speed in pattern generation

$$\left(\frac{l_p^2}{\tau} \right)$$

Both raster scan and vector control pattern generators face basic limitations in operating speed. The rate at which a pattern is usually generated can be limited by two factors:

- The bandwidths of the systems that control the electron beam place limits on how fast the pattern can be transmitted.
- Resist sensitivity can limit the writing rate of the electron beam.

4.9. Ion-Beam Lithography

4.9.1. Introduction

Ion beams will become increasingly more important for submicron lithography. Not only can they be used for direct writing onto wafers[68] but also for mask fabrication.[69,70] In these operating modes the ion-beam systems can directly replace the electron-beam machines. They can also be used for ion implantation, direct ion milling, or for producing bombardment damage that will enhance a follow-up sputtering or etching step. Figure 4.39 shows the dependence of the etch rate factor on the ion (proton) dose. Oxide layers can

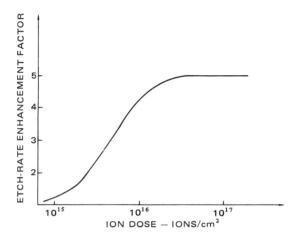

FIGURE 4.39. Etch rate enhancement factor as a function of the ion dose.

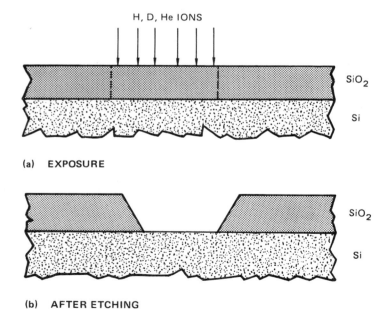

FIGURE 4.40. Radiation damage patterning of SiO_2 with H, D, and He ions.

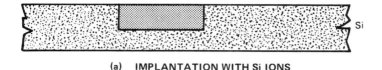

(a)   IMPLANTATION WITH Si IONS

(b)   THERMAL OXIDATION (~950°C) ENHANCED LINEAR OXIDATION RATE
IN THE ION-IMPLANTED AREA

FIGURE 4.41. Variable-thickness SiO₂ pattern generation on silicon.

be patterned without any resist because of the enhanced etch rate initiated by radiation damage. The etch rate dependence on the ion dose shows a saturation for SiO_2 with a dose of 2×10^{16} proton/cm². Figure 4.40 depicts the etching behavior of SiO_2. The increase in the linear oxidation rate of implanted silicon can be used to generate oxide layers of variable thickness (see Figure 4.41). After silicon-ion implantation and thermal oxidation the SiO_2 thickness increases in the irradiated area. Nitrogen implantation into Si retards the oxidation rate of the irradiated areas and the nitrogen-ion bombardment of the thin films changes the resistivity of the film.

Ion irradiation can generate patterns directly in metal layers (Ni or Mo) in a resistless process. Thus thin Ni or Mo layers can be used as metallic ion-sensitive resists. The pattern generation process for a metallic resist is shown in Figure 4.42.

The resolution of ion-beam lithography is inherently good because the secondary electrons produced by an ion beam, being of low energy, have a short diffusion range and practically no backscattering occurs.

Ion-projection machines with or without mask demagnification can be used for the resistless patterning of semiconductors (Si, GaAs, poly-Si), insulating layers (SiO_2, Si_3, N_4), metals (Al, Ni, Mo, Au), and also patterning organic thin films (resists). A further possible approach is shadow printing through thin free-standing and self-supporting mask patterns. The use of such masks is limited to cases where no free-standing opaque areas are needed. Employing a thin amorphous carrier membrane sufficiently transparent for

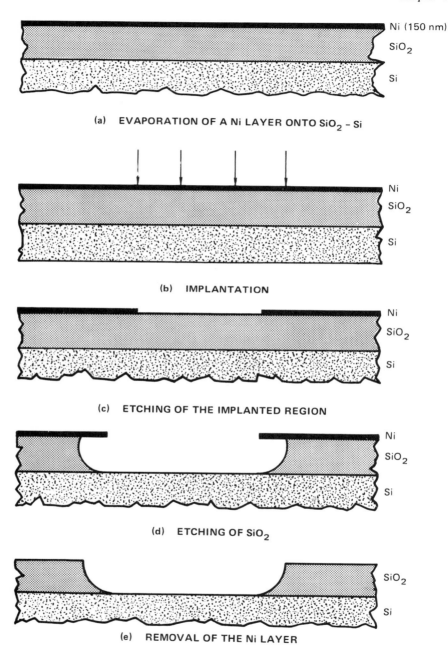

FIGURE 4.42. Pattern generation with metallic resist.

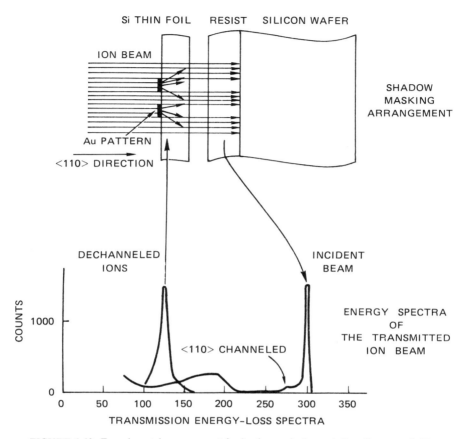

FIGURE 4.43. Experimental arrangement for ion-beam shadow printing. Source: ref. 71.

light ions can solve this problem, but the trade-off is the wide-angle scattering of the energetic ions in the membrane. This angular dispersion makes the replication of patterns by means of proximity exposure difficult. Contact exposure, on the other hand, results in a decreased yield and reliability.

A different experimental approach for masking is based on the use of a ⟨110⟩ silicon membrane 3–6 μm thick and a thin (1000 Å Au) metal pattern to be duplicated.[71] The resist film is exposed through the single-crystal silicon foil in the channeling direction, using a collimated helium beam with an initial energy of 300–2000 keV. The masking technique employs the dechanneling effect of a thin metal layer as well as the difference in energy loss for random and channeling directions (see Figure 4.43).

Microfocused ion beams can be used to expose resists, machine materials, and change the electrical and mechanical properties of semiconductors by implantation. All of these can be accomplished directly without masks.

Until recently the limited intensity of ion-beam sources made these applications impractical, but the development of high-intensity and stable liquid-metal (LM) ion sources[72,73] are highly encouraging for future applications.

4.9.2. Ion-Beam Resists

Chemical changes of resist materials can be induced effectively also by ion-beam exposure. The amount of energy deposited per unit path length of ion beam is much greater than that of electron beam. Therefore sensitivity to the ion beam measured in terms of the incident charge per unit area is expected to be higher. High resolutions can be also expected, since the lateral spread of ion beams in resist films is much less than in the case of electron beams. Using ion beams, the energy dissipation of ions in the resist is caused by electronic and nuclear collisions (see Sections 2.7.2.1 and 2.7.2.2). Since the effects of the two components are considered to be independent of each other, the radiation yields corresponding to both components can be estimated separately.[74] The two components are different in their contribution to chemical events, and energy deposition rates corresponding to nuclear and electronic collision losses are also different in their dependence on the penetration depths, as shown in Figure 4.44; thus the radiation yield is separated into two components and the number of scissions is given by the summation of both contributions. Therefore the average molecular weight M_f after ion exposure is given as (see Section 4.8.4.1 for positive electron resists)

$$M_f = \frac{M_n}{1 + KQ[G_e(z)S_e(z) + G_n(z)S_n(z)]}$$

where $G_e(z)$ and $G_n(z)$ are the radiation yields for chain scission caused by electronic and nuclear collisions, respectively.

The solubility rate of the polymer can be expressed (see Section 4.8.4.1) as

$$R = R_0 + \beta M_f^{-\alpha}(z)$$

where R_0, β, and α empirically determined constants. When the resist film is very thin, $S_e(z)$ and $S_n(z)$ can be regarded as constants and the resist thickness removed, d, after development is given as

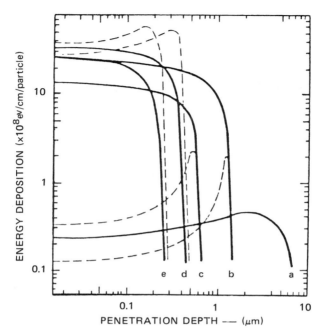

FIGURE 4.44. Calculated curves of the energy deposition rate in PMMA resist as a function of the penetration depth: (a) 20-kV electron, (b) 200-kV He$^+$, (c) 60-kV He$^+$, (d) 250-kV Ar$^+$, and (e) 150-kV Ar$^+$. For the curves of the ions, solid and broken lines are of electronic and of nuclear collision loss, respectively. Source: ref. 74.

$$d = RT$$

where T is the development time. For thick films the relationship between T and the dissolved thickness d becomes

$$T = \int_0^d \frac{dz}{R_0 + \beta M_f^{-\alpha_1}(z)} \qquad (4.30)$$

for PMMA the best fit between the theoretical curves calculated from Eq. (4.30) and the experimental results is found when[74]

$$R_0 = 84 \text{ Å/min}$$
$$\beta = 3.9 \times 10^8 \text{ Å/min}$$
$$\alpha = 1.41$$

$$G_e = 1.7$$
$$G_n = 0.9$$

Then the radiation yields for scission caused by the electronic collisions of ions are nearly equal to those in an electron exposure, where $G = 1.9$.

The radiation yields caused by nuclear collisions are one-half in PMMA in comparison with the radiation yields in electron-beam exposures.

4.9.3. Liquid-Metal (LM) Ion Sources

LM ion sources have been developed[73-75] for a fairly wide variety of metallic elements (Ga, Bi, Ge, Hg, Au, Woods metal, and alkali metals).

The mechanism of the ion source operation (see Section 2.4.5) appears to involve the formation of a field-stabilized cone of the liquid metal ("Taylor cone") from which field evaporation and ionization occur. The supply of material to the cone apex occurs via viscous flow along the emitter surface driven by the capillary forces and the electrostatic field gradients near the emitter tip.

In a scanning ion-beam machine[68] a LM Ga source has been utilized in the optical column to focus 0.1 nA of Ga^+ current into a 0.1-μm beam diameter. The experimentally measured brightness value is

$$\beta = \frac{1}{\pi \alpha^2 A_s} = 3.3 \times 10^6 \text{ A/cm}^2 \cdot \text{sr}$$

at the image plane of the 57-kV Ga^+ beam. Here A_s is the emitting area, that is, $A_s \approx \pi \alpha_0^2 \, r_a^2$, where $\alpha_0 \approx 0.2$ rad is the emission half-angle and $r_a \approx 600$ Å. Extracted currents of 1–10 μA can be achieved for gallium beams.

Energy distribution measurements[76] have shown that the energy spread (ΔE) increases with both current and particle mass. The energy spread for Ga, In, and Bi are 5, 14, and 21 eV, respectively, at an angular intensity of 20 μA/ sr.

Thus a small acceptance angle ($\backsim$ 1 mrad) still allows a few nanoamperes of ion-beam current to be realized.

4.9.4. Scanning Systems

Figure 4.45 shows a high-intensity scanning ion probe with a liquid metal source.[68] The imaging of the source is carried out by a unit-magnification accelerating lens with a post lens deflector to form the scanned beam. The electrostatic deflector enables the beam to be scanned in the x and y directions.

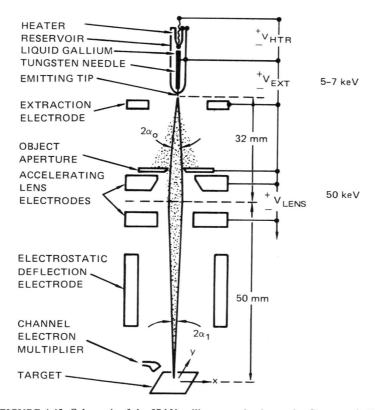

FIGURE 4.45. Schematic of the 57-kV gallium scanning-ion probe. Source: ref. 68.

The deflector under computer control can move the beam in a line or raster scan at a speed of $5 \times 10^4 \ \mu m/s$.

The spot diameter is limited by chromatic aberration and the lens acceptance angle α_0. The spot diameter is given by

$$d = C_c \, \alpha_0 \, \frac{\Delta E}{E_0} \approx 1000 \ \text{Å}$$

where

$$\alpha_0 = 1.2 \ \text{mrad}$$
$$C_c = 33 \ \text{mm}$$
$$\Delta E \sim 14 \ \text{eV}$$
$$E_0 \sim 5.7 \times 10^4 \ \text{eV}$$

This fine microbeam can be used for micromachining, doping, or resist exposure. It was found, however, that the exposure properties of resists are different from what would be expected on the basis of their electron exposure properties.[77]

In conclusion, one can say that the liquid ion sources have excellent properties for applications in microfabrication. Ion currents of 1–100 μA may be obtained, corresponding to source angular intensities of up to 10^{-4} A/sr, which is approximately 100 times greater than the gas-phase field-ion source.

The uv, X-ray, electron-, and ion-beam systems have resolutions below one micron. The ion-beam spot size could be inherently even smaller, with a resolution as low as 100–500 Å. From the discussions in this chapter it is clear that there are several avenues for the realization of submicron lithographies and fabrication processes. The actual method used in a given submicron fabrication process will ultimately depend on the characteristics of the fabricated device itself.

5

Special Processes Developed for Microcircuit Technology

Since we are illustrating the principles of microfabrication with application to microelectronics, we devote this chapter to discussing in some detail those special processes that have been developed for planar silicon technology, namely, epitaxy, oxidation, doping, and annealing. This is not to imply that analogous or similar processes could not or have not been applied to other materials and devices, but it reflects the fact that the major thrust in the microdevice area over the past two decades has been associated with silicon, and hence more work has been done in this area. We have included some special epitaxial processes developed for devices utilizing III–V compound semiconductors, since these are of emerging importance for high-speed microcircuits.

5.1. Epitaxy and Oxidation

5.1.1. Silicon Epitaxy

Epitaxial techniques[1,2] have been credited with increasing the yield of ICs because they eliminate the long diffusion times required to produce uniformly doped layers (see Section 1.3.5). The focus of this section is on the epitaxial growth process and its simple physical modeling.

5.1.1.1. Growth Processes

The growth mechanisms can be divided into direct and indirect processes. In the direct process silicon atoms are transferred from the source to the sub-

349

strate (Si wafer) surface and stick there. Under proper deposition conditions the Si atoms are mobile on the heated surface and align themselves according to the substrate crystal structure. Epitaxial growth occurs because of the formation of nucleation sites and subsequent lattice matching as two-dimensional islands grow across the surface (see Figure 5.1).

The rate of formation R_n of nucleation sites depends on the concentration of the silicon in the vapor and on the free energy of formation of the site, that is,

$$R_n \approx n_0 \exp\left(-\frac{\Delta G}{kT}\right)$$

where n_0 is the silicon concentration in the vapor, ΔG is the free energy of formation of the nucleation sites, T is the temperature of deposition, and k is Boltzmann's constant.

In addition to evaporation, other examples of the direct deposition method are sublimation and sputtering (see Chapter 3). Although epitaxial films can be obtained by these vacuum methods, the indirect techniques are better suited and controlled for the fabrication of doped layers.

5.1.1.2. Indirect Processes

Indirect processes are those in which silicon atoms are obtained by the decomposition of a silicon compound at the heated substrate surface. Figure 5.2 shows the schematic of a rf-heated horizontal epitaxial reactor.[4] Table 1 describes the various source gases and their functions in the reactor.[5]

Silane ($SiCl_4$) is the most common deposition chemical because it is readily available and its high temperature of deposition makes it less sensitive to oxidation, which (at a few ppm) can lead to surface defects. The minimum deposition temperature and the maximum growth rate are limited by the trapping of structural defects, which occurs if the surface diffusion rate is lower than the arrival rate of the silicon atoms.

At the silicon surface silane decomposes according to the chemical equation

$$SiH_{4(g)} \xrightarrow{1000\,^\circ C} Si_{(s)} + 2H_{2(g)}$$

and the silicon growth rate is proportional to the partial pressure of silane. For cases where the reaction is surface controlled, epitaxial growth proceeds

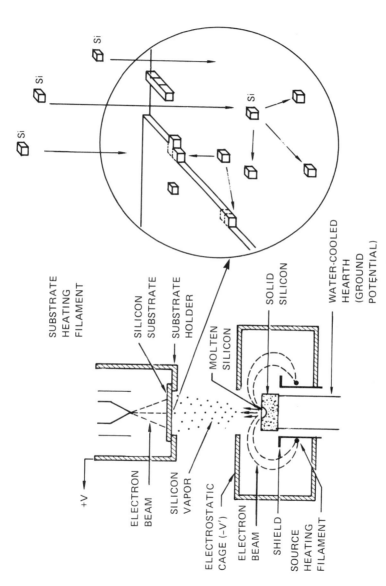

SUBSTRATE HEATING FILAMENT

SILICON SUBSTRATE

SUBSTRATE HOLDER

+V

ELECTRON BEAM

SILICON VAPOR

ELECTROSTATIC CAGE (−V′)

ELECTRON BEAM

SHIELD

SOURCE HEATING FILAMENT

MOLTEN SILICON

SOLID SILICON

WATER-COOLED HEARTH (GROUND POTENTIAL)

Si Si Si Si

FIGURE 5.1. Evaporation and substrate heating by electron bombardment. Source: ref. 3.

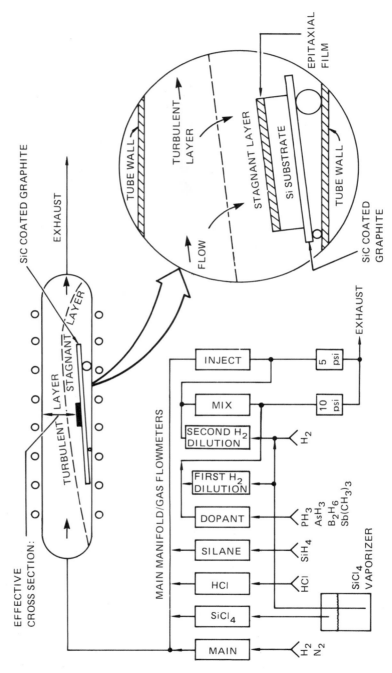

FIGURE 5.2. Rf-heated horizontal epitaxial reactor.

TABLE 1. Source Gases and Their Functions as Used in Epitaxial Deposition

Gas	Function	Comments
N_2	Main flow	Purges out explosive/poisonous gases prior to opening the reactor tube to air
H_2	Main flow	Most common ambient for the growth of epitaxial layers
$SiCl_4$	Si source	Common liquid Si source; vaporized in a H_2 bubbler; corrosive vapor; temperature range, 1150–1200°C; growth rate, 0.2–10 μm/min
SiH_4	Si source	Common gaseous Si source; pyrophoric gas; temperature range, 1000–1050°C; growth rate 0.2–1.0 μm/min
HCL	Si etchant	Most common Si etchant used for substrate preparation; corrosive poison gas
PH_3	Si dopant	Most common phosphorus source for doping epitaxial silicon; flammable poison gas
AsH_3	Si dopant	Behaves like PH_3
$Sb(CH_3)_3$	Si dopant	A liquid antimony source used as a vapor at a concentration of a few hundred ppm in H_2; used because SbH_3 is unstable; poisonous vapor
B_2H_6	Si dopant	Behaves like PH_3

according to the following steps[5]:

- Mass transfer of the reactant molecules (SiH_4) by diffusion from the turbulent layer across the boundary layer to the silicon surface
- Adsorption of the reactant atoms on the surface
- The reaction or series of reactions that occur on the surface
- Desorption of the byproduct molecules
- Mass transfer of the byproduct molecules by diffusion through the boundary layer to the main gas stream
- Lattice arrangement of the adsorbed silicon atoms

The overall deposition rate is determined by the slowest process in the list above. Under steady-state conditions all steps occur at the same rate and the epitaxial layer grows uniformly. One of the major problems of silicon epitaxy is the treatment of dopant inclusion.[6–10] Doping is normally accomplished by the incorporation, in hydride form, of B, As, or P in the main gas flow. The amount of dopant incorporated is approximately proportional to its partial pressure. The other sources of impurities (besides those intentionally introduced in the main gas flow) are the reactor (walls, structure) and the wafer itself.

Autodoping is the term used to describe the dopant contribution from the wafer to the epitaxial layer. Autodoping can be divided into two categories[4]: macroautodoping, where unwanted dopant is transported within the reactor from one wafer to another, and microautodoping, where unwanted dopants are incorporated into a localized region of the same wafer. The trends in silicon epitaxy are toward reduced growth temperatures to avoid wafer distortion and provide better control for autodoping and growth rates.

An interesting application of silicon epitaxy is the silicon-on-sapphire (SOS) technology, where a standard bulk silicon component is constructed within a silicon island that is electrically isolated from other such islands within the same IC.[11]

These silicon "islands" represent the basic differences between SOS components and bulk silicon devices (Section 1.2).

For this process a 0.5–1 μm silicon layer is growth epitaxially on the sapphire substrate in which these islands are etched, and p and n wells fabricated for further silicon processing. The result is a very densely packed IC without the parasitic devices obtained with bulk silicon processing.

5.1.2. Liquid-Phase Epitaxy

The fabrication of devices from the III–V compound semiconductors is based totally on epitaxial layers grown on bulk single-crystal substrates. For many materials (e.g., GaAs) epitaxial layers are grown from liquid solutions. In liquid-phase epitaxy (LPE) the substrate is thermally equilibrated with the solution and growth takes place upon supersaturation of the solution brought about by a temperature decrease.

In current-controlled LPE[12,13] (electroepitaxy) an electric current is made to flow through the growth interface while the temperature of the system is maintained constant. Layers of semiconductors, including InSb, GaAs, InP, and GaAlAs as well as garnet layers, have been successfully grown in this way. The growth behavior is related to the Peltier cooling of the liquid–substrate interface, which leads to supersaturation and electromigration[14,15] (caused by electron–momentum exchange).

Standard LPE reactors are easily modified to permit current flow through the solution–substrate interface [see Figure 5.3(a)]. Since the substrate and solution have different thermoelectric coefficients, a current flow across their interface is accompanied by Peltier cooling or heating, depending on the current direction. The magnitude of this heat, Q, is equal to the difference in Peltier coefficients π_p times the current density j. For III–V semiconductors Q is

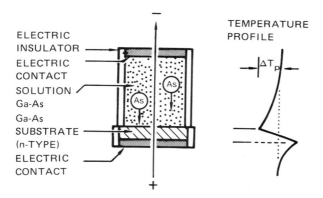

(a)   **GROWTH CELL USED FOR ELECTROEPITAXIAL GROWTH OF GaAs FROM Ga-As SOLUTION**

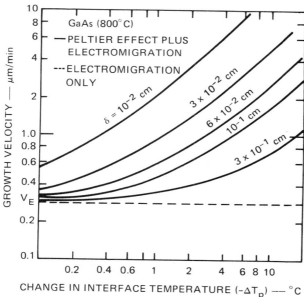

Source: Ref. 12.

(b)   **GROWTH VELOCITY OF GaAs FROM Ga-As SOLUTION AS A FUNCTION OF ΔT_p**

FIGURE 5.3. Electroepitaxial growth.

on the order of 1 W/cm^2 and the current density is about 10 A/cm^2. ΔT_p is typically on the order of 0.1–1 °C.

The growth rate is strongly influenced by the formation of a solute boundary layer of thickness δ, as shown in Figure 5.3(b), where the growth velocity is plotted as a function of the interface temperature difference ($-\Delta T_p$). From Figure 5.3(b) it is evident that the contribution of electromigration to growth is dominant for large boundary-layer thicknesses, but the Peltier effect dominates the growth for small δ.

5.1.3. Molecular Beam Epitaxy (MBE)

Molecular beam epitaxy is essentially the coevaporation of elemental components onto a heated single-crystal substrate, but it is so named to emphasize the control of the crystal composition. (See Sections 1.3.1 and 3.3.3 on this subject.) The technique consists of directing controlled "beams" of the required atoms from effusion-cell ovens (which can be shuttered to change from the growth of one type of crystal to another) toward the heated substrate in a vacuum of 10^{-10} Torr. The technology of this technique has recently been extensively reviewed.[16] An important feature of MBE is that its growth rate is relatively low (60–600 Å/min), which permits the precise control of the thickness of the epitaxial layers. In fact, it is possible to grow AlAs and GaAs in alternate monolayers, permitting the growth of structures with tailored transport properties.

Probably the most important advantage of the MBE growth technique is that it is possible to do several different types of analyses *in situ* during the growth of the epitaxial layers. Most commercial MBE growth systems include reflection electron-diffraction equipment, a mass spectrometer, and an Auger spectrometer with ion-sputtering capability. Although an ultrahigh vacuum is required, most commercial systems also incorporate a sample-exchange interlock so that atmospheric contamination of the growth chamber is prevented and rapid throughput exchange can be accomplished. Most of the work on MBE has been directed toward GaAs and AlGaAs to fabricate novel single-crystal structures. Single-crystal multilayered structures having component layers that differ in composition but are lattice-matched throughout form the basis for semiconductor devices in which both photons and charge carriers (holes and electrons) can be manipulated. The fabrication of the double-heterostructure (DH) laser is an important example of this type of device.[16]

The GaAs–Al$_x$Ga$_{1-x}$As DH laser in its simplest version is a small, rectangular single-crystal parallelepiped consisting of a n-type GaAs substrate with at least three layers (n-Al$_x$Ga$_{1-x}$As, p-GaAs, and p-Al$_x$Ga$_{1-x}$As) grown

epitaxially onto it, as illustrated in Figure 5.4(a). The alignment of the conduction and valence bands of the composite structure when forward biased (n
side negative) with a voltage of about the width of the GaAs energy gap is
shown schematically in Figure 5.4(b). As the result of the forward bias, electrons are injected into the conduction band of the p-GaAs layer, where they
recombine with the holes and emit radiation with approximately the energy of
the GaAs energy gap, Eg_{GaAs}.

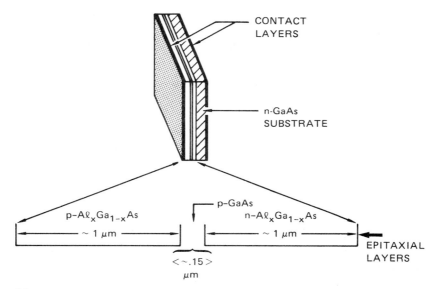

(a)   **THE GaAs-Aℓ_xGa$_{1-x}$As DOUBLE HETEROSTRUCTURE LASER**

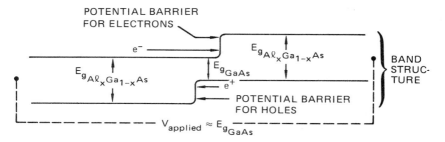

(b)   **SCHEMATIC OF THE ALIGNMENT OF THE CONDUCTION
AND VALENCE BANDS OF THE COMPOSITE STRUCTURE**

FIGURE 5.4. A double-heterostructure laser. Source: ref. 16.

Note in Figure 5.4(b) that at the heterojunction there are potential barriers that prevent holes (e^+) and electrons (e^-) from diffusing beyond the GaAs region. The injected electrical carriers are then confined to the GaAs layer. In addition, because GaAs has a higher refractive index than does $Al_xGa_{1-x}As$, the $Al_xGa_{1-x}As$–GaAs–$Al_xGa_{1-x}As$ three-layer sandwich is a waveguide so that the generated light tends to be confined to the GaAs layer. The cleaved ends of the parallelepiped act as partial mirrors. Thus light of energy approximately Eg_{GaAs} is generated by an electronic transition in a waveguide within a Fabry–Perot cavity formed by the mirrors. With a sufficiently high current through the device, stimulated emission and lasing result.

The DH laser provides an excellent illustration of how heterostructures are used to manipulate light and electrical carriers in a single solid-state device. MBE techniques also can be used to fabricate monolayer structures and quantum wells, which consist of alternating layers of Ga, As, and Al, with each layer about 50–400 Å thick. The steps joining the conduction and valence band edges at the heterojunctions form the boundaries of potential wells that tend to contain electrons in the conduction band and holes in the valence band. These quantum wells modify the conduction properties of the multilayer structures so that they differ from the bulk semiconductors. Since the wells act as one-dimensional containers for the electrons, the energy of the electrons is quantized, giving rise to a band structure for these confined carrier quantum states.[16]

MBE, unlike other crystal growth methods, also permits the growth of complex epitaxial structures with controlled lateral dimensions by using shadow masking. These developments have resulted in the fabrication of very thin (5–500 Å) structures and novel optical and microwave devices.

5.1.4. Oxidation

The oxidation of single-crystal silicon to form a passive and protective SiO_2 layer is one of the most important processes in microelectronics. The thermal oxidation of silicon is commonly performed in a water vapor or oxygen atmosphere over the temperature range 700–1250°C. The oxidation process consists of the diffusive transport of oxygen through the growing amorphous SiO_2 layer, followed by the reaction with Si at the interface (see Section 1.3.2.). The kinetic model of silicon oxidation is well known[17] and fits H_2O oxidation data very well. However, present trends in oxidation technology are quickly outstripping the usefulness of this simple model.[18,19] For example, the crystal orientation, the dopant concentration, the partial pressures of oxygen and HCl, and other ambient parameters can change the oxide growth parameters,[5] and these are not treated in the first-order model. In addition, there are new oxide-charge models of the SiO_2–Si interface[18,20] that can predict oxide-charge densities for all process conditions during oxidation. After a short discussion of the basic model the orientation, concentration, high-pressure effects, and oxidation in HCl–O_2 mixtures will be reviewed in the following sections.

5.1.4.1. The Basic Deal–Grove Model

Silicon oxidation kinetics are based on the general oxidation relationship (see Section 1.3.2)

$$\frac{z_0^2}{B} + \frac{z_0}{B/A} = t + \tau \tag{5.1}$$

where z_0 is the oxide thickness, t is the oxidation time, and A, B, and τ are constants that are functions of the oxidation conditions. The quadratic equation [Eq. (5.1)] for z_0 can be easily solved in the following limiting cases:

$$\text{If } t + \tau \ll \frac{A^2}{4B}, \quad \text{then } z_0 \approx \frac{B}{A}(t + \tau) \tag{5.2}$$

$$\text{If } \tau \gg \frac{A^2}{4B}, \quad \text{then } z_0^2 \approx B(t + \tau) \tag{5.3}$$

B/A is called the "linear" rate constant and B is called the "parabolic" rate constant. These limiting forms indicate that the oxidation process in the parabolic domain is diffusion limited and in the linear region it is surface-reaction limited (see Section 1.3.2).

At the present time B, B/A, and τ are known[5] only for $\langle 111 \rangle$ oriented lightly doped conditions, in which case, using the nomenclature of Section 1.3.2, we have

$$B = C_1 e^{-E_1/kT} = 2D_{\text{eff}} \frac{C^*}{N_i} \tag{5.4}$$

$$\frac{B}{A} = C_2 e^{-E_2/kT} = \frac{C^*}{N_i(1/K + 1/h)} \tag{5.5}$$

and

$$\tau = (z_i^2 + Az_i)/B \tag{5.6}$$

where z_i is the effective value of the oxide thickness at $t = 0$. The parameters for dry oxidation[20] are the following:

$$N_i = 2.2 \times 10^{22}/cm^3$$
$$C_1 = 7.72 \times 10^2 \ \mu m^2/hr$$
$$C_2 = 6.23 \times 10^6 \ \mu m/hr$$
$$E_1 = 1.23 \ eV/molecule$$
$$E_2 = 2.0 \ eV/molecule$$
$$z_i = 0 \ \text{Å}$$

For the wet oxidation process the parameters are

$$N_i = 4.4 \times 10^{22}/cm^3$$
$$C_1 = 2.24 \times 10^2 \ \mu m^2/hr$$
$$C_2 = 8.95 \times 10^7 \ \mu m/hr$$
$$E_1 = 0.71 \ eV/molecule$$
$$E_2 = 1.97 \ eV/molecule$$
$$z_i = 200 \ \text{Å}$$

The crystal orientation, the oxidant pressure, the dopant concentration, and the addition of chlorine to the oxidant have all been shown to affect B, B/A, or both. In the following sections these effects will be briefly discussed.

5.1.4.2. Crystal Orientation Effects

Orientation differences in the oxidation rate appear only after the oxide thickness exceeds about 100 Å. The growth rate difference is largest (40%) at relatively low temperatures (700°C) and decreases gradually at higher temperatures (2% at 1200°C). Generally the $\langle 111 \rangle$ orientation oxidizes faster than the $\langle 100 \rangle$ orientation. The orientation dependence should appear[17] in the linear rate constant B/A, since it involves the reaction rate constants (K, h) (see p. 31) via the constant A.

5.1.4.3. Impurity Doping Effects

The effects of impurity doping on thermal oxidation rates are intimately connected to impurity redistribution. As a thermal oxide is grown over a doped

Si substrate, there results a redistribution of the impurity (as seen in Figure 5.5). Phosphorus, arsenic, and antimony dopant atoms tend to pile up at the interface ($C_s > C_B$), but in boron a surface depletion ($C_s < C_B$) takes place.

It has been observed that the oxidation rates are generally faster for heavily doped ($C_B \cong 10^{19}$ cm^{-3}) p and n substrates.[5] The two concentrations that are involved in this increased oxidation are the dopant concentration of the SiO$_2$–Si interface (C_s) and the impurity concentration in the oxide (C_{ox}). From recent experiments[5] it has been concluded that the pileup of n-type dopants at the SiO$_2$–Si interface affects the oxidation kinetics far more than does the increased impurity concentration (C_{ox}) in the oxide. It is the linear rate constant B/A that is changed by a high impurity dopant concentration. Figure 5.6 shows the linear and parabolic rate constants as functions of the dopant concentration at 900°C.[5]

5.1.4.4. High-Pressure Oxidation

The advantage of high-pressure (20-atms) oxidation is its ability to oxidize at lower temperatures, thereby decreasing the redistribution of dopants and decreasing the generation of silicon defects. The relationship between oxide thickness and steam pressure is shown in Figure 5.7.

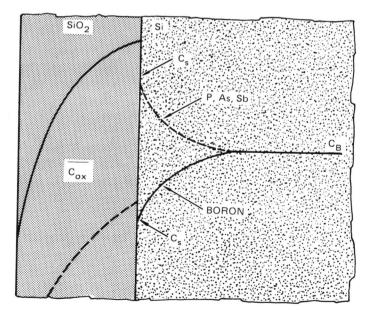

FIGURE 5.5. Redistribution of the impurity doping profiles near the SiO$_2$–Si interface. Source: ref. 5.

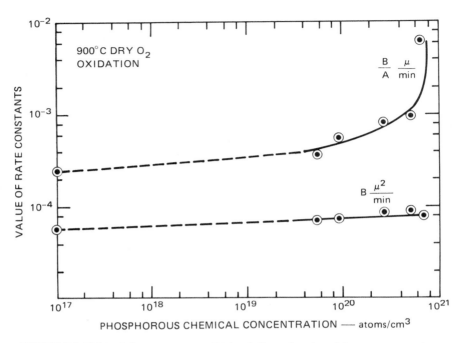

FIGURE 5.6. Value of the rate constants B/A and B as a function of dopant concentration at 900°C dry O_2 oxidation. Source: ref. 5.

The relation for the parabolic rate constant B should be modified to include the effect of the steam pressure, since according to Henry's law the equilibrium solid concentration is proportional to the bulk gas particle steam pressure P_g, and $C^* = HP_g$, but, using Eq. (5.4), $B = 2D_{eff} C^*/N_1$. A is independent of the steam pressure and affected by the Si crystal orientation and the doping concentration. A rule of thumb is that for every increase of 1 atm the temperature of oxidation may be reduced by approximately 30°C.[22] As a consequence of the increased pressure, processes operating at 1100°C may be reduced to about 800°C without incurring a time penalty. For a 25-atm system 1 μm of oxide may be grown in about 15 min at 920°C, as opposed to approximately 10 h at 1 atm.

5.1.4.5. Oxidation in HCl–O_2 Mixtures

A significant development in thermal oxidation has been the addition of chlorine during oxidation.[5] Such additions result in an improved threshold stability for MOS devices and an increased dielectric strength. In addition, chlorine increases the rate of silicon oxidation. According to recent experiments,[20] the monotonic increase in B with HCl concentration (Figure 5.8) is related to the effective diffusion coefficient D_{eff}. The increased chlorine concentration may cause the SiO_2 lattice to be strained, thereby allowing the diffusion of the oxidant to occur more easily. Figure 5.8 shows the parabolic rate constant as a function of the percentage of HCl for different orientations and oxidation temperatures.

5.1.4.6. Oxide Charges

The densities of the charges and states associated with the Si–SiO_2 system are shown in Figure 5.9.[18,20] They consist of charges trapped in the interface of charge density Q_{it} (C/cm²), charges fixed in the oxide close to the interface Q_f, mobile impurity ions in the oxide Q_m, and space-charge trapped in the bulk

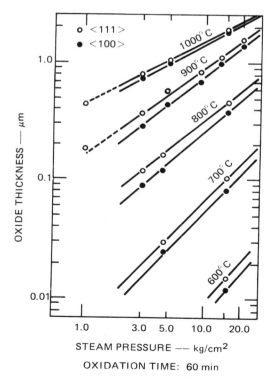

FIGURE 5.7. Oxide thickness as a function of steam pressure for $\langle 111 \rangle$ and $\langle 100 \rangle$ silicon. Source: ref. 21.

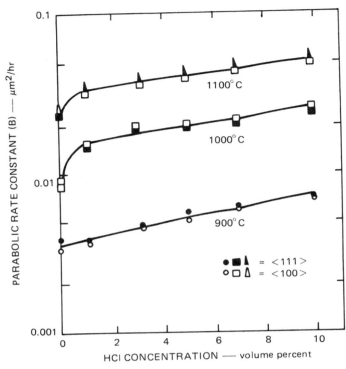

FIGURE 5.8. Parabolic rate constant (B) as a function of percent HCl for $\langle 111 \rangle$ and $\langle 100 \rangle$ oriented *n*-type silicon at 900, 1000, and 1100°C. Source: ref. 20.

of the oxide Q_{ot}. Each charge density has a corresponding number density given by $|Q|/q = N$ (number/cm²), where q is the electronic charge. The sign of Q is either positive or negative, depending on whether the majority of the charge is positive or negative. By definition, however, N is always positive.[23] The interface trap states are located at the oxide–silicon interface and are introduced into the forbidden gap near the semiconductor surface because of the disruption of periodicity of the lattice. It is believed that there should be approximately one surface state for every surface atom, resulting in a number density $N_{it} \sim 10^{15}$/cm².[18,20] The charges Q_f are fixed in the oxide near the interface between the oxide and the silicon. Located within 200 Å of the SiO$_2$–Si interface, they cannot be discharged or moved, but the value of Q_f is a strong function of oxidation, annealing conditions, and the orientation of the Si crystal. The dependence of Q_f on the ambient temperature of the final heat treatment suggests excess ionic silicon as a possible oxide in the origin of Q_f, although recently bond coordination defects have been shown to be a more likely cause.[24] In addition, there may be mobile charges Q_m present within the oxide layer caused by impurity ions (Na^+, Li^+, K^+). These charges Q_m are responsible for the voltage drift in MOS device characteristics. Finally, a positive space-charge buildup Q_{ot} can be generated in the oxide from external ionizing radiation (X ray, electron beams).[25]

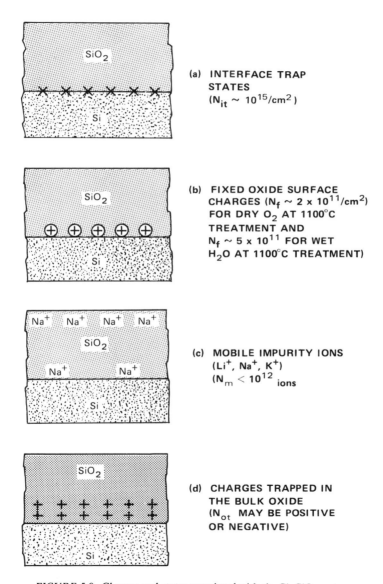

(a) **INTERFACE TRAP STATES**
$(N_{it} \sim 10^{15}/cm^2)$

(b) **FIXED OXIDE SURFACE CHARGES** $(N_f \sim 2 \times 10^{11}/cm^2)$ **FOR DRY** O_2 **AT 1100°C TREATMENT AND** $N_f \sim 5 \times 10^{11}$ **FOR WET** H_2O **AT 1100°C TREATMENT)**

(c) **MOBILE IMPURITY IONS** (Li^+, Na^+, K^+) $(N_m < 10^{12}$ **ions**

(d) **CHARGES TRAPPED IN THE BULK OXIDE** (N_{ot} **MAY BE POSITIVE OR NEGATIVE)**

FIGURE 5.9. Charges and states associated with the Si–SiO₂ system.

This radiation-induced oxide space charge can be annealed out at relatively low temperature ($>300\,°C$). Figure 5.10 shows the location of these oxide charges in thermally oxidized silicon structures.

So far we have assumed that the oxidizing species proceeding through the oxide are uncharged. However, experiments[26] studying the effect of an electric field applied across the oxide on the oxidation rate have indicated that the oxidizing agent (O_2) is negatively charged. It is possible that the molecular oxygen dissociates upon its entry into the oxide as

$$O_2 \rightleftarrows 2O^- + 2\ holes$$

and both the oxygen ion and the hole diffuse through the oxide coupled by an electric field (ambipolar diffusion). Another theory[27] predicts that as long as the oxide thickness is smaller than some critical distance, oxidation will be rapid. At greater oxide thicknesses the growth rate will slow down. Estimates of these critical thicknesses yield 150 Å for silicon oxidation in O_2 and 5 Å for oxidation in H_2O. These values are very close to the z_i values used in the Deal–Grove theory of oxidation (Section 5.1.4.1).

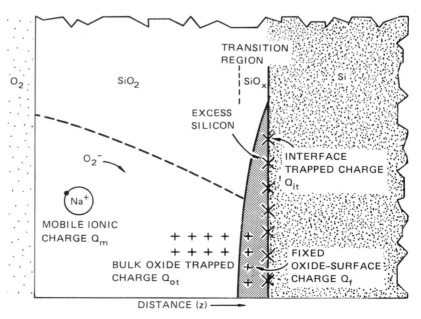

FIGURE 5.10. Location of oxide charges in thermally oxidized silicon structures. Source: ref. 20.

5.2. Doping

5.2.1. Diffusion

The principal purpose of doping semiconductors is to control the type and concentration of impurities within a specific region of the crystal to modify its local electrical properties. Diffusion has been found to be a highly practical way of doing this and has been extensively studied.[28–30]

Diffusion describes the process by which atoms (ions) move in a crystal lattice (see Section 1.3.5.1). The movement of atoms (impurities) in a lattice takes place in jumps. These jumps occur in all three dimensions, their net flux being the statistical average over a period of time. The mechanism by which jumps can take place is as follows: The atoms of the crystal form a series of potential hills [see Figure 5.11(a)] that impede the motion of the impurities. The height of the potential barrier is typically on the order of 1 eV in most materials. The distance between successive potential barriers is on the order of the lattice spacing, which is 1–3 Å. If a constant electric field is applied, the potential distribution as a function of distance will be tilted, as shown in Figure 5.11(b).[6] This will make the passage of positively charged particles to the right easier and their passage to the left more difficult. Let us now calculate the flux F at position z. This flux will be the average of the fluxes at positions $z - a/2$ and $z + a/2$. In turn, these two fluxes [Figure 5.11(b)] are given by $F_1 - F_2$ and $F_3 - F_4$, respectively.

Consider the component F_1. It will be given by the product of (1) the density per unit area of impurities in the plane of the potential valley at $z - a$, (2) the probability of a jump of any of these impurities to a valley at z, and (3) the frequency of attempted jumps ν. Thus we can write

$$F_1 = [aC(z - a)] \exp\left[-\frac{q}{kT}\left(W - \frac{1}{2}aE\right)\right]\nu \qquad (5.7)$$

where $aC(z - a)$ is the density per unit area of particles situated in a valley at $z - a$, and the exponential factor is the probability of a successful jump from the valley at $z - a$ to the valley at z. Note the lowering of the barrier due to the electric field E.

Similar formulas can be written for F_2, F_3, and F_4. When these are combined to give a formula for the flux F at position z, with the concentrations $C(z \pm a)$ approximated by $C(z) \pm a(\partial C/\partial z)$, we obtain

$$F(z) = -(\nu a^2 e^{-qW/kT})\frac{\partial C}{\partial z}\cosh\frac{qaE}{2kT} + (2a\nu e^{-qW/kT})\,C\sinh\frac{qaE}{2kT} \qquad (5.8)$$

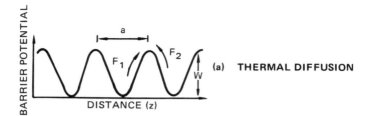

(a) THERMAL DIFFUSION

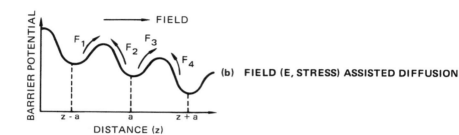

(b) FIELD (E, STRESS) ASSISTED DIFFUSION

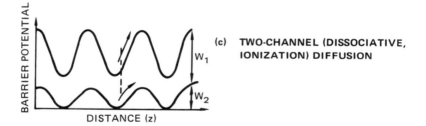

(c) TWO-CHANNEL (DISSOCIATIVE,
     IONIZATION) DIFFUSION

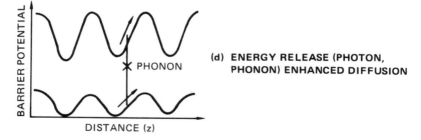

(d) ENERGY RELEASE (PHOTON,
     PHONON) ENHANCED DIFFUSION

FIGURE 5.11. Model of diffusion mechanisms in crystals.

An extremely important limiting form of this equation is obtained for the case when the electric field is relatively small, that is, $E \ll kT/qa$. In this case we can expand the cosh and the sinh terms in the Eq. (5.8). Noting that cosh $z = 1$ and sinh $z = z$ for $z \to 0$, this results in the limiting form of the flux equation for a positively charged species

$$F(z) = -D\frac{\partial C}{\partial z} + \mu EC \tag{5.9}$$

where

$$D = \nu a^2\, e^{-qW/kT} \tag{5.10}$$

and

$$\mu = \frac{\nu a^2\, e^{-qW/kT}}{kT/q} \tag{5.11}$$

Note that the mobility μ and the diffusivity D are related by

$$D = \frac{kT}{q}\mu \tag{5.12}$$

This is the well-known Einstein relationship.

It is customary to identify the contribution to the flux, which is proportional to the concentration gradient, as the diffusion term, while the contribution that is proportional to the concentration itself is referred to as the drift term. The migration of the diffusing species follows Fick's law:

$$\frac{\partial C}{\partial t} = D\frac{\partial^2 C}{\partial z^2} - \mu_+ E\frac{\partial C}{\partial z} \quad \text{for positive charge}$$
$$\frac{\partial C}{\partial t} = D\frac{\partial^2 C}{\partial z^2} + \mu_- E\frac{\partial C}{\partial z} \quad \text{for negative charge} \tag{5.13}$$

Interstitial diffusion takes place when the impurity atoms move through the crystal lattice by jumping from one interstitial site to the next. The diffusion of substitutional impurities, that is, impurities that occupy sites in the silicon lattice, usually proceed by the impurities jumping into silicon vacancies in the lattice. Here the activation energy includes the energy required to form a Si vacancy as well as the energy required to move the impurity. The diffusion

of interstitial species usually proceeds more rapidly than does the corresponding species located on a substitutional site.

In some systems an atom, which is normally substitutional, has a very
small energy difference between the substitutional and the interstitial site. In
this case another diffusion mechanism, called dissociative diffusion, can occur.
In this mechanism the substitutional atom is excited to become an interstitial
atom, leaving behind a vacancy. The interstitial atom diffuses until it encoun-
ters another vacant site, at which point it becomes substitutional again. The
one-dimensional diffusion equations now become

$$\frac{\partial}{\partial t} S = D_S \frac{\partial^2}{\partial z^2} S - R_S + Q_I$$

$$\frac{\partial}{\partial t} I = D_I \frac{\partial^2}{\partial z^2} I + R_S - Q_I$$

(5.14)

with S denoting the concentration of substitutional atoms and I the concentration of interstitials. That is, the diffusion equation is augmented by reaction
terms, R_S denoting the rate of loss per unit time from the S channel and Q_I
the rate of loss from the I channel. Unless $R_S = Q_I$ (i.e., the channels are in
equilibrium), the diffusion is non-Fickian. There are several impurity–host systems that yield a non-Fickian diffusion.[31]

In addition to the normal thermally activated diffusion process, the diffusion mechanisms can be enhanced by particle bombardment. Particle bombardment creates both ionization and recoil atoms, both of which can enhance
diffusion. If the energy of the recoil atom is high enough to create vacancies
and interstitials, these defects can enhance diffusion (see Section 5.2.3.).
(Defect-enhanced diffusion is often referred to as radiation-enhanced diffusion.) Vacancies and interstitials can each create one or more diffusing species
in the system, so that the proper description of the diffusion process requires a
number of coupled equations such as those for dissociative diffusion. Moreover,
the reaction terms in the equations are frequently not in thermal equilibrium,
so that the diffusion is non-Fickian.

5.2.1.1. Field-Enhanced Diffusion

Field-enhanced diffusion occurs in the presence of a field, whether electric,
stress, or other. As shown in Figure 5.11(b), the potential-energy barriers tra-
versed by the diffusing species through the lattice now experiences a bias,
resulting in a net drift in the diffusion process. The diffusion equation for the

field-enhanced case is given by

$$\frac{\partial C}{\partial t} = \frac{\partial}{\partial z} D(z,\, t) \frac{\partial}{\partial z} C - v \frac{\partial}{\partial z} C$$

where the last term describes the drift due to the field. Field-enhanced diffusion can readily happen at surfaces. For example, particle-induced electron emission at the surface of a low-conductivity material creates an electric field, while the evaporation of one constituent of an alloy could create a strain field at a surface. Recoil-enhanced diffusion occurs when a high-energy collision imparts a recoil energy to the diffusing species so that its net probability of diffusing is enhanced (see Section 5.2.4.).

Clearly, if there is a directionality to the bombarding particle beam, then there will be a net direction for the drift of the recoil atoms. The diffusion equation for recoil-enhanced diffusion is essentially the same as that for field-enhanced diffusion with a different origin for the coefficients. Recoil-enhanced diffusion in fact occurs in the form of recoil implantation, for example, the recoil implantation of oxygen atoms from a SiO_2 layer on top of a silicon lattice. This process is also responsible for the alteration of an alloy composition at a sputtered surface, since the sputtering collisions not only favor the "emission" of one constituent from the surface but they also favor driving the other constituent(s) into the bulk.

5.2.1.2. Ionization-Enhanced Diffusion

Ionization-enhanced diffusion occurs when the diffusion species possesses charge states, one of which has a lower barrier to migration than does the other. Figure 5.11(c) shows the potential-energy curve as a function of distance through the lattice of two channels for the ionization-enhanced diffusion mechanism.

5.2.1.3. Energy Release-Enhanced Diffusion

For the energy release-enhanced diffusion mechanism we can distinguish several cases. The creation of an electron–hole pair in a material system with a forbidden gap results in the temporary storage of energy, since the energy of the electron is substantially higher than that of the hole. Recombination of the electron and the hole then provides energy that may enhance diffusion, particularly if that recombination occurs at the site of the diffusing species, for example, at a recombination center as shown in Figure 5.11(d). Of course, if the energy is released in the form of a photon, then the photon can escape the vicinity of the diffusing species and will not contribute to an enhancement of the diffusion.

If, on the other hand, some or all of the energy is released locally in the form of phonons, these phonons can enhance the diffusion. This energy-release mechanism has been observed to occur frequently in III–V compounds.[32] If the energy is released in the form of heat in the vicinity of the defect, it may enhance diffusion. The released energy contributes to the jump probability and can be incorporated into the diffusion coefficient of the simple (one-channel) Fick equation.

5.2.2. Ion Implantation

The use of ion beams in microelectronics has grown rapidly over the past 10 years. Ion beams have been used for many years in a variety of scientific experiments. More recently they have emerged into use in engineering applications and even in production processes, such as sputtering deposition and etching in semiconductor device manufacture, fine surface polishing, and ion-implantation doping of semiconductor materials and devices. Ion beams clearly offer certain unique advantages, which has resulted in their increased use for semiconductor and other microelectronic applications.[33-35]

Ions are heavy—they have 10^3–10^5 times the mass of an electron. Thus for a given energy an ion carries approximately 10^2–10^4 times more momentum than an electron. As a result of this characteristic, ions can produce quantitatively different effects in their interactions with crystal lattices (see Section 2.7 on ion interactions).

An ion retains most of the chemical properties of the original atom. After it has been injected into the lattice it may recombine or remain ionized, but it usually behaves chemically just like an atom introduced into the lattice by diffusion, alloying, or during epitaxial growth. Thus ion implantation can be used to "dope" a semiconductor. In addition, because an ion is electrically charged, it can be accelerated to any desired velocity and electrically deflected and focused. Thus an ion beam can be positioned at will and caused to dope only a tiny volume of material.

In comparison to diffusion, ion implantation has a number of advantages, the more important of which are listed here:

- Doping levels can be controlled precisely, since the incident ion beam can be measured accurately as an electrical current.
- Doping uniformity across a surface can be accurately controlled.
- The depth profile can be regulated by choice of the incident ion energy.

- It is a low-temperature process. This feature is not necessarily important with silicon, but some compound semiconductors are unstable at high temperatures.
- Extreme purity of the dopant can be guaranteed by mass analysis of the ion beam.
- Particles enter the solid as a directed beam, and since there is little lateral spread of the beam, smaller and faster devices can be fabricated.
- Dopants can be introduced that are not soluble or diffusable in the base material.

Because of these features, circuits of identical characteristics can be made across a large wafer. In addition, because of the precision of doping levels, devices can be made that are either very difficult or even impossible to make by other means. Together with this process control, implantation provides improved yields. By far the greatest use of ion implantation to date has been in the fabrication of devices with a turn-on, or threshold, voltage less than 1.5 V. By the more old-fashioned diffusion techniques it is very difficult to get the turn-on voltage as low as 1.5 V in a routine and controlled manner, because the amount of dopant that has to be introduced is so small by normal diffusion standards. Using ion implantation, the process becomes simple. It is largely because of ion implantation that hand calculators and digital watches can be operated by standard 1.5-V batteries. The disadvantages of ion implantation in comparison to diffusion are that expensive and complicated equipment is required, the junctions are not automatically passivated, and the crystal structure may be damaged.

5.2.2.1. Qualitative Features of Ion Implantation

In the process of ion implantation[36] atoms of the desired doping element are ionized and accelerated to high velocities and then made to enter a substrate lattice by virtue of their kinetic energy (or momentum). After the energetic ion comes to rest and equilibration has occurred the implanted atom may be in a position in which it serves to change the electronic properties of the substrate lattice; that is, doping occurs. Lattice defects caused by the energy-loss process of the ions may also change the electronic properties of the substrate.[37]

Some ions from a well-collimated beam that is directed toward a channel (certain crystallographic directions in which open spaces exist among the rows of atoms) are able to penetrate deeply into the crystal lattice before coming to rest in interstitial or substitutional sites (Figure 1.34). The directions that allow large amounts of this ion channeling ions are limited. When the crystal is examined from directions other than along a channel or plane, the atoms appear more randomly oriented, roughly as in a dense atomic gas. However, even when ions are injected along a nonchanneling direction, it is difficult to completely prevent channeling.

The penetration depth and final distribution of the ions in the crystal depend on the ion energy, the crystal and the ion species, and the angular alignment of the ion beam with the crystal axis. These characteristics do not depend strongly on the crystal temperature. One can distinguish two primary classes of ion–crystal interaction that yield quite different penetration depths and ion density distributions. Ions penetrating a target that is either amorphous or crystalline, but with the ion path misaligned from any crystal axis, will suffer collisions that reduce their inward motion in such a way that the resulting density profile with depth is roughly Gaussian. This case is illustrated in Figure 1.34, where it is shown as the "amorphous" distribution near the surface.

If the ion dose is low and the trajectories directed precisely in an open crystallographic direction, the ions can penetrate deeply into the crystal and will stop rather abruptly at the end of their range. The resulting profile is also shown in Figure 1.34, by the curve marked "channeled peak." Some fraction of the ions traveling along a channel leave the channel prematurely. Channeled profiles cannot be maintained at high dose levels, since energetic ions striking the surface of a crystal will displace the crystal atoms, resulting in a near-amorphous condition close to the surface. Although the ion dose required to create this amorphous layer varies somewhat with the ion and crystal species, it is usually about 10^{14} ion/cm^2 or more. Thus, with high dose levels, the implanted distribution will always be near Gaussian. Typical parameters for an ion-implanting machine are shown below:

- Dopants: P, As, Sb, B
- Dose: 10^{11}–10^{16}/cm^2
- Energy: 10–400 keV
- Depth of implant: typically 1000–5000 Å
- Reproducibility and uniformity: $\pm 3\%$
- Temperature: normally room temperature
- 50-μA current/3-in. wafer corresponds to a dose rate of 6×10^{12} atom/ cm$^2 \cdot$s^{-1}

5.2.2.2. Range Theory—Lindhard, Scharff, Schiott (LSS)[38]

It is customary to assume that there are two major forms of energy loss for an ion entering a target (see Section 2.7). These are interactions of the ion with the electrons in the solid and collisions of the ion with the nuclei of the target. Therefore the total energy loss can be written as the sum [Eq. (2.217)]

$$-\frac{dE}{dz} = N[S_n(E) + S_e(E)] \qquad (5.15)$$

where $-dE/dz$ is the average rate of energy loss with distance along R (eV/ cm), E is the energy of the ion at point z along R (eV), $S_n(E)$ is the nuclear stopping power (eV·cm²), $S_e(E)$ is the electronic stopping power (eV·cm²), and N is the target atom density $= 5 \times 20^{22}/\text{cm}^3$ for Si. The total distance that the ion travels before coming to rest is called its range (R); the projection of this distance onto the direction of incidence is called the projected range (R_p). The distribution of the stopping points for many ions in space is the range distribution.

If $S_n(E)$ and $S_e(E)$ are known, then the total range can be calculated [Eq. (2.218)]:

$$R = \int_0^R dz = \frac{1}{N} \int_0^{E_0} \frac{dE}{S_n(E) + S_e(E)} \qquad (5.16)$$

R is the *average* total range and one should expect a distribution of R in the direction of incidence (ΔR_p) around R_p, the projected range, and a transverse distribution (ΔR_T) around R_T, the transverse range. Generally, the shape of the distribution depends on the ratio between the ion mass M_1 and that of the substrate atoms, M_2. The relative width $\Delta R_p/R_p$ of the distribution depends on the ratio of M_1 to M_2. For light ions such as B, $\Delta R_p/R_p$ is large, and for heavy ions such as As and Sb $\Delta R_p/R_p$ is small.

5.2.2.3. Useful Approximations

The expression for the nuclear stopping power can be written[37] in the form [Eq. (2.208)]

$$S_n = 2.8 \times 10^{-15} \frac{Z_1 Z_2}{Z^{1/3}} \frac{M_1}{M_1 + M_2} \text{ eV·cm}^2 \qquad (5.17)$$

where S_n is the nuclear stopping power (independent of E), Z_1 is the ion atomic number, M_1 is the ion atomic mass, Z_2 is the substrate atomic number (14 for Si), M_2 is the substrate atomic mass (28 for Si), and $Z^{1/3} = (Z_1^{2/3} + Z_2^{2/3})^{1/2}$. The electronic stopping power can be approximated as [Eq. (2.213)]

$$S_e(E) = kE^{1/2} \qquad (5.18)$$

Here E is the energy of the ion and k is a constant that depends on both the ion and the substrate and is given by

$$k = Z_1^{1/6} \frac{0.0793 Z_1^{1/2} Z_2^{1/2} (M_1 + M_2)^{3/2}}{(Z_1^{2/3} + Z_2^{2/3})^{3/4} M_1^{3/2} M_2^{1/2}} \frac{C_R}{C_E^{1/2}} \tag{5.19}$$

where

$$C_R = \frac{4\pi a^2 M_1 M_2}{(M_1 + M_2)^2}$$

and

$$C_E = \frac{4\pi \epsilon_0 a M_2}{Z_1 Z_2 q^2 (M_1 + M_2)}$$

where ϵ_0 is the permittivity of free space $= 8.85 \times 10^{-15}/\text{F}\cdot\text{cm}^{-2}$, a is the Bohr radius $= 0.529 \times 10^{-8}$ cm, and $q = 1.602 \times 10^{-9}$ C. For an amorphous Si substrate k is independent of the type of ion and Eq. (5.19) reduces to

$$k = 0.2 \times 10^{-15} \text{ eV}^{1/2}\cdot\text{cm}^2$$

Note that S_n is independent of E [Eq. (5.17)] and S_e increases with E. At some critical energy E_c the electronic and nuclear stopping power are equal; that is, at $E = E_c$

$$S_n(E_c) = S_e(E_c)$$

and

$$\sqrt{E_c} = \frac{S_n^0}{k}$$

hence

$$\sqrt{E_c} = 14 \frac{14 Z_1}{(14^{2/3} + Z_1^{2/3})^{1/2}} \frac{M_1}{M_1 + 28} \tag{5.20}$$

For amorphous Si the critical energy is given as

$$E_c \cong 10 \text{ keV for B } (Z = 5, M = 10)$$

$$E_c \cong 200 \text{ keV for P } (Z = 15, M = 30)$$

$$E_c > 500 \text{ keV for As and Sb}$$

Thus B tends to be stopped by electronic interactions; P, As, and Sb tend to be stopped by nuclear collisions. If $E \ll E_c$, then

$$\frac{dE}{dz} = kS_n$$

and from Eq. (5.16)

$$R = (0.7 \text{ Å}) \frac{Z^{1/3}}{Z_1 Z_2} \frac{M_1 + M_2}{M_1} E_0 \tag{5.21}$$

Equation (5.21) is useful for heavy ions such as As, Sb, and sometimes P. If $E \gg E_c$, then

$$\frac{dE}{dz} = NkE^{1/2}$$

$$R = 20\sqrt{E_0} \text{ Å} \qquad \text{for } E_0 \text{ in eV} \tag{5.22}$$

Equation (5.22) is useful for boron if channeling does not take place. In Eqs. (5.21) and (5.22) R is the total range, not the projected range R_p. For the general case when the above limiting cases do not apply,

$$R = \frac{1}{N} \int_0^{E_0} \frac{dE}{S_n + kE^{1/2}} \tag{5.23}$$

5.2.2.4. Projected Range

The previous equations have been in terms of R, the total range, but we really require R_p, the projected range. It can be shown that[36]

$$\frac{R}{R_p} = 1 + b\frac{M_2}{M_1} \tag{5.24}$$

where $b \cong \frac{1}{3}$ for nuclear stopping and $M_1 > M_2$, that is, for Sb and As. Although b is smaller for electronic stopping (B, P), $\frac{1}{3}$ is still correct for the first order. For example, As : $M = 75$, therefore $R/R_p \cong 1.12$. For other mass and energy combinations the projected range can be estimated using the R_p/R calculations in Table 2.

TABLE 2. *Projected Range Corrections* R_p/R
(Si Substrate)

Ion	R_p/R Values				Rule of Thumb Value $(1 + M_2/3M_1)^{-1}$
	20 keV	40 keV	100 keV	500 keV	
Li	0.54	0.62	0.72	0.86	0.4
B	0.57	0.64	0.73	0.86	0.54
P	0.72	0.73	0.79	0.86	0.77
As	0.83	0.84	0.86	0.89	0.89
Sb	0.88	0.88	0.89	0.91	0.93

The dopant concentration distribution as a function of the distance in the target is shown in Figure 5.12, where

$$R_p = \frac{R}{1 + bM_2/M_1}$$

and the standard deviation

$$\Delta R_p \cong \frac{2}{3} \frac{\sqrt{M_1 M_2}}{M_1 + M_2} R_p = \frac{\text{half-width at } (1/2) N_{max}}{(2 \ln 2)^{1/2}}$$

$$N_{max} \cong \frac{N_D}{2.5 \Delta R_p}$$

and N_D is the number of implanted atoms per square centimeter. The ion concentration $N(z)$ will be normally distributed about R_p; that is,

$$N(z) = N_{max} \, e^{\dfrac{-(z - R_p)^2}{2\Delta R_p^2}} \tag{5.25}$$

dropping by one decade at

$$z = R_p \pm 2\Delta R_p, \qquad N(z) = \tfrac{1}{10} N_{max}$$

two decades at

$$z = R_p \pm 3\Delta R_p, \qquad N(z) = \tfrac{1}{100} N_{max}$$

and five decades at

$$z = R_p \pm 4.8\Delta R_p, \qquad N(z) = 10^{-5} N_{max}$$

In practice, R_p and ΔR_p are tabulated for most common impurities as a function of the ion energy.

Thus to find an implantation profile one needs to do the following:

- Look up R_p and ΔR_p in Table 3.
- Assume that the distribution is Gaussian.
- Calculate N_{max} from

$$N_{max} = \frac{N_D}{2.5\Delta R_p}$$

The profile is then given by Eq. (5.25).

The doping profile is usually characterized by the projected range R_p and the standard deviation in the projected range, ΔR_p. Figure 5.13 shows R_p, ΔR_p and the transverse distribution ΔR_T for boron, phosphorus, and arsenic ions implanted into silicon.[39] The penetration depth is roughly linear with energy and is shallower the more massive the ion.

5.2.2.5. Masking

Spatially selective doping is achieved by means of "masking." Being basically a mechanical process, ion implantation only requires some physical barrier of a sufficient thickness in order to be masked. There is actually a considerably wider choice of masking materials usable with ion implantation than

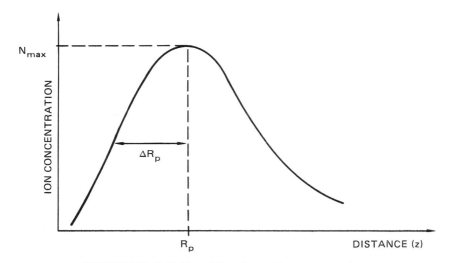

FIGURE 5.12. Definition of Gaussian profile for implanted ions.

TABLE 3. Projected Range and Standard Deviation in Projected Range for Various Ions in Silicon (Ranges and Standard Deviations Quoted in Å)[a]

Ion	(keV)	20	40	60	80	100	120	140	160	180	200
B	R_p	714	1413	2074	2695	3275	3802	4289	4745	5177	5588
	(ΔR_p)	276	443	562	653	726	793	855	910	959	1004
N	R_p	491	961	1414	1847	2260	2655	3034	3391	3728	4046
	(ΔR_p)	191	312	406	479	540	590	633	672	710	745
Al	R_p	289	564	849	1141	1438	1737	2036	2335	2633	2929
	(ΔR_p)	107	192	271	344	412	476	535	591	644	693
P	R_p	255	488	729	976	1228	1483	1740	1998	2256	2514
	(ΔR_p)	90	161	228	291	350	405	495	509	557	603
Ga	R_p	155	272	383	492	602	712	823	936	1049	1163
	(ΔR_p)	37	64	88	111	133	155	176	197	218	238
As	R_p	151	263	368	471	574	677	781	885	991	1097
	(ΔR_p)	34	59	81	101	122	141	161	180	198	217
In	R_p	133	223	304	381	456	529	601	673	744	815
	(ΔR_p)		23	38	51	63	75	86	97	108	119
Sb	R_p	132	221	300	376	448	519	590	659	728	797
	(ΔR_p)	22	36	49	60	71	82	92	102	112	122

[a]Source: ref. 36.

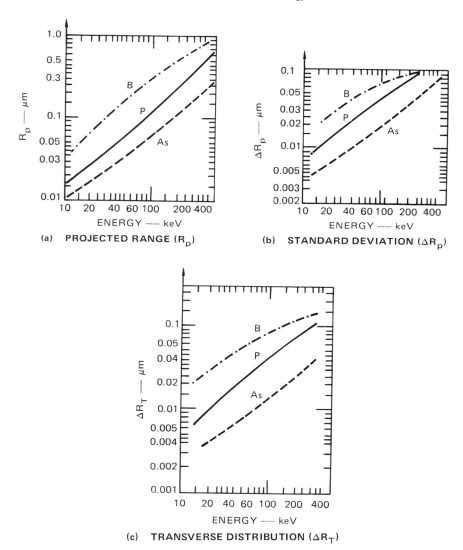

FIGURE 5.13. Projected range (R_p), standard deviation (ΔR_p), and transverse distribution (ΔR_T) for boron, phosphorus, and arsenic as functions of the implanted ion energy. Source: ref. 36.

with diffusion. Figure 5.14 shows some of them and offers a comparison of different masking layer thicknesses required to prevent all but about 0.1% of an implant from penetrating. SiO_2 makes a good mask, Si_3N_4 is even better, Al is good, and photoresist (KTFR) in thicknesses commonly applied is adequate for all but high-energy boron implants.

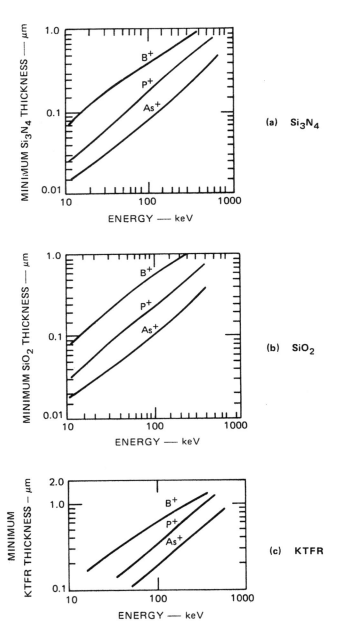

FIGURE 5.14. Thickness required to mask implants. Source: ref. 20.

5.2.2.6. Channeling

The range distributions in single-crystal targets are different if the ion beam is aligned along the crystal axis (see Section 2.7.2.4). In this case, because projectiles can channel along open directions, a deeper penetration can occur. The channeling angle ψ is given as (see Figure 5.15)

$$\psi \cong \left(\frac{a}{d} \psi_1 \right)^{1/2}$$

where $\psi_1 = [(2 Z_1 Z_2 q^2)/ Ed]^{1/2}$, $a \cong 0.5$ Å, and d is the atomic spacing (2.5 Å for Si).

Critical angles as a function of energy for several different projectiles are also given in Figure 5.15 for the three major orientations of a Si target. The channeling effect can result in any of the several types of profiles shown in Figure 5.15. The amorphous peak A or A' is generated by ions A' entering a crystal at random directions to the lattice direction, or by ions A in well-aligned trajectories, but that impact substrate atoms at the surface end of the lattice rows. This amorphous distribution exhibits a Gaussian shape characterized by a mean projected range R_p and a standard deviation ΔR_p. The channeling peak C comprises ions whose trajectories are well within the critical angle of the crystal direction. These ions lose energy by electronic collisions and channel to approximately the maximum range. Ions with higher values of electronic stopping produce more-pronounced (higher) channeling peaks. A channeling peak is not exhibited for all ions incident at room temperature. Channeling is purposely avoided in most devices because it is difficult to control accurately. Usually, implantation is done with the wafer tilted to get a random, "amorphous" profile. For example, a 7° tilt off the $\langle 100 \rangle$ axis results in greater than 99% of the ions being stopped, as if the silicon were amorphous.

5.2.2.7. Implantation

To summarize the previous results,

- For light ions (B) and high energies the electronic stopping is the major energy-loss mechanism. For heavy ions (As, Sb) and low energies the dominant energy loss is due to nuclear collisions.
- The ion distribution in the target is Gaussian in both the z and x directions (see Figure 5.16), or

$$N(z, x) = N_{max} \exp \left(\frac{-(z - R_p)^2}{2\Delta R_p^2} \right) \exp \left(\frac{-x^2}{2\Delta R_T^2} \right) \qquad (5.26)$$

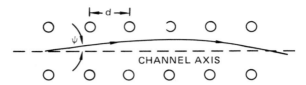

(a)   **SCHEMATIC TRAJECTORY OF A CHANNELED PARTICLE**

Ion	Energy (keV)	Channel Direction		
		$<110>$	$<111>$	$<100>$
Boron	.30	4.2'	3.5	3.3'
	50	3.7'	3.2'	2.9°
Nitrogen	30	4.5°	3.8'	3.5°
	50	4.0'	3.4°	3.0'
Phosphorus	30	5.2'	4.3'	4.0'
	50	4.5°	3.8°	3.5'
Arsenic	30	5.9°	5.0'	4.5'
	50	5.2'	4.4°	4.0'

(b)   **CRITICAL ANGLES FOR CHANNELING OF SELECTED IONS IN SILICON**

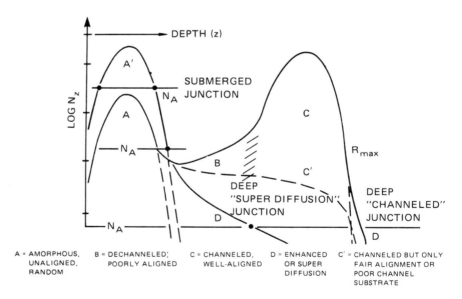

(c)   **DEPTH PROFILE**

FIGURE 5.15. Channeling effect. Source: ref. 40.

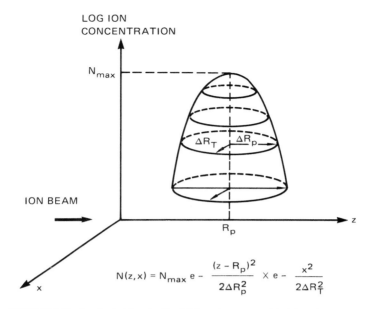

FIGURE 5.16. Implanted longitudinal (z) and transversal (x) ion distribution.

LSS theory allows the prediction of R_p, ΔR_p, and ΔR_T for an arbitrary ion and substrate.[38] Channeling can be a problem if the beam is aligned along a crystal axis, but this is normally avoided in devices by tilting the wafer with respect to the beam. Deviations from the Gaussian shape are usually caused by the unwanted channeling effect or by the enhanced diffusion during annealing as a result of damage (see Section 5.2.3). Figure 5.17 shows two actual measured profiles along with a Gaussian fit.[39]

5.2.2.8. Damage

For range determination both electronic and nuclear interactions are important. However, concerning radiation damage, only the atomic or nuclear collisions should be considered because only these interactions can transfer enough energy to the target atoms.[37]

For crystalline substrates atoms are bound together with energy E_d. The necessary condition to cause damage is that $E_{transf} \geq E_d$. Heavy ions (Sb, As) stop primarily by nuclear collisions; therefore they cause more damage than B or P, which stop mainly by electronic interactions. Nuclear stopping dominates at low energies; therefore one should expect that most of the damage for B occurs near where the ions stop.

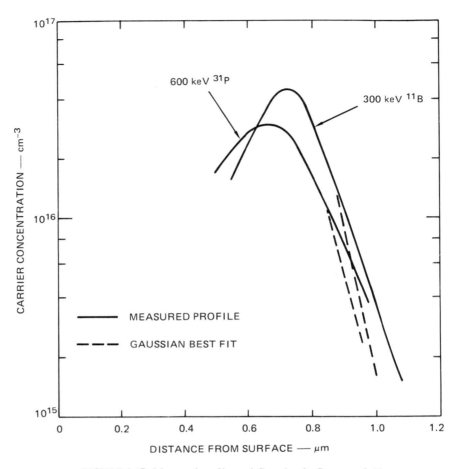

FIGURE 5.17. Measured profiles and Gaussian fit. Source: ref. 39.

5.2.2.8.1. Heavy Ions—Sb, As   For heavy ions stopping is primarily due to nuclear collisions ($E_c > 500$ keV for As and Sb and $E_c \cong 200$ keV for P), therefore one would expect a great deal of damage.

The typical situation is understood by observing that the nuclear stopping power S_n of a 100-keV Sb projectile in a silicon target is about 0.2 keV/Å over the entire trajectory.[37] Since the spacing between the lattice planes in silicon is about 2.5 Å, this means that the Sb projectile loses about 500 eV per lattice plane on the average. One can expect the majority of this energy to be given to one primary silicon recoil atom, as the average recoil will have an energy approaching 500 eV; it would then have a range of roughly 25 Å. For comparison, the range of the 100-keV Sb ion is 500 Å, which is not substantially larger than the possible range of the average primary recoil atoms it creates.

The collision cascade of each primary recoil will contain roughly 15 displaced target atoms, since the number of recoils $\cong E/2E_d$ for heavy ions, where E_d is the bonding energy of the target atom, $\cong 15$ eV for Si. The damage volume $V_D \cong \pi(25 \text{ Å})^2$ (500 Å) (see Figure 5.18). Within this volume there are roughly 15 displacements per lattice plane, or 3000 total displacements. The average vacancy density is about $3000/V_D \frown 10^{22}/\text{cm}^3$, or about 20% of the total number of atoms in V_D. As a consequence of the implantation, the material has turned essentially amorphous.

5.2.2.8.2. Light Ions—B Much of the energy loss is a result of electronic interactions, which do not cause damage. There is more damage near the end where nuclear stopping dominates. For boron $E_c \cong 10$ keV. As a consequence, most of the damage is near the final ion position. For example, 100-keV B has $R_p \cong 3400$ Å, which equals 500 recoils or vacancies created in the last 1700 Å, or about 200 over the first 1700 Å. In the damage volume ($V_D \frown 1.6 \times 10^{-18}$ cm³) there are 500 displacements, and with this the average vacancy density is $500/V_D \frown 3 \times 10^{20}/\text{cm}^3$, which is less than 1% of the atoms; higher doses are needed to create amorphous material with light ions.

5.2.2.8.3. The Creation of Amorphous Material To create amorphous material an energy deposition of $\cong 10^{21}$ keV/cm³ is required, the same as needed for melting. To estimate the dose for converting a crystalline material to an amorphous form by heavy ion bombardment, consider a 100-keV Sb ion that has a range $R_p \cong 500$ Å (5×10^{-6} cm). For this ion beam the dose to make amorphous Si is

$$D = \frac{(10^{21} \text{ keV/cm}^3) R_p}{E} = 5 \times 10^{13}/\text{cm}^2$$

For 100-keV B $R_p \cong 3300$ Å $= 3.3 \times 10^{-5}$ cm. The dose to make amorphous Si $D \cong 10^{21}(3.3 \times 10^{-5})/100 = 3.3 \times 10^{14}/\text{cm}^2$.

In practice, higher doses are required for boron because the damage is not uniformly distributed along the path of the ion.

5.2.2.8.4. Damage Profiles To the first order the damage profile may be assumed to be Gaussian. It is generally shallower than the ion profile (see Figure 5.19). Based on LSS theory, it is possible to calculate the range values for the damage profile. For example, for boron

$$\frac{M_{\text{Si}}}{M_{\text{B}}} = \tfrac{12}{5} = 2.4$$

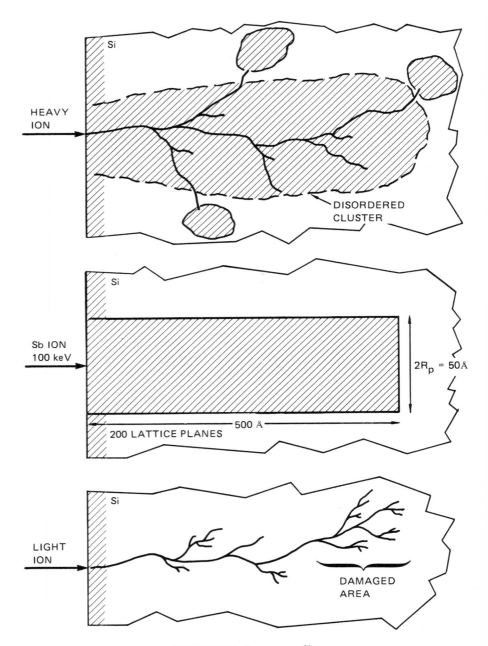

FIGURE 5.18. Damage profiles.

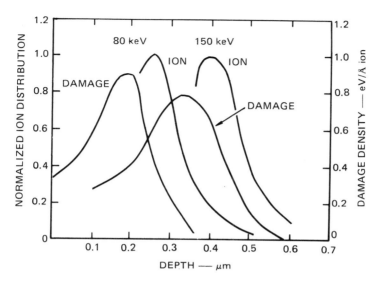

(a) DAMAGE PROFILE

(b) RANGE DISTRIBUTION PARAMETERS

$\dfrac{M_{target}}{M_{ion}}$	$\dfrac{\langle z \rangle}{E/NC_1}$	$\dfrac{\langle \Delta z^2 \rangle}{\langle z \rangle^2}$	$\dfrac{\langle x^2 \rangle}{\langle z \rangle^2}$	$\dfrac{\langle zx^2 \rangle}{\langle z \rangle \langle x^2 \rangle}$	$\dfrac{\langle \Delta z^3 \rangle}{\langle z^3 \rangle}$
1/10	0.842	0.058	0.018	1.07	0.007
1/4	0.577	0.125	0.044	1.16	0.021
1/2	0.453	0.195	0.089	1.20	0.043
1	0.369	0.275	0.176	1.20	0.079
2	0.297	0.409	0.343	1.16	0.135
4	0.229	0.710	0.674	1.12	0.221
10	0.153	1.684	1.671	1.07	0.345

(c) DAMAGE DISTRIBUTION PARAMETERS

$\dfrac{M_{target}}{M_{ion}}$	$\dfrac{\langle z \rangle_D}{E/NC_1}$	$\dfrac{\langle \Delta z^2 \rangle_D}{\langle z \rangle^2_D}$	$\dfrac{\langle x^2 \rangle}{\langle z \rangle^2_D}$	$\dfrac{\langle zx^2 \rangle_D}{\langle z \rangle_D \langle x^2 \rangle_D}$	$\dfrac{\langle \Delta z^3 \rangle_D}{\langle z \rangle^3_D}$
1/10	0.692	0.434	0.192	1.80	0.580
1/4	0.489	0.437	0.181	1.71	0.433
1/2	0.376	0.386	0.152	1.47	0.218
1	0.295	0.380	0.157	1.41	0.172
2	0.241	0.457	0.257	1.42	0.272
4	0.198	0.623	0.485	1.31	0.391
10	0.143	1.215	1.153	1.16	0.567

FIGURE 5.19. Damage profile and associated range and distribution parameters. Source: ref. 41.

Using the Sigmund–Sanders tables,[41] one can estimate the damage distribution by noting that

$$\langle z \rangle_D = R_D$$

and

$$\langle \Delta z \rangle_D = \Delta R_D$$

The damage range can be calculated by taking ratios from the second column of the tables in Figure 5.19. Thus

$$\langle z \rangle_D = \frac{0.22}{0.26} R_p \approx 0.8\, R_p \tag{5.27}$$

The standard deviation in the damage distribution is given from ratios in the third column of the table as

$$\langle \Delta z \rangle_D = 0.75 (\Delta R_p) \tag{5.28}$$

which can be used together with Eq. (5.27) to construct an actual damage distribution.

5.2.2.8.5. The Location of Implanted Ions For low doses many of the ions end up on lattice sites; they are substitutional and also electrically active. For high doses most of the ions end up as interstitials and are not electrically active. In general B shows a higher percentage of ions in interstitial or nonelectrically active sites.

5.2.3. Radiation-Enhanced Diffusion

All the impurities used in fabricating silicon devices diffuse either by a substitutional or an interstitial mechanism. As a consequence, the diffusion process is sensitive to the vacancy concentration and can be readily influenced by the supply of excess vacancies. One way of generating the vacancies is by means of displacement reactions caused by nuclear collisions in the ion-implantation process. This scheme is known as radiation-enhanced diffusion.[42] Figure 5.20 shows the experimental results for B, P, As, and Si under the intrinsic condition $N_A < n_i$. From the theory at low temperatures and high fluxes the

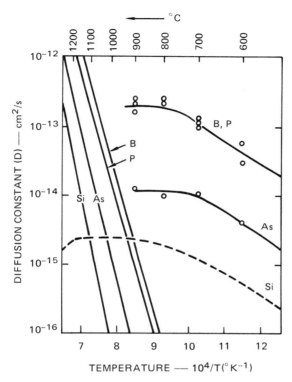

FIGURE 5.20. Temperature dependence of enhanced diffusion (5×10^{11} proton/cm$^2 \cdot$s^{-1}).
Source: ref. 42.

enhanced self-diffusion coefficient in Si depends on the flux by a square-root law

$$D_s \backsim k(\text{flux})^{1/2}$$

and at low fluxes $D_s \backsim k'$ (flux) depends linearly on the flux.

5.2.4. Recoil Phenomena

Nuclear collisions are responsible not only for the radiation damage in the target materials, which leads to the radiation-enhanced diffusion of impurity profiles, but also for the modification of impurity distributions by the introduction of additional impurity species by recoil from surface layers (see Figure

5.21). This effect occurs when dielectric films are applied to the surface of a silicon wafer for selective doping. The total number of recoil atoms from the surface film that reach the silicon substrate depends on the type of bombarding ions, their energy (E_0), the type of atoms in the film, and the film thickness (W).

The depth distribution for the dielectric–silicon interface is shown in Figure 5.22 and is given by[44]

$$N(D) = BN \left[\left(\frac{E_0}{W} \right)^{1/3} - \left(\frac{E_0}{D} \right)^{1/3} \right] \left[1 + \left(\frac{W}{W_0} \right)^{2/3} \right] \text{ atom/volume}$$

where B is a coefficient that contains the atomic parameters of the film substrate and the projectile and W_0 is an empirical factor that characterizes the formation of the collision cascade. Figure 5.22 shows the oxygen profile obtained from the 50-keV recoil collisions of 10^{13} ions/cm^2 traversing an oxide layer of 10 Å.

Recoil, range, and damage distributions following ion implantation are of considerable importance in device design. Calculations based on the theory of

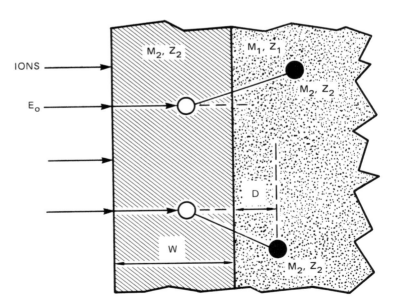

FIGURE 5.21. Diagram of recoil collisions.

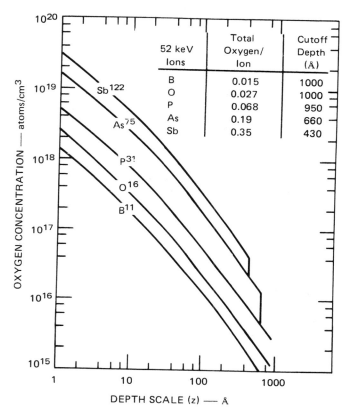

FIGURE 5.22. Oxygen profile obtained from 50-keV recoil collision at 10^{13} ion/cm². Source: ref. 42.

Lindhard, Scharff, and Schiott (LSS) have been developed for range and damage profiles for ions implanted in semiconductors and they give a satisfactory accounting of the primary-ion range distributions in semi-infinite substrates.[43] However, practical device processing involves implantation into targets with one or more thin films. In these cases we are interested in the spatial distributions of not only the primary ions but also the recoil-implanted atoms in each layer.

It has been shown recently that numerical integration of the extended Boltzmann equation can be used to obtain estimates of the primary-ion and recoil range distributions, the energy deposition profiles, and the energy and angular distributions.[44]

5.2.5. *High-Dose Implantation Limits*

One of the major advantages of ion implantation is the external control of the number of implanted ions. In high-dose implantation this advantage is lost owing to target sputtering. The maximum achievable concentration of an implanted species is determined in the limit by sputtering or erosion of the implanted surface.[45] In the implantation process ion sputtering transfers energy from the incident ion to the target material, which results in the ejection of atoms from the surface.

For a dose of 10^{17} ion/cm^2 (about 100 monolayers) it is possible to remove 100–1000 layers of target material, which corresponds to a thickness change of 500–5000 Å. Then during implantation the surface profile is a result of erosion and sputtering of both the target and the implanted ions. The simplest estimate gives the concentration of implanted species to be proportional to $1/S$, where S is the sputtering yield of the target.

Recent experiments on the sputtering of compounds[45] have indicated that preferential sputtering effects can influence the maximum achievable ion concentration. The influence of preferential sputtering (r) is generally to reduce the concentration of higher-mass atoms at the surface of the target. According to this model, the maximum concentration is proportional to r/S, where S is the total sputtering yield and r is the preferential sputtering factor ($\frac{1}{2} < r < 2$). Since lower-mass elements tend to be preferentially sputtered, one can achieve a higher concentration of heavy elements than of lighter elements in the substrate.

5.2.6. *Device Applications*

The older predeposition and drive-in technique "automatically" passivates junctions with a SiO$_2$ layer (see Figure 5.23); however, ion implantation must normally be followed by oxidation.

5.2.6.1. *Bipolar Devices*

For bipolar applications high-dose implantations are used for buried layers, emitters, and base contact doping, while low- or moderate-dose implants are employed to dope active base regions and to make resistors.

Figure 5.24 shows an example of how a low-resistivity As buried layer can be formed under a high-quality epitaxial layer. In this application a high-dose As implant is made through an oxide cut. A high-temperature drive-in and oxidation step serves as the anneal. The oxide grown is sufficient to consume

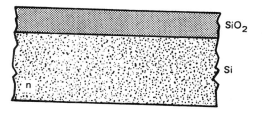

(a)   OXIDE GROWTH

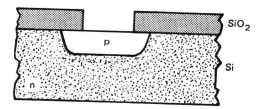

(b)   DIFFUSION

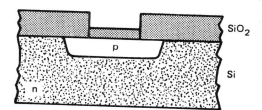

(c)   OXIDE REGROWTH

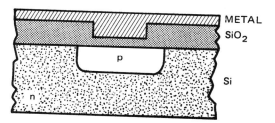

FIGURE 5.23. Passivation process.

(d)   METALIZATION

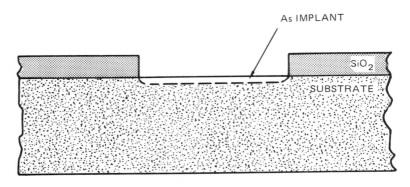

(a)   IMPLANT As ($10^{15} - 10^{16}$ cm^{-3})

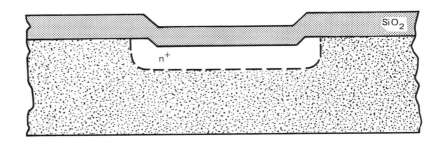

(b)   ANNEAL AND OXIDIZE

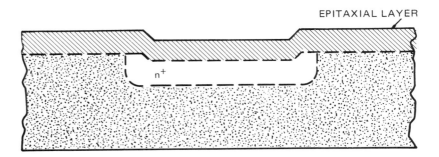

(c)   STRIP OXIDE AND GROW EPITAXIAL LAYER

FIGURE 5.24. Buried-layer formation.

the silicon in which the implanted layer originally resided. This eliminates residual damage typical of very-high-dose implants and allows the growth of high-quality epitaxial silicon over the buried layer. Sheet resistivities of less than 10 Ω per square can be achieved by this process.

5.2.6.2. MOS Devices

Ion implantation was first applied on a large scale in MOS devices, and its use continues to grow. It has provided a reproducible means of introducing small amounts of impurities near the surface and has led to many applications. Most MOS applications of ion implantation involve threshold voltage control. Adjustment of the active device threshold voltage by placing a controlled low dose close to the gate oxide–substrate interface was the first of these applications and remains the most widely used. Figure 5.25 shows some data for a *p*-channel MOS in which a boron implant is made through the gate oxide with the peak of the distribution at the interface so that approximately 50% of the dose

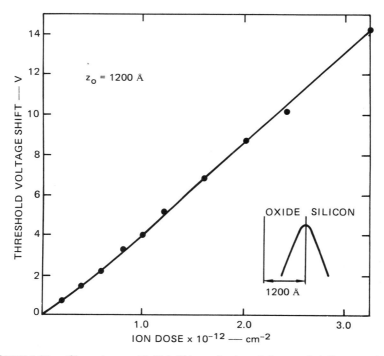

FIGURE 5.25. *p*-Channel MOS with 30-keV boron implant, fully annealed. Source: ref. 39.

goes into the silicon. For the ideal case of the sheet charge placed right at the interface the threshold voltage will shift linearly with the implanted dose.

5.2.6.3. Resistors

Ion-implanted resistors can be used for any IC technology that has a need for them as well as for monolithic precision resistor networks. Dose control is also a key factor, as the higher sheet resistivities are very difficult to obtain reproducibly with diffusion. A wide variety of conditions are possible; some of these are illustrated in Table 4 for the case of boron-implanted resistors. Sheet resistivities of from less than 100 Ω per square to many thousands are achievable, although surface conditions make control difficult for the very high resistivities. Of great interest is the temperature coefficient of resistance (TCR). For a fully annealed implant, represented by the last column of Table 4, the TCR is positive and higher the lower the dose. At lower annealing temperatures, however, combined damage and doping effects can cause a negative TCR for all but very low doses. There exists an annealing condition that yields a zero TCR, representing partial annealing of the damage. This is shown in the fourth column of Table 4. This partial annealing has found use in precision resistor networks, which are stable over some desired operating temperature range. The second column of Table 4 represents the range of observed TCR values in the annealing range 400–950°C. For all but the very low doses the TCR starts out negative, crosses to zero, is maximum at about 600–700°C, and falls back to its final value shown for 950°C. This final value represents a fully annealed implant with no residual damage effects.[39]

5.2.7. Neutron Transmutation Doping

Irradiating semiconducting materials in a nuclear reactor was initiated in 1950 at the Oak Ridge National Laboratory.[46] The primary interest at that time was the study of radiation damage effects, although the transmutation of silicon was used to determine isotropic cross sections. This work led to the direct use of neutron transmutation doping (NTD) to convert pure single-crystal ingots of high-resistivity silicon into homogeneously and precisely phosphorus-doped n-type material[47] (see Section 1.3.5). NTD, however, cannot be used for spatially selective doping as can diffusion and ion implantation.

The resistivity variation of NTD material can be as low as 1%, corresponding to doping variations of less than 1% across the wafer.[48]

Nuclear reactors provide the source of thermal ($E_n = 0.025$ eV) neutrons for NTD. However, the thermal neutron flux is always accompanied by a fast

TABLE 4. Boron-Implanted Resistors

Ion Dose (cm^{-2})	Parameter Range for 400°C–950°C Anneals		Anneal Temperature for TCR = 0 (°C)	TCR for 950°C Anneal (%/°C)
	$R_\square$ (Ω/□)	TCR (%/°C)		
10^{12}	100,000–25,000	~+0.7	None	~+0.7
10^{13}	13,000–2500	−0.1–+0.45	475	+0.45
10^{14}	3000–400	−0.15–+0.3	475	+0.15
10^{15}	2000–50	−0.2–+0.2	540	+0.10

neutron component that is not useful in providing transmutations but which produces unwarranted displacements of atoms that must be repaired by annealing after the irradiation.

Thermal neutrons interact only weakly with the atomic electrons through their magnetic moments, but, being neutral, they can impact the target nuclei and be captured by it. This interaction is described in terms of a capture cross section σ_c. The number of captures per unit volume for a fluence (flux exposure time) of neutrons ϕ is given by

$$N = N_T \sigma_c \phi$$

where N_T is the number of target nuclei per unit volume. At low energies

$$\sigma_c \sim E^{-1/2} \sim \frac{1}{v}$$

where E is the neutron energy and v the corresponding velocity. For a given nuclear radius $1/v$ is proportional to the interaction time, so that σ_c may be considered as representing the probability of interaction between the nucleus and the thermal neutron.

After neutron capture the target nucleus differs from the initial nucleus by the addition of one nucleon forming a new isotope in an excited state. The excited state must relax by the emission of energy, which is usually in the form of high-energy gammas. The time for the decay of this excess energy by gamma radiation can be very short (prompt gammas) or can take an appreciable time, in which case a half-life, the time for the number of particles emitted per second to decrease by a factor of 2, can be measured. The gamma emission spectrum is used to characterize the nuclear energy levels of the transmuted target nuclei.

The absorption of a neutron and the emission of gammas is represented by the notation A_X (n, γ) $(A + 1)_X$ where (n, γ) represents (absorption, emission), A is the initial number of nucleons in the target element X before neutron absorption, while $A + 1$ is the number after absorption. It is possible for the product isotope $(A + 1)_X$ to be naturally occurring and stable. In many cases, however, the product isotope is unstable. Unstable isotopes further decay by various modes involving the emission of electrons (β decay), protons, and alpha particles and K-shell electron capture or internal conversion until a stable isotopic form is reached. These decays produce radioactivity characterized by their half-lives $T_{1/2}$.

In the case of silicon, absorbing the thermal neutron, three stable target isotopes are formed by (n, γ) reactions as in Table 5.[49]

The first two reactions produce no dopants and only slightly redistribute the relative abundances. The third reaction produces ^{31}P, the desired donor dopant, at a rate of about 3.355 ppb per $\phi = 10^{18}$ n_{th}/cm^2. This production rate is calculated from the rate equation for ^{30}Si capture ($\sigma_c = 0.11 \times 10^{-24}$, $N_T = 5 \times 10^{22}$ $Si/cm^2 \times 0.031$).

In addition to the desired phosphorus production reaction and its relatively short half-life for β-decay, the reaction

$$^{31}P(n, \gamma)^{32}P \rightarrow {}^{32}S + e^- \ (\sigma_c + 0.19 \text{ b}, \ T_{1/2} = 14.3 \text{ days})$$

occurs as a secondary undesirable effect. The decay of ^{32}P is the main source of residual radioactivity in NTD Si.

Once the dopant phosphorus has been added to the silicon ingot by transmutation of the ^{30}Si isotope, this radiation-damaged and highly disordered material has to be annealed to make it useful for electronic device manufacture. Several radiation-damage mechanisms contribute to the displacement of the silicon atoms from their normal lattice positions. They are the following:

- Fast neutron knock-on displacements
- Fission gamma-induced damage
- Gamma recoil damage
- Beta recoil damage
- Charged particle knock-ons (n, p), (n, α), etc. reactions

Estimates can be made of the rate at which Si atom displacements are produced by these various mechanisms, once a detailed neutron energy spectrum is known.[48]

NTD is now quite well established as a method of introducing a uniform distribution of phosphorus dopant into high-purity silicon in order to obtain a

TABLE 5.

Reaction	Cross Section σ_c (Barns) (1 b $= 10^{-24}$ cm^2)	Isotope Abundance
1. ^{28}Si(n, γ) ^{29}Si	0.08	92.3% ^{28}Si
2. ^{29}Si(n, γ) ^{30}Si	0.28	4.7% ^{29}Si
3. ^{30}Si(n, γ) ^{31}Si $\rightarrow$ ^{31}P $+ e^-$		
The half-life of ^{31}Si is 2.62 h.		

uniform n-type starting material. This is particularly interesting for power devices where n-type starting material is required for high-power device design and performance, including a more precise control of avalanche breakdown voltages, a more uniform avalanche breakdown, that is, a greater capacity to withstand overvoltages and a more uniform current flow in the forward direction. In addition to the area and spatial uniformity of the dopant distribution, the NTD silicon offers three clearly identifiable advantages over silicon doped by more traditional techniques:

- Precise control of the doping level
- Elimination of dopant segregation at grain boundaries in polycrystalline silicon
- Superior control of heavy-atom contaminants

The precise control of doping levels is important in any application such as avalanche or infrared detectors, which require a high-resistivity material. The transmutation doping of epitaxial layers deposited on low-resistivity n-type or p-type silicon substrates has also been demonstrated.[50]

5.3. Laser-Assisted Processes

5.3.1. Laser Annealing

The distribution of implanted ions in the solid substrate depends on both the acceleration voltage (ion energy) and mass relations between the implanted ion and the atoms of the target. Stopping of an energetic ion occurs as a result of collisions between the ion and the electrons and atoms of the target. If these collisions are sufficiently energetic, host atoms will be displaced from lattice sites, thus leaving vacancies behind. These displaced atoms may in turn displace other atoms in the lattice, forming as a result a disorder cluster around the ion track (see Section 5.2.2.8). Furthermore, since the process just described is far from thermal equilibrium, only some of the implanted impurities will be located on substitutional sites after the implantation. In general, a large fraction will be found in interstitial positions and still other impurity atoms will form complexes with simple lattice defects (lattice vacancies and interstitials). Moreover, the doping characteristics of the impurity may be overwhelmed by the damage. As a result, the as-implanted substrate will usually exhibit an electrical carrier concentration far below that which would be predicted simply from the implanted-impurity concentration profile. In other words, only a small percentage of the implanted impurities contribute to electrical carriers. Hence an annealing process is required to restore crystal perfection.

In conventional furnace-annealing procedures the entire wafer is heated to temperatures of 1000°C or higher for periods of about half an hour. This process is not always able to restore crystal perfection in the implanted region and often leads to deleterious effects on both the crystalline perfection and the chemical purity of the silicon in the bulk of the wafer. Moreover, such thermal treatments often result in a high concentration of electrically inactive dopant atoms in the form of precipitates close to the surface of the wafer, leading in turn to short carrier lifetimes within this region.

A new but still experimental method for obtaining high-quality anneals of near-surface displacement damage uses the heat generated by the absorption of radiation from high-powered lasers.[51-59] Several advantages are claimed for the laser annealing technique[51]:

- Fine spatial control of the area that is recrystallized on the wafer is possible.
- Control of the dopant diffusion depth is possible by varying the conditions of pulse length and pulse intensity.
- Because of extremely high recrystallization velocities (10^2–10^3 cm/s) of the melted surface layer, substitutional solid solutions are obtained with dopant impurity levels well in excess of conventionally observed values.
- Only the surface region of the wafer is exposed to elevated temperatures, so that bulk crystallinity is not degraded.
- Since laser annealing is an extremely rapid event, there is no need for vacuum conditions or special inert atmospheres to prevent oxidation or contamination during the annealing procedure.

In annealing ion-implanted damage in semiconductors by means of pulsed lasers, the defect layer, generally amorphous, must be melted by the laser; the melt depth must extend slightly into the single-crystal layer to achieve optimum annealing and substitution of the implanted ions in the lattice.

For semiconductor materials such as Si and Ge implanted with As, B, or P, and using a Q-switched Nd, glass: Nd, YAG, or ruby laser, the laser pulse is in the duration range 10–100 ns. The energy density of irradiation on the material is in the range 0.5–10 J/cm^2. The resulting duration of the melt is generally a few hundred nanoseconds. (Because the absorptive properties of the host materials vary with wavelength, these parameters will vary also.) In addition, the absorptive properties of the liquid differ from those of the solid.

These conditions can create a problem in coupling adequate energy into the material during the short pulse. A way around this problem was found by using a Q-switched Nd: glass laser at both its basic wavelength of 1060 nm and a frequency-doubled wavelength of 530 nm.[60] The shorter-wavelength light couples well into the solid silicon and initiates the melt. The 1060-nm-wavelength light then couples well into the melted material and completes the process of annealing.

The laser annealing of ion-implanted damage in semiconductors basically completely restores the crystalline structure by epitaxial regrowth, with the implanted ions taking up substitutional positions in the lattice structure.[52] Pulsed-laser annealing results in none of the defects found in thermal annealing, that is, dislocations, stacking faults, and loops in the crystal. In application, the laser beam is rastered, with the scan rate coupled with the spot size and interpulse delay so that the irradiated spots overlap.

Another approach to laser annealing is the use of a continuous-wave (cw) or CO_2 laser. In this technique the laser beam is scanned continuously over the surface of the semiconductor material at a rate that allows the surface to heat to a temperature just below the melting point. As in the case of the pulsed laser, the restoration of the crystal structure is essentially complete, with the implanted ions taking up substitutional positions in the lattice structure. In this case, however, there is no melting and there is a solid-phase epitaxial regrowth. With no melting, there is no redistribution of the dopant ions, so the crystal retains its high dopant density.[61]

The scan rate of the cw beam is usually about 0.5–10 cm/s, giving a dwell time of about 10–100 ms. The typical flux is about 200 J/cm^2. In application, the scanning can be done by moving either the laser beam or the table on which the semiconductor material is mounted. An experimental arrangement[62] of the former type is shown in Figure 5.26. In cw laser annealing it has generally been necessary to preheat the substrate to a temperature in the 200–400°C range to reduce lateral thermal stress. This also allows the use of a lower-pow-

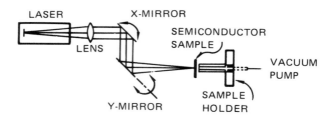

FIGURE 5.26. Laser-annealing apparatus. Source: ref. 62.

ered laser than would otherwise be necessary. Each mode of laser operation (pulsed, cw) has its advantages and disadvantages, depending in part on the materials and dopants used and on the size of the wafer to be processed.

5.3.2. Thermal Analysis of Laser Annealing

In pulsed laser annealing recrystallization is a result of the formation of a thin $(0.1-1.0 \ \mu m)$ molten layer on the surface of the silicon wafer. Since the duration of the annealing pulse is $\leq 10^{-7}$ s, and from experiments we know that the migration of dopant atoms is several thousand angstroms, we can calculate the diffusion coefficient required to cause a diffusion length of l cm in a time τ by using the relationship

$$l = (D\tau)^{1/2}$$

or

$$D = \frac{l^2}{\tau}$$

With typical experimental values, $l = 10^{-5}$ cm and $\tau = 10^{-7}$ s, this formula gives $D \cong 10^{-3}$ cm^2/s. This value is more typical of diffusion constants in liquid materials than in solid-state materials, where $D_s = 10^{-12}-10^{-14}$ cm^2/s. Therefore both experimental (recrystallization) and theoretical evidence support the melted-layer model.

In typical experimental configurations (the diameter of the incident laser beam is $\cong 1$ cm and its duration is $< 1 \ \mu$s) the diffusion length for heat is

$$l \approx (k\tau)^{1/2} \approx 1 \ \mu m$$

where k is the thermal diffusivity and τ is the duration of the diffusion. Since l is much less than the beam diameter, the radial (x, y) diffusion can be neglected. The temperature distribution $T(z, t)$ can be calculated from the one-dimensional diffusion equation[51]

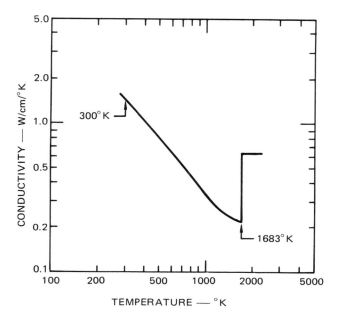

FIGURE 5.27. Thermal conductivity of silicon as a function of temperature. Source: ref. 51.

$$\frac{\partial T}{\partial t} = \frac{1}{\rho C}\left[\frac{\partial}{\partial z}\left(K(T)\,\frac{\partial T}{\partial z}\right) + A(z, t)\right] \qquad (5.29)$$

where ρ is the density and C is the specific heat, which are temperature independent. However, this equation is nonlinear, since the normal conductivity K is temperature dependent (see Figure 5.27):

$$k = K/\rho C$$

$A(z, t)$ represents the energy density absorbed from the incident laser pulse. The heat losses (radiation, conduction) are very small (1 W/cm²) compared to the input pulse energies (10^6 W/cm²) and they can be neglected. Equation (5.29) can be solved for the following limiting cases.

5.3.2.1. Adiabatic Limit

The distribution of the laser light intensity on the wafer is

$$I = (1 - R)I_0\, e^{-\alpha z} \; \text{W/cm}^2 \qquad (5.30)$$

where $R(T)$ is the reflectivity of the silicon surface. The generation of heat can be calculated as

$$A(z) = -\frac{dI}{dz} = \alpha(1 - R)I_0\, e^{-\alpha z}\ \text{W}/\text{cm}^3 \qquad (5.31)$$

If diffusion is negligible during the pulse (adiabatic limit), the temperature distribution is given by

$$T(z) = \frac{A(z)\tau}{\rho C} \qquad (5.32)$$

In this limit, $l = (k\tau)^{1/2} \ll \alpha^{-1}$, the temperature distribution in the wafer is determined by the initial distribution of the energy from the pulsed laser beam. Here one has to remember that the reflectivity is a function of the temperature.[51]

5.3.2.2. Diffusion Limit

If $l = (k\tau)^{1/2} \gg \alpha^{-1}$, the temperature distribution in the silicon is determined by the heat diffusion into the bulk. In this case the temperature distribution is given as a solution of Eq. (5.29) in the following form[51]:

$$T(z, t) = \frac{2(1 - R)I_0}{K}\left[\left(\frac{kt}{\pi}\right)^{1/2} \exp\left(\frac{-z^2}{4kt}\right)\right.$$
$$\left. - \frac{z}{2}\, \text{erfc}\left(\frac{z}{2(kt)^{1/2}}\right)\right] \qquad (5.33)$$

and the temperature rise at the surface, $z = 0$, is

$$T(0, \tau) = \frac{2(1 - R)I_0}{K}\left(\frac{k\tau}{\pi}\right)^{1/2} \qquad (5.34)$$

In this diffusion limit the surface temperature rise depends on the absorbed flux $(1 - R)I_0$ and is independent of the absorption coefficient α.

Figure 5.28 shows the measured dependence of the threshold ($T = 1683°\text{K}$ at the surface) incident power density I_0^T on the absorption coefficient α and pulse duration τ. Figure 5.28 shows that in the adiabatic region

$$I_0^T \approx \frac{1}{\alpha}$$

Using Eq. (5.31), the energy required to elevate a surface layer with a thick-

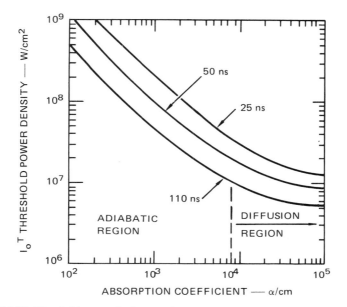

FIGURE 5.28. Threshold power destiny as a function of the coefficient $\alpha(\lambda)$. Source: ref. 51.

ness δz to the melting point T_M during the pulse duration τ can be obtained from

$$A\tau = \tau\alpha(1 - R)I_0^T \, \delta z = C\rho \, \delta z T_M$$

so the threshold incident power density is given as

$$I_0^T = \frac{C\rho T_M}{(1 - R)\tau} \frac{1}{\alpha} \tag{5.35}$$

and in diffusion limited case

$$I_0^T = \frac{KT_M}{2(1 - R)}\left(\frac{\pi}{k\tau}\right)^{1/2} \tag{5.36}$$

Equation (5.36) is independent of α, as can be seen from Figure 5.28 if $\alpha \geq 3 \times 10^4$. The threshold energy also can be calculated for the two limits, using Eqs. (5.35) and (5.36), as

$$E^T = I_0^T\tau = \frac{C\rho T_M}{(1 - R)} \frac{1}{\alpha} \tag{5.37}$$

for the adiabatic limit and

$$E^T = I_0^T \tau = \frac{kT_M}{2(1 - R)} \left(\frac{\pi}{k} \right)^{1/2} (\tau)^{1/2} \tag{5.38}$$

for the diffusion limit. Here less energy is required for shorter pulses because the absorbed energy does not diffuse far into the silicon and the surface can reach the melting point with less material being heated. The results of this one-dimensional thermal diffusion model[51] are in reasonable agreement with experiments over a wide range of conditions for λ, α, and τ.

The application of lasers has been extended beyond laser annealing for material processing in semiconductors. The most important areas are crystalline conversion, laser-assisted diffusion, and graphoepitaxy.

5.3.3. Crystalline Conversion

The conversion of amorphous or polycrystalline materials to single-crystal or large-grain structures can be accomplished by laser processing without affecting the underlying characteristics of the substrates. In polycrystalline materials grain boundaries act as barriers to the movement of charge carriers across junctions. Impurities may diffuse more rapidly along grain boundaries than they do in single crystals, catastrophically shorting out a junction. After conversion the resistivity is reduced by a factor of about 3. Applications of this crystalline conversion are found in the formation of interconnect structures on sophisticated IC devices and deposited polycrystalline Si on recrystallized metallurgical substrates.[61]

Laser radiation also can be used advantageously to alter the surface characteristics of a thin layer of polysilicon (5000 Å) by localized transient melting of the surface without producing deleterious heating effects at either interface of an underlying silicon dioxide layer on a silicon crystal substrate. The growth of high-quality oxides over polysilicon has long been a major problem in double-layer polysilicon structures, such as floating-gate memory devices (FAMOS) and charge-coupled devices (CCD).

5.3.4. Laser-Assisted Diffusion

Both n-type and p-type junctions can be obtained by the laser-assisted diffusion of dopants, with potentially higher activation densities and sharper profiles than those created by normal solubility techniques. In application, the dopant (such as As) may be spun onto the surface of the base material (such as Si) and irradiated by a laser at an energy density of about 10 J/cm² at a 1.06-μm wavelength, or 0.7 J/cm² at a 0.53-μm wavelength, providing sufficient mobility to allow the dopant ions to migrate to substitutional lattice sites at a desired depth. The resulting profile is sharp and an essentially uniform distribution of dopant is achieved in the melt.[63]

5.3.5. Graphoepitaxy

Graphoepitaxy[69] (see Section 1.4.3) of silicon offers the possibility of forming single-crystal films over amorphous artificial microstructures. This film-formation method yields polycrystalline films on smooth amorphous substrates, and oriented films if the film formation is over a surface relief structure. It has been reported[64] that laser crystallization of amorphous silicon over a surface grating in fused silicon also yields a polycrystalline film with grains several tens of micrometers in diameter. These new techniques could lead to new combinations of substrates and films and new three-dimensionally integrated devices.

5.4. Electron-Beam-Assisted Processes
5.4.1. Electron-Beam Annealing

Pulsed electron-beam processing for annealing ion-implantation damage offers advantages over laser annealing in that it is independent of surface optical properties [$\alpha(T, \lambda)$, $R(T, \lambda)$] and it can irradiate large areas. These properties can be utilized for the low-cost automated manufacture of silicon solar cells[65] and the large-scale annealing of integrated circuits. The electron-beam pulse-annealed wafers also show better electrical recovery than the furnace-annealed wafers. The furnace-annealed implant regions typically exhibit high residual defect densities, while the electron-beam pulsed-annealed implanted layer is found to be free of dislocations.[66]

Pulsed electron beams can also anneal through narrow (5–50 μm) windows in oxide films. The insulating coating does not affect electron-beam parameters at moderate doses. However, at the very high processing dose rates (10^{15} rads/s) even the best insulators become conducting. The trapped charges can be dissipated with a subsequent low-temperature (500°C) processing step.

Finally, focused electron beams also can be used for local annealing.[67] The principle of this new experimental technique is that areas of the ion-

TABLE 6. Comparison of Electron- and Laser-Beam Processing

Characteristic	Electron Beams	Lasers
Pulse width	10–200 ns	10–130 ns
Absorption and reflection dependence	Material density, electron-beam energy	Wavelength, temperature, crystal structure, doping level, dopant, surface preparation
Beam control	Electric and magnetic fields	Optics
Diameter of anneal area	To 76 mm	30 μm–20 mm
Uniformity		
Macroscopic	$\pm 5\%$	Gaussian beam profile
Microscopic	Improves by self-interaction	Hot spots, diffraction patterns
Energy flux	≥ 1 J/cm^2	1–10 J/cm^2
Processing depth	Controllable through electron energy spectrum	Maximum melt depth ≥ 1 μm because of damage mechanisms
Environment	Vacuum	Air or vacuum
Possible limitations	Residual charge (depends upon subsequent processing); charge carried across surface of insulators through radiation-induced conductivity	Patterned oxide surface films cannot be pulsed (interface pattern); perturbations in beam grow in terms of heating; and surface is not flat after pulsing

implanted silicon chip are locally annealed to form conducting regions while the unannealed regions remain of high resistivity and form the isolation between adjacent conducting areas. Using this localized annealing technique, diode arrays can be fabricated to make integrated devices such as vidicon targets and optical sensors. To summarize, one can say that laser- and electron-beam-annealing techniques complement ion implantation, especially in the area of submicron device fabrication. Table 6 shows a comparison[66] of pulsed electron-beam and pulsed laser processing.

5.4.2. Thermal Analysis of Electron-Beam Annealing

Electron-beam annealing models are similar in mathematical structure to laser models. But in the case of electron-beam heating the source is a volume source[68] and not a surface source, as it is for laser heating. This is because the electron penetration is typically much larger than the thickness of the layer to be annealed. The temperature distribution as the function of depth (z) is given as a solution of the diffusion equation in the following form [see Eq. (5.33)] [68].

$$T(z, t) = \frac{I_0}{K}\left\{ \left(\frac{kt}{\pi}\right)^{1/2} \left[\exp\left(\frac{-(z - z_0)^2}{4kt}\right) + \exp\left(-\frac{(z + z_0)^2}{4kt}\right) \right] \right.$$
$$\left. - \frac{|z - z_0|}{2} \operatorname{erfc} \frac{|z - z_0|}{2(kt)^{1/2}} - \frac{|z + z_0|}{2} \operatorname{erfc} \frac{|z + z_0|}{2(kt)^{1/2}} \right\}$$

$$(5.39)$$

where the plane source is at a depth z_0 and the decrease by a factor of 2 takes place because the heat is now conducted away in both z directions. Because it is assumed that no heat is lost from the boundary ($z = 0$) through convection and radiation, an image source of equal sign and strength I_0 is added, located at $z = -z_0$. Using this image method, temperature profiles can be calculated for stationary and moving sources.[68]

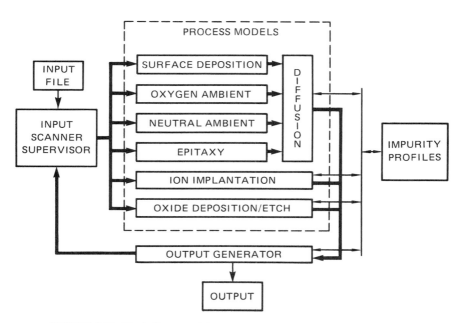

FIGURE 5.29. Block diagram of the SUPREM process simulator. Source: ref. 69.

5.5. Computer Simulation of Fabrication Processes

Earlier in this book we have shown how a set of geometrical patterns can be transformed into an integrated system by the use of simple physical fabrication processes. This integrated system consists of a number of conducting and insulating layers with dimensions in the micron and submicron domains. In planar technology each step is carefully devised to produce features with the minimum possible deviation from the ideal device behavior. The complex fabrication sequence requires accurate physical and computer models for process structure[69] and system modeling.[70] The Stanford University Process Engineering Models (SUPREM) program is a computer simulation capable of simulating most of the IC fabrication steps.

The fabrication-step simulation is based on the process models of oxidation/drive-in, predeposition (gaseous or solid), epitaxial growth, ion implantation, etching, and oxide deposition. These various process models are implemented in SUPREM as subprograms, each consisting of a number of subroutines. Figure 5.29 shows the block diagram of the SUPREM process simulator.[69] The input specifications have been designed as process run sheet data and documentation, which includes a series of process steps. The output of the program consists of the one-dimensional profiles of all dopants present in the silicon and Si–SiO$_2$ materials.

6

Submicron Microscopy and Microprobes

6.1. Introduction

Having built a device consisting of elements made to submicron tolerances, it is often necessary to examine it to see whether the device was built as specified and whether the components are of the required materials with the desired physical properties.

In the period 1830–1880 optical microscopes using light in the visible region were brought to a remarkable level of perfection—close to that dictated by the wavelength of light. We assume that the reader is familiar with the principles of optical microscopy and will simply observe that in general we are mainly concerned with a few important aspects, namely,

- Field of view
- Magnification (M)
- Resolving power (δ)
- Depth of focus (D)

The theory of the compound microscope gives

$$\delta = \frac{\lambda}{\text{NA}} \qquad (6.1)$$

where NA is the numerical aperture of the objective lens and

$$\text{NA} = \frac{n \times \text{radius of objective lens}}{\text{focal length}}$$

Here n is the refractive index of the material between the object and the lens. For the best modern objective lenses NA may be as great as 0.95 in air, and by filling the space between the object and the objective lens with oil, the NA may be increased to as much as 1.5. Since the useful range of the visible spectrum is only 0.4–0.7 μm, even with the best optical microscopes we cannot observe details much smaller than 0.3 μm (or 3000 Å).

The depth of focus is the distance, measured along the optical axis, in which the resolving power is unaffected by defocusing, and it is given by

$$D = \frac{\delta}{\sin \alpha} \qquad (6.2)$$

where 2α is the angle of divergence of the imaging rays. Since $\sin \alpha \cong 1$ in a high-powered objective, we see that the depth of field in optical microscopes is approximately equal to the resolution.

Clearly, optical microscopes are barely adequate for viewing device elements with tolerances much less than a micron. But as was indicated in Chapter 2, electrons obey optical laws similarly to photons, and for energies above 100 eV they have wavelengths much smaller than visible photons. Electron lenses have not approached the same degree of perfection as optical lenses and are largely limited to paraxial rays (or narrow angles of 10^{-2} rad or less) by the spherical aberration of electron lenses.[1] Nevertheless, resolving powers to 10 Å are routinely obtained, and with certain types of electron microscopes 5 Å can be obtained with the greatest care. Even atoms ($\delta < 3$ Å) can be resolved under special circumstances.[2,3] Although the angle of divergence usable by electron lenses is small, this is compensated to some extent by the fact that their depth of field D [Eq. (6.2)] is correspondingly large, and the low resultant particle flux is compensated by the high energy per particle, which makes them easy to detect.

Another fundamental limit to microscopy lies in the radiation damage caused by the illuminating particle.[4] The energy required to break the chemical bonds that hold an atom to its position within the crystal or molecular structure of which it forms a part lies in the range of 1 to 10 eV. Thus, if an inertial energy substantially greater than this is transferred to the atom by the irradiating particle, the valence bonds can be broken and the atom displaced from its original position. Even relatively low-energy ultraviolet light can cause irreversible changes in the bonding of polymers, and although electrons have a small mass for momentum transfer in elastic collisions with atoms, their initial energy is usually sufficiently high (10^4–10^6 eV) so that energies of several electron volts can be transferred to the atomic particle in an elastic collision. Energy transfer may also take place by way of secondary events following inelastic collisions.

Despite these limitations, electron microscopes have emerged as the principal tool for viewing devices with submicron tolerances. Although there is a necessity for putting the sample in a vacuum chamber, commercial electron microscopes are not difficult to use and can be supplemented with additional analytical capabilities.

Electron microscopes can be divided into three categories, namely transmission, reflection, and emission, and they can be utilized in two modes, namely, projection and flying spot (raster scan) modes.

In the flying spot mode the beam is deflected by electrostatic or electromagnetic systems (see Section 2.5.6) so that the focused spot on the specimen describes a series of parallel lines spaced a spot width apart. This is the familiar raster scan seen on television screens. The amplitude of the line, the spot size, the number of lines, and the writing rate can all be varied. A detector with a wide field of view is used to measure the intensity of a chosen radiation, such as electrons or photons, which originates only at the illuminated spot. The intensity of a synchronous CRT display is modulated by a signal that is a suitable function of the radiation intensity measured by the detector, thus forming a viewable image. The magnification is given by the linewidth on the display divided by the linewidth on the specimen.

Each of the six alternatives has its special area of usefulness, but the scanning electron microscope (SEM), which utilizes (secondary electron) emission in the raster scanning mode, has found the most general usage for microdevice viewing, despite its relatively poor resolving power (30–100 Å). This has been because of its extraordinary ability to record the surface structure of three-dimensional objects and its ability to view large areas at low magnifications and "zoom in" on small features of interest at high magnifications. Furthermore, the energy spectrum of the secondaries can be used to obtain information about the composition of the surface layers, and X-ray fluorescence can be used to obtain information about the bulk composition.

The optimum useful magnification of any microscope (M_0) is that which will enlarge the smallest resolvable image detail in the microscope (δ) to a size resolvable by the unaided eye (0.2 mm). That is, for lengths measured in meters

$$M_0 = \frac{2 \times 10^{-4}}{\delta} \tag{6.3}$$

For the optical microscope

$$M_0 = 1000 \qquad (\delta = 2 \times 10^{-7} \text{ m})$$

for the electron microscope

$$M_0 = 100,000 \qquad (\delta = 2 \times 10^{-9} \text{ m})$$

and to view atoms

$$M_0 = 1,000,000 \qquad (\delta = 2 \times 10^{-10} \text{ m})$$

Magnifications smaller than these can be utilized by exposing a photographic surface at a magnification that is comfortably resolved by silver halide film ($\delta_f = 10^{-5}$ m) and then optically projecting or enlarging the developed image to obtain the total magnification necessary for viewing.

Ion microscopes have also been developed using the same six alternatives producing submicron resolutions. However, the field-ion emission microscope as developed by Müller and co-workers[5] is the most remarkable in its clear resolution of arrays of atoms on the surface of metal crystals of sufficiently small radius.

X-ray microscopes rely on simple projection enlargement. A highly focused electron beam on thin metallic foil can produce a tiny X-ray source. The main limitation to date is in the X-ray intensity, and for this reason the radiographs appear to have a practical resolution limit of about 0.1 μm, far less than their theoretical limit and marginally better than for optical microscopes.

In the following sections we will describe those microscopes currently of interest for examining structures with tolerances in the range of 0.5 μm to atomic dimensions (1–3 Å). We will also cover the various types of "microprobes" that are used for the chemical analysis of surfaces and films.

6.2. Transmission Electron Microscopes

The transmission electron microscope (TEM) (see Figure 6.1)[6,7] is essentially the electron analog of the optical compound microscope. TEMs were developed in the 1930s, using the technologies gained in building CRT displays. In concept, a thin specimen is uniformly irradiated on one side by an electron beam of a given energy formed by an illuminating system. On the

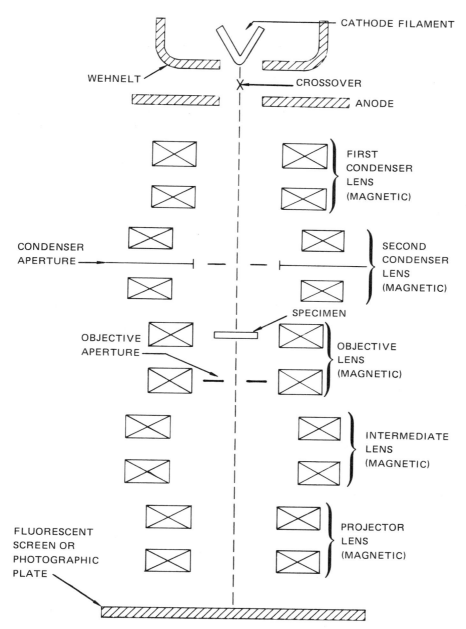

FIGURE 6.1. Schematic representation of a transmission electron microscope (TEM). Source: ref. 6.

other side a powerful objective lens images the plane containing the specimen. The objective lens is followed by a projection system that displays the final magnified image of the specimen on a fluorescent screen to a size that can be comfortably viewed by the eye. The fluorescent screen can be replaced by a photographic film or plate to record the image.

The specimen must be thinner than the mean free path for elastic collisions of the electrons illuminating it, and in any case it must be as thin as possible, otherwise a vast amount of overlapping detail appears in the image because of the high resolving power and great depth of focus of the electron microscope. As a rule of thunb, the thickness should not exceed 10 times the resolving power required. Ultramicrotomes have been developed for biological specimens that can prepare sections from 1000–100 Å thick, and microetching techniques have been developed to obtain films of metals, alloys, and semiconductors in the same range, so that the full resolution capabilities of the microscope can be utilized.

Contrast in the magnified image is obtained because some electrons traveling through the specimen collide with atomic particles, resulting in scattering or energy loss, which prevents their being imaged by the objective lens. To collect the scattered electrons an aperture stop is placed in the image focal plane of the objective lens, thus preventing them from forming a background fog on the fluorescent screen. This aperture also serves the function of reducing the spherical aberration of the objective lens, and its size is optimally reduced to the point where the image blurring due to spherical aberration is equal to that due to diffraction. Further reduction would increase the diffraction blurring and would therefore not be useful. The higher the electron energy, the smaller its wavelength and the smaller the aperture stop that can be used. At the usual working voltages of 50–100 kV the optimum aperture angle turns out to be $\frac{1}{2}°$, as compared with over 80° for optical microscopes. For a focal length of the objective lens of 2 mm this corresponds to an aperture diameter of 40 μm.

To view the image at the fluorescent screen or to record the image on a photographic film in the same plane in a reasonable time, the illuminating electron brightness at the specimen must be as intense as possible. The illuminating system is designed with this and other features in mind. It usually consists of an electrostatic triode gun (see Chapter 2) followed by two magnetic condenser lenses. A crossover is formed between the cathode and the anode of the triode gun, with a divergence after leaving the anode aperture of about 10^{-2} rad ($\frac{1}{2}°$). The first condenser lens, which is always operated at full strength, forms a demagnified image of the crossover inside its lower pole piece. This image is magnified by the second condenser lens by a small factor in the range 1–2 to form an illuminating spot at the specimen. The diameter of the spot illuminating the specimen is controlled by varying the strength of the second condenser lens and need not be larger than the area to be viewed. The beam current is determined by the size of the condenser aperture, the voltages on the gun electrodes, and ultimately by the maximum emission brightness of which the cathode is capable. Typically, the first condenser lens demagnifies the crossover by a factor of 10–15. The diameter of the illuminating spot on the specimen can be reduced to 1 μm, using standard V-shaped filaments, and it can be made even smaller with point electron sources if required.

The objective lens has a focal length that can be made as small as 1 mm, and the specimen is positioned essentially at the focal point. The division of magnification between the objective, intermediate, and projection lenses is approximately 100 : 20 : 100, respectively, for a magnification of 200,000 times. The function of the intermediate lens is to shorten the microscope column length over that required for a single projection lens, given the difficulty of making magnetic lenses with focal lengths any smaller than 1 mm.

The condensing and magnifying systems are made using magnetic lenses (see Section 2.5.2), because of the smaller aberrations obtained in these lenses in comparison with electrostatic lenses. In TEMs the axial positions of the specimen and all the lenses are fixed. Focusing and magnification changes are accomplished by simply varying the focal lengths of the objective, intermediate, and projection lenses by changing the currents through their magnetizing coils. The stability of these currents and of the gun voltages is therefore of prime importance in obtaining the best performance. In addition, the alignment of the optical axes of the lenses and apertures must be maintained to a high degree of precision.

For viewing living biological matter a greater range of electrons is required to produce greater penetration and less radiation damage. A 100-keV beam will not penetrate even the smallest living organism sufficiently to produce a good image. Hence microscopes utilizing up to 3-MeV beams have been developed for this purpose. While the TEM can be made to resolve atoms under special circumstances (and improvements are continually being made), its range of general usefulness can be considered at this time to take over from optical microscopes for resolutions below 1 μm–10 Å for specimens that can be obtained in the form of very thin films.

Surface structures may be viewed with the TEM by the techniques of shadowing in the case of a thin sample, and by replicating in the case of a thick sample. In shadowing, an electron-dense material is evaporated on the surface at a shallow angle so that it condenses thicker on one side than the other, as shown in Figure 6.2(a). An enhanced contrast is then obtained in normal viewing. In replicating [Figure 6.2(b)], a thin layer of coating material is deposited on the surface. The substrate is then removed, usually by dissolving it, leaving the replica behind. Coating materials may be organic, evaporated inorganic materials, electrolytically deposited metals, or oxides of the specimen itself. Replicas formed by these techniques are usually of constant thickness (in the range 100–1100 Å) and although they accurately follow the contours of the surface, replicas are usually "shadowed" before viewing in the electron microscope.

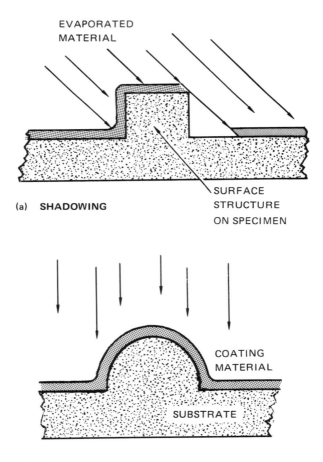

EVAPORATED
MATERIAL

SURFACE
STRUCTURE
ON SPECIMEN

(a) **SHADOWING**

COATING
MATERIAL

SUBSTRATE

(b) **REPLICATING**

FIGURE 6.2. Methods of viewing surface structures with the TEM.

6.3. Scanning Electron Microscopes

The SEM (see Figure 6.3) requires an illumination system in which the spot at the specimen can be made as small as possible and can be deflected to raster over the surface.[7,8] In its conventional mode of operation a fraction of the secondary electrons generated by the primary beam are collected and accelerated onto a cathodoluminescent surface situated at one end of a glass rod, which acts as a "light pipe." At the other end of the light pipe, usually external to the vacuum, is placed a photomultiplier tube, which is essentially a low-noise, high-sensitivity, high-gain, high-speed amplifier. The output from the multiplier is used to modulate the intensity of the display CRT. The display is a map of the number of electrons collected by the detection system at each point of the scan. This depends not only on the variation of the secondary electron emisson coefficient over the surface but also on the solid angle subtended to the collector system of the particular area on which the spot is impinging. Thus the topography of even a surface of uniform secondary-electron coefficient is effectively displayed with an astonishing three-dimensional appearance when the specimen is tilted slightly from the plane of the normal to the optical axis of the system. Because the spot size does not change appreciably over a few microns, the depth of field is large, making it especially useful for viewing microcircuits and devices, not to mention insects, cells, and fossils. The ability of SEMs to view surface structures, almost completely lost in optical microscopy (because of the small depth of field) and transmission microscopy (due to the thin sections used), has proved to be of great value, even at low magnifications.

The key electron-optical element in the SEM is the illuminating system, which differs somewhat from that of the TEM in that as small a spot as possible is required and lower beam energies in the spot are necessary (5–25 kV). This latter requirement is because high-energy secondaries generated deep in the specimen can emerge from the surface a distance away from the spot where the primary beam was incident (see Section 2.6.2), resulting in a loss of resolution. The electron gun produces a crossover close to the anode. The first weak condensing lens contains an aperture and, combined with the second strong lens, enables a demagnified image of the crossover to be focused down to the limit set by spherical aberration in the gun optics, the energy distribution of the electrons leaving the cathode, and the Boersch effect. Using point-field-emission sources or point-thermionic sources of lanthanum hexaboride, spots down to 10 Å in diameter can be obtained with sufficient brightness.

If very thin samples and higher beam energies are used, the microscope can be used in the transmission mode (STEM). In this case the elastically scat-

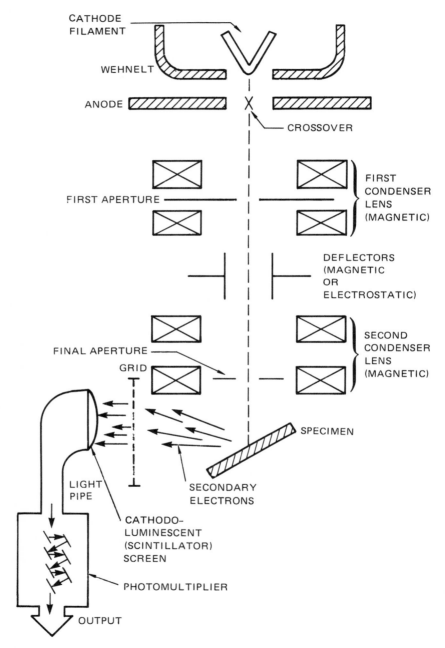

FIGURE 6.3. Schematic representation of a scanning electron microscope (SEM). Source: ref. 7.

tered electrons are collected on an annular ring scintillator formed at the end of a glass tube light pipe, and the unscattered primary beam and small-angle inelastically scattered electrons are lost down the center. In this way resolutions of 3–5 Å have been obtained and heavy single atoms have been viewed.[2] Alternatively, an energy spectrum of the entire transmitted beam can be obtained with an electron energy analyzer. Electrons that have suffered inelastic collisions lose a certain amount of energy characteristic of the atomic species involved in the collision, so that an analysis of the material through which the electrons have passed can be obtained. This is called electron loss spectroscopy (ELS).[9]

In the conventional SEM we utilize all the types of secondary electrons. If, however, we first pass them through an energy analyzer, we can select only a particular energy range for image modulation. Now an electronic vacancy in an atom, created by collision of the primary beam, can be filled by an orbital electron from a less tightly bound state, with the simultaneous emission of a second electron from another less tightly bound state. The emitted electron has an energy characteristic of the atom concerned. This is the Auger effect (see Section 2.6.2). Since Auger electrons have relatively low energies, they can only escape from the surface layer of atoms. Thus, by selecting an Auger peak of a given element, we can obtain a map of that element over the surface. Because of the low intensity of the Auger peaks, it is necessary to use slow scanning times, a large scanning spot, and an efficient electron energy analyzer to obtain a signal of measurable amplitude above the noise in a reasonable time. Nevertheless, fractional micron resolutions can be obtained. Because the Auger electrons only arise from the surface atomic layer, analysis can only be carried out in ultrahigh vacuums after the surface contaminants have been removed, usually by argon sputtering. This device is known as a scanning auger electron microscope (SAEM).

When the scanning spot interacts with the surface, it also generates X rays characteristic of the material being bombarded (X-ray fluorescence; see Section 2.6.2). When electronic vacancies are filled by electrons from less tightly bound states, the emission of photons characteristic of the atom concerned occurs. Since X-ray photons can escape from much deeper in the material than electrons of the same energy, they are more representative of the bulk composition, and hence poorer vacuums than necessary for Auger spectroscopy can be used. The X rays are detected by solid-state counters that count individual photons and measure their energy. If only photons of a given energy characteristic of a given material are counted, a map of that material over the scanned area is obtained. Intensity considerations again limit the resolution to barely submicron dimensions. X-ray crystal spectrometers operating on the principle of Bragg reflections are used instead of counters when more precision is required for lower-energy photons originating from low-atomic-number materials. This device is known as the scanning X-ray microprobe.

Of course, in these latter variations of the SEM we could hold the beam at a single point of interest on the specimen, first identified by using it in the standard detection mode, to measure the composition of a particular point of interest. Such instruments are sometimes called microprobes.

If the scanning beam strikes the surface of a semiconductor device, currents and voltages can be induced that alter the paths of the secondary electrons. Such effects are used to analyze the mechanism of operation and locate faults in microcircuits. In the voltage-contrast technique[10-13] the change of secondary-electron signal gives a measure of the local potentials on the surface of an IC. In voltage-contrast SEM micrographs positive-biased components of an IC appear dark compared to areas of lower potential. This contrast results from the existence of retarding fields above positive-biased specimen areas, which cause a decrease in the secondary-electron signal.

However, this voltage-contrast measurement gives only qualitative results in determining the local surface voltage because of the following restrictions:

- The retarding fields close to the surface of ICs depend not only on the geometry and voltages of the spot, but also on the voltage distribution of the whole specimen surface.
- A nonlinearity exists between the signal and the specimen voltage, caused by the energy distribution of the secondary electrons.
- The voltage contrast in SEM micrographs is superimposed on the material and topography contrast.

Electron-beam-induced current technique (EBIC) in semiconductors is another important family of measurements for microelectronic devices.[14-16] In this technique a high-energy (E_B) electron beam is focused on a small area of a microcircuit and can penetrate through several layers of the device structure. In the semiconducting regions electron–hole pairs are generated by incident electrons of energy E_B of number $E_B/3E_G$ per incident electron, where E_G is the band gap of the semiconductor. With appropriate external voltages applied to the microcircuit, the currents generated by the newly created charge carriers are measured. This technique permits the observation and study of inversion and depletion layers, breakdowns, carrier multiplication effects, and the phenomena of impact ionization. It also provides a method for determining carrier diffusion lengths, evaluating doping profiles near a junction, and measuring surface recombination velocities.

The combined application of EBIC with other techniques (X-ray, Auger) permits analysis of the chemical nature of the precipitates that generate the localized breakdown. It can be used jointly with cathodoluminescence to relate light-emitting regions to microplasmas or to evaluate the doping distribution of *p–n* junctions. Furthermore, the EBIC technique allows the study of dislocations and other defects in semiconductors.

The SEM is the single most useful instrument for microcircuit analysis. Table 1 lists the different kinds of operational modes of SEMs for the characterization of semiconductor devices, showing the various physical effects, obtainable information, and expected spatial resolutions.

6.4. Electron-Reflection Microscopes

In electron-reflection microscopes the surface to be viewed is held at a potential such that all or a large fraction of the incident electrons do not physically strike the sample surface. Those that do strike the surface do so with low energies (in a range up to only a few volts) so that the surface is barely perturbed by the illuminating electron beam. Reflection principles can be utilized in both the projection and flying-spot modes.

The projection device[17,7] is called the electron-mirror microscope (EMM), and the principle of its operation is shown in Figure 6.4. A collimated electron beam is directed normal to the sample surface. As electrons pass through the final lens aperture, they are rapidly decelerated, with a turning point determined by the potential of the sample surface with respect to the cathode and the field strength at the sample surface. After turning, the electrons are accelerated back through the lens and an enlarged image is projected onto a cathodoluminescent screen. Additional magnification may be obtained by separating the outgoing beam from the incoming beam with a weak magnetic field and inserting an additional magnifying lens in the path of the outgoing beam.

Contrast in the outgoing beam is obtained from several sources, namely,

- Surface topography
- Variations in electric potential over the surface
- Variations in magnetic fields over the surface

Any deviation from flatness of the sample surface will influence the shape of the equipotentials near the surface. Thus lateral components of velocity will be given to reflected electrons by a bump on the surface, depending on the detailed shape of the turning equipotential. These electrons will be deflected and give rise to a decrease in the intensity of the enlarged image corresponding

TABLE 1. *Operational Modes of SEM for the Characterization of Semiconductor Devices*

Mode of Operation	Physical Effect	Information and Applications	Spatial Resolution
Secondary emission (SE)	Dependence of SE on angle of electron-beam incidence and materials	Pattern structure, magnetic domain imaging	0.01 μm
Backscattered emission (BE)	Dependence of backscattered coefficient on the angle of incidence	Surface structure, measure of mean atomic number, chemical crystal orientation	0.01–1 μm
	Material dependence	Material contrast, depth information	
Current mapping (SE) + (BE)	Current change in the probe is due to the changes in SE	Topological contrast, material contrast, and BE on the surface	0.1–1 μm
Channeling	Dependence of SE and BE on crystal orientation	Crystal structure	1–10 μm
Voltage contrast	Influence of probe potential and probe–detector geometry on SE signal	Mapping of potential distribution in IC surface structures, location and height of barriers, resistivity changes	1–10 μm
Induced current (EBIC)	Incident electrons create excess current, which induce current in an external circuit	Detects built-in barriers in inhomogeneous semiconductors, diffusion length determination, depth and thickness of $p-n$ junctions, defect location, resistivity	0.1 μm
Cathodoluminescence (CL)	Emission of photons (infrared to uv) from the sample after electron-beam irradiation	Characteristics of local band gaps, doping distribution, relaxation times for radiative process	0.1 μm
X-ray analysis	Emission of characteristic radiation	Qualitative and quantitative elemental analysis	1–10 μm
Auger electrons	Measurement of Auger electrons (100–1000 eV) with an energy spectrometer; displacement of Auger peaks resulting from local fields	Material analysis (light elements) element mapping, local measurement of potential	Depth 2–10 Å 0.1 μm
Transmission	Elastic scattering	Atomic movements, nucleation studies, energy-loss spectroscopy, nanolithography	10–100 Å

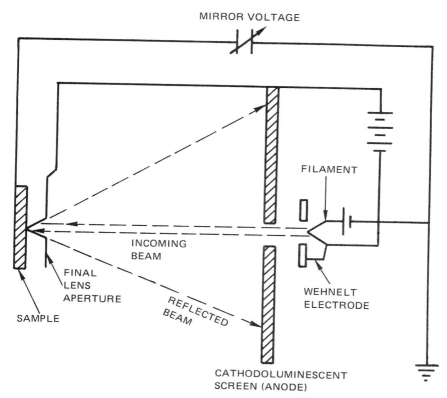

MIRROR VOLTAGE

FILAMENT

INCOMING
BEAM

FINAL
LENS
APERTURE

REFLECTED
BEAM

WEHNELT
ELECTRODE

SAMPLE

CATHODOLUMINESCENT
SCREEN (ANODE)

FIGURE 6.4. Principle of the electron-mirror microscope. Source: ref. 17.

to the position and size of the projection. In practice, resolution steps about 24 Å high can be comfortably observed.[17] However lateral resolutions of only about 500 Å are observed, despite the fact that a theoretical limit of about 40 Å is predicted.

When a conductive polycrystalline material is polished, each exposed crystallite surface may exhibit a contact potential difference with its neighboring crystallites. This is because of differing exposed crystalline planes if they are of the same material, or differences in crystal structure if they are of different materials. Contact potential differences in the range of tens to thousands of millivolts are typically obtained. These cause abrupt changes in the equipotentials at the lines of contact analogous to the case for geometric perturbations. Resolutions down to 50 mV have been observed. The voltage applied to the sample may be adjusted so that one set of crystallites reflects the incident electrons before they reach their surfaces, while another set of crystallites is at a potential so as to just collect the incident electrons. This gives rise to a very high contrast. Insulating particles on the surface may reach high negative potentials, as some energetic electrons in the beam are able to reach them. This again gives rise to high contrast. A charge pattern distributed over a flat insulating surface will similarly change the path of the reflected electrons, giving rise to high contrast in the image.

Changes in the local magnetic field over a sample surface will also give rise to perturbations in the paths of the reflected electrons. The problem of estimating the magnetic contrast is more complex, since the radial component of the electron velocity interacts with the normal magnetic field in accordance with the Lorentz equation (see Section 2.5.2). However, the effect is large enough so that information stored on magnetic tapes or disks may be read out with the EMM.

The electric field E in the space between the final lens aperture and the sample is limited by vacuum breakdown and lies typically in the range 10^5–10^6 V/cm. If the sample is held at $-V$ volts with respect to the cathode, then the electrons will turn at a distance $d = V/E$ cm from the sample. This assumes that all electrons leave the lens with the same energy. In fact, an average energy spread δV exists owing to factors discussed in Section 2.2. Thus the turning distance d lies in a range $\delta d = \delta V/E$. For $E = 10^6$ V/cm and $\delta V = 1$ V, then $\delta d = 10^{-6}$ cm $= 100$ Å. The actual step-height resolution is somewhat better than this at the center of illumination (see ref. 17, p. 221).

Reflection principles can also be used in the scanning mode. The device is similar to a conventional scanning electron microscope except that at the spot focal point the beam is rapidly decelerated and reflected before the spot has time to grow appreciably. In one form, called SLEEP (scanning low-energy electron probe),[18] this device is used to investigate the work-function distribution over the surface of thermionic cathodes in various stages of activation and life.

While not usually considered a "microscope," the vidicon television camera tube operates on very similar principles (see Figure 6.5). In this device one surface of a photoconductor is charged by the electron beam to about 10^{11} electronic charges/cm^2. The other surface interfaces with a transparent conductive layer on glass held at ground potential. A light image is focused on the interface, thereby partially discharging the photoconductor and leaving a charge pattern corresponding to the light pattern. The electron beam scans the surface at a potential such that in undischarged areas the beam is completely reflected, but in discharged areas part of the beam is utilized in recharging the surface to its reflection potential and the remainder is reflected. Readout may be accomplished by measuring the displacement current in the transparent conductive layer or by monitoring the current in the return beam, using an electron multiplier. While the resolution of this device is currently quite poor by optical

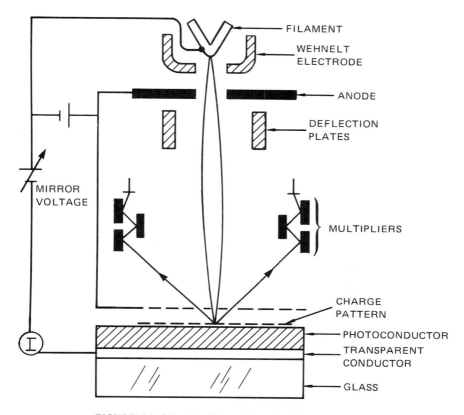

FIGURE 6.5. Principle of the vidicon television camera tube.

microscopy standards, it may be possible to view charge patterns on a submicron scale by using similar principles. Ion implantation could be used to fabricate such submicron photosensitive surfaces (see Section 5.2).

6.5. Electron Emission Microscopes

Under special circumstances a sample surface can be made electronically self-luminous. That is, it emits electrons directly from its surface, such as arises from heating the sample (thermionic emission), applying a strong field to the sample surface (field emission), or by particle bombardment of the surface (secondary emission). Such electrons are usually emitted in random directions, with low energies and with a small energy spread. In the first two cases the electrons can be directly focused into an enlarged image by electron lenses, without the complications of an external source of electrons or other particles. In the third case illumination of the sample surface can, with some ingenuity, be accomplished without interfering with the electron-optical system.

In the TEM, shown in Figure 6.6, [19] the sample surface is made an electrode of the system, forming an immersion lens with the Wehnelt electrode and the anode. A homogeneous electric field E is assumed to exist in the cathode-to-anode region. To improve the resolution an aperture stop is placed in the focal plane of the lens to intercept electrons with high thermal lateral velocities. It has been shown[20] that the resolution of this device, δ, is given by

$$\delta = \frac{\phi}{|E|} \left(\frac{r}{f}\right)^2 \tag{6.4}$$

where f is the focal length of the lens, ϕ is the potential at the focal plane, and r is the radius of the aperture stop.

With $\phi = 40$ kV, $|E| = 10^5$ V/cm, $r = 10$ μm, and $f = 5$ mm, we obtain $\delta = 160$ Å. The aperture cannot be made too small because of loss in intensity in the image, and, of course, need not be made smaller than the electron diffraction limit for the aperture.

The TEM is restricted to materials with a low work function and low volatility that can be heated to emit reasonable current densities. Use of the photoelectric effect vastly increases the range of materials that can be used.[21] Intense ultraviolet sources are focused over the area of interest and the photoelectronic image is projected onto the fluorescent screen. Since the energy distribution of the photoelectrons is typically much larger than that for thermionic electrons, the use of the aperture stop is imperative even for modest resolutions (fractions of a micron). Similarly, the surface can be bombarded with high-energy electrons and only the low-energy secondaries will be imaged. Secondary electrons can also be generated by ion bombardment. This tends to etch the surface by sputtering, producing beautiful patterns for metallography.

The TEM can only be used in the projection mode. However, in principle, both scanning electron and scanning ion microscopes are emission microscopes when they utilize the secondary electrons generated at the surface by the scanning beam. While a scanning photon microscope is conceptually possible, it is not used because of the relatively large minimum size of the focal spot (on the order of 1 μm) and the small depth of focus.

The field emission microscope (FEM), shown in Figure 6.7,[22] must be the simplest microscope ever devised. It consists of a needle of tip radius r a distance R from a cathodoluminescent screen held at a potential V. However,

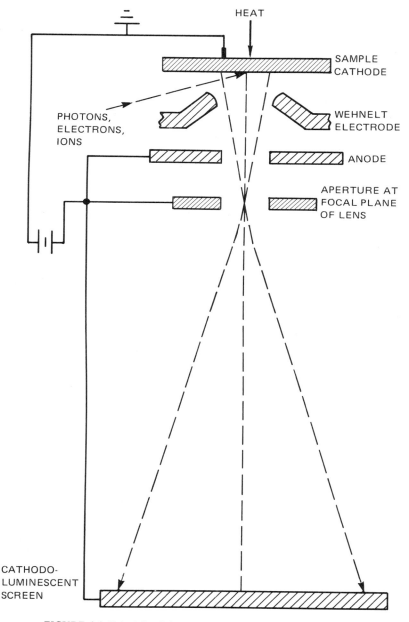

FIGURE 6.6. Principle of the emission microscope. Source: ref. 19.

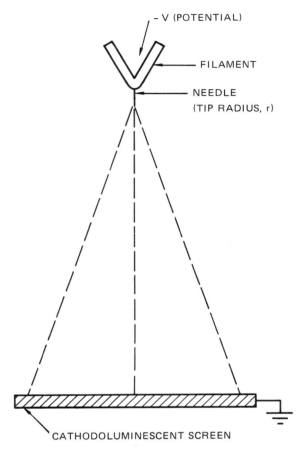

FIGURE 6.7. Principle of the field emission microscope (FEM).

it requires an ultrahigh vacuuum ($p < 10^{-10}$ Torr) to operate for useful periods of time. The field at the tip is given by

$$E = \frac{V}{kr} \tag{6.5}$$

where k is a constant that depends on the shape of the needle and a value of about 7 is generally used. To obtain electrons from the tip in substantial quantities (10^4–10^8 A/cm^2) by field emission, fields in the range 10^7–10^8 V/cm must be applied. The needles are usually annealed by the application of heat and electric fields into single crystals of tip radius of about 1 μm, so voltages of about 10 kV have to be applied. The electrons simply travel on radial paths and the magnification is given by

$$M = \frac{R}{r} \tag{6.6}$$

for $R \sim 10$ cm, $r \sim 1$ μm, and $M \sim 10^5$.

The resolution is dependent on the velocity distribution of the emitted electrons, and the momentum uncertainty and is given by[23]

$$\delta = \left(\frac{2\hbar\tau}{mM}\right)^{1/2} (1 + 2m\tau v_0^2/\hbar M)^{1/2} \tag{6.7}$$

or

$$\delta = 2.72 \left(\frac{r}{\sqrt{V}}\right)^{1/2} \left(1 + 0.222\frac{r}{\sqrt{V}}\right)^{1/2} \text{Å}$$

for r in angstroms and V in volts, where τ is the transit time from tip to screen, v_0 is the average transverse velocity, m is the mass of electron, and the other symbols are as conventionally or previously defined.

Studies of the variation of the work function of single crystal faces and the effects of adsorbed species have been made using this microscope. Resolutions down to 10 Å are conventionally obtained and large adsorbed organic molecules can be made visible.[22] For tip radii below 2000 Å it is difficult to obtain a crystal structure. However, the evidence strongly suggests, in agreement with the theory, that for tip radii below 100 Å in radius[3] atomic resolution can be observed, but the conditions are such that only relatively few (1–20) atoms can be observed in this way owing to the small tip radius. Such small radius tips exist in metallic whiskers formed by various methods including electrolytic etching.

6.6. Field-Ion Microscopes

The field-ion microscope (FIM)[5] is unique in that it is the only microscope to date in which arrays of atoms are routinely observed. The principle, as shown in Figure 6.8, is similar to that of the field electron emission device except that electrons are replaced by positive ions.

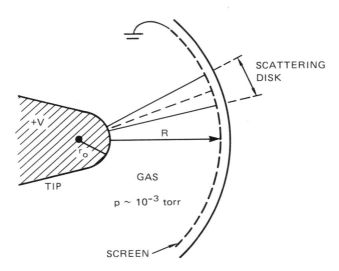

FIGURE 6.8. Schematic diagram of the field-ion microscope.

If an atom or molecule is placed in a field approaching 10^8 V/cm (or 1 V/Å) it first becomes polarized and, as a critical field is exceeded, an electron tunnels out of its orbit to the vacuum region, leaving a positively charged ion. Around a sharp tip, with appropriate voltages applied, there is both a high field and an increasing field gradient as the tip is approached. The action of the field gradient pulls the polarized particle toward the tip and, at a critical distance, emits an electron forming an ion. The electric field now forces the ion away from the tip along a radial path toward the screen. Since the ionization usually takes place within 1 Å or so from the surface, perturbation of the local electric fields by the surface atoms plays an important role, since the position of a molecule when it is ionized is more likely to be close to that of a surface atom than to the space between them. This effect is made stronger by cooling the tip to liquid-hydrogen temperatures. The inert gas helium is preferred, since it is monatomic, chemically inert, and has a low sputtering coefficient, causing less wear on the screen. Pressures of about 10^{-3} Torr are used. The equations governing resolution are the same as for the FEM, except that the mass of the helium ion is 7.4×10^{-3} greater than that of the electron and a substantially smaller resolution is obtained.

The ion current densities reaching the screen are quite small, and to view the image in a reasonable time a "channeltron" multiplier is used. This device consists of a two-dimensional array of tiny capillaries, with an electric field applied parallel to the axis of the capillaries. When an ion strikes a capillary, it generates secondary electrons, which in their passage through the tube are amplified by a chain reaction of collisions with the capillary walls. These electrons emerging from the other side of the capillary are then accelerated to a cathodoluminescent screen, where a bright amplified image is formed.

A problem with the FIM is that the electric fields at the tip necessary for field ionization are close to values that will pull the surface atoms away from their crystals, that is, field evaporation. Many materials field evaporate at lower fields than those at which field ionization occurs and hence cannot be viewed. However, the effect can be used to advantage to identify surface atoms. A short voltage pulse is applied to the system to exceed the evaporation field of some surface atoms. The time these atoms take to reach the screen is then measured so that an estimate of the mass of the particle can be made. This device is called an atom probe[24,25] and has proved to be of value in the study of impurities in metal crystals.

6.7. X-Ray and Ion Microprobes

A number of corresponding methods for chemically analyzing surfaces and films are available that utilize X-ray photons and ions rather than electrons.

A collimated beam of X rays impinging on a surface is more penetrating than an electron beam of the same particle energy and can therefore provide more information concerning the constitution of the material at greater depths, although the depth is still limited by the range of the photoelectrons generated. X rays also produce photoelectrons from the inner shells that carry information concerning the chemical bonding state of the excited atom. Thus analysis of the energy distribution of the electrons emitted from a surface bombarded by X rays is useful for subsurface chemical analysis. Instruments operating on these principles are known as ESCA (electron spectroscopy for chemical analysis) or XPS (X-ray photoelectron spectroscopy) instruments. Soft X-ray sources are used for ESCA, typically using the Al $K\alpha$ (1486.6 eV) or Mg $K\alpha$ (1253.6 eV), and are of quite low intensities compared to electron beams. To read out the data in a reasonable time larger spots must be used than in electron devices and longer integrating times must be used in the detection electron energy analyzers. However, ESCA does not have the noise from the strong secondary background to contend with as does Auger electron spectroscopy (AES), and surface charging is not as severe a problem as with electron bombardment. These properties have made ESCA particularly useful for the analysis of organic and polymeric materials.

Ion beams can now be obtained with relatively high intensities, focused into a small spot on the surface of interest with appropriate lenses and deflected in ways similar to those used for electron beams. At low energies (10^2–10^3 eV) ions are scattered from the surface molecules principally by billiard-ball-type elastic scattering. In ion-scattering spectroscopy (ISS) the energy loss measurements are made on the scattered beam ions at a specific scattering angle. Most of the incident ions are neutralized, particularly if they penetrate the first atomic layer. However, the few percent that are reflected exhibit energy peaks mainly characteristic of elastic scattering at the surface atoms. The incident ion beam is held to a narrow energy range and the reflected intensity is measured as a function of the ratio of the scattered energy to the incident energy. Noble gas ions are mostly used, since they give sharper peaks.

The elastic backscattering of high-energy light ions, usually called Rutherford backscattering spectrometry (RBS), has also proved to be a very useful tool in the study of implantation damage and annealing, the location of impurity atoms in crystal, and the investigation of surfaces and thin films.[26–28]

A well-collimated beam of singly ionized 1–3 MeV helium ions is used. The specimen to be analyzed is placed on a goniometer and exposed to the beam in a target chamber at a pressure of 10^{-6} Torr. Information about the specimen is contained in the energy spectrum of the ions backscattered through an angle of 150°. These ions are detected by a silicon surface barrier detector. The signal is amplified and analyzed, stored, displayed, and recorded.

A silicon sample, for example, exposed to a beam of energy E_0 gives rise to a spectrum consisting of a sharp edge at $0.59 E_0$ corresponding to helium ions elastically scattered by surface silicon atoms and deflected into the detector. The sharp edge is followed by a smoothly varying yield at lower energies. Helium ions not scattered at the surface penetrate into the target, losing energy through inelastic collisions and being scattered elastically in the bulk. They eventually leave the target, to be detected after losing more energy inelastically on the way out. If the stopping power of the target and the geometry of the system are known, one can convert the energy scale into a depth scale.

If the ion beam is incident in a low index direction of a high-quality crystal specimen, it would appear to the beam to be a very open structure consisting of hollow "tubes" or "channels," with lattice atoms forming the walls of the tubes. Channeling minimizes collisions of incident ions with target atoms and reduces the observed backscattered yield for all depths in the crystal. A small surface-damage peak observed only in the channeled spectrum is due to localized surface disorder. Even high-quality crystal samples show this surface-disorder peak. If the crystal has been damaged, for example, by ion implantation, and the displaced atoms occupy sites in the channel areas, the channeled ions suffer elastic collisions with these atoms, resulting in a higher scattered yield.

The ratio of the yield in the channeling and nonchanneling directions is a measure of the quality of single-crystal specimens. For good-quality silicon this ratio, often referred to as K_{min}, is 0.03.

RBS, in conjunction with channeling, can also be used to determine the concentration and position of impurity atoms in the crystal, the technique being more sensitive for heavier elements. This technique has also been applied in a variety of surface and thin-film problems.

In secondary-ion mass spectrometry (SIMS) the ions sputtered from the surface by the primary-ion beam are monitored in a mass spectrometer. Typically, quadrupole mass spectrometers are used at resolutions of 1 amu. The imaging of surface inhomogeneities can be effected by projection images of the secondary ions or by raster scanning. Resolutions of the order of 1 μm have been reported in both cases. The primary beam is also used to remove surface layers for depth profiling, and depth resolutions to 50 Å have been reported.

Figure 6.9 shows the schematic of a SIMS in which secondary ions are extracted from a sample by an electric field and analyzed according to their

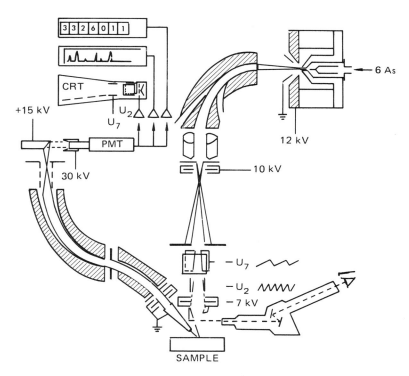

FIGURE 6.9. Schematic diagram of a SIMS system. Source: ref. 27.

mass-to-charge ratio by a mass spectrometer. The SIMS instrument can operate in three modes:

- An ion microprobe may be used, in which local analysis is performed by microfocusing of the primary-ion beam.
- A true image may be formed by geometrical extraction of the secondary ions and aperturing within this image.
- Broad-beam instruments may be used, in which analysis over areas greater than 100 μm is performed instead of microanalysis.

SIMS offers the following capabilities:

- All the chemical elements can be observed, including hydrogen.
- Separate isotopes of an element can be measured, offering the possibility of determining self-diffusion coefficients, isotopic dating, and isotopic labeling.
- The area of analysis can be as small as 1 μm in diameter.
- The secondary ions are emitted close to the surface of the sample.
- The sensitivity of this technique exceeds that of AES or electron probe microanalysis in many but not all situations, with detection limits in the 10-ppm range under favorable conditions.
- The microanalytical part of this technique can be used to prepare ion images (ion microscope) or ion area scans (ion microprobe), which can give pictorial information about the distribution of species of interest.
- Material can be eroded away in a regular and carefully controlled fashion, and a profile in depth can be prepared.

6.8. Laser Scanning

Laser scanning is a new and very versatile way of investigating the inner workings of active semiconductor devices. The scanner does not damage the device and can be used to map dc and high-frequency gain variations in transistors to reveal areas of the device operating in a nonlinear manner, to map temperature within devices, to determine internal logic states in ICs, and to selectively change these states.[29] Figure 6.10 shows the light and signal paths of a dual-laser scanner.

Two lasers are used for a greater measurement flexibility. Visible or near-infrared radiation incident on silicon creates electron–hole pairs with a generation rate that decreases exponentially with distance into the material. The penetration depends on the wavelength of the incident radiation. The visible light from the 0.633-μm laser has a characteristic penetration depth of about 3 μm in silicon. Because most modern silicon devices have their active regions

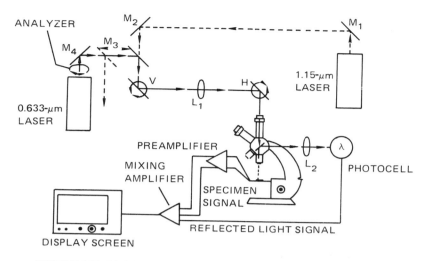

FIGURE 6.10. Light and signal paths of a dual-laser scanner. Source: ref. 29.

within a few micrometers of the surface, the 0.633-μm laser is effective in exciting active regions of such devices. The intensity at the specimen can be varied to produce junction photocurrents over a range from about 10 pA to about 0.1 mA.

Silicon at room temperature is almost transparent to the 1.15-μm infrared radiation from the second laser; the characteristic penetration depth of this radiation is about 1 cm. The infrared laser is used for three classes of measurement:

- Examination of the silicon–header interface
- Device temperature profiling
- Examination of the device through the backside of the silicon chip

Each of these applications makes use of the penetrating nature of the radiation. In the first application the reflected-light circuit is used to look through the silicon wafer and observe irregularities at the silicon–header interface in the "flying-spot microscope" mode. The second application made use of the temperature sensitivity of silicon absorption: A larger signal is produced on the display screen for areas that are warmer than others. Utilizing this sensitivity, one has an electronic technique for the thermal mapping of devices, which appears to have a number of advantages over the more traditional methods. The third application uses the penetrating infrared radiation to photogenerate carriers deep within silicon devices. This capability allows the operation of face-down bonded components, such as beam-lead devices to be examined.

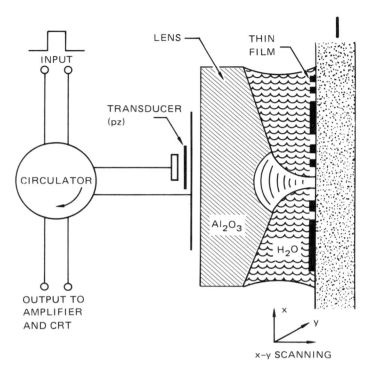

FIGURE 6.11. Arrangement for a reflection scanning acoustic microscope (SAM) system. Source: ref. 22.

6.9. Scanning Acoustic Microscope

The scanning acoustic microscope operates at gigahertz frequencies and has a greater resolution than optical microscopes. It can look beneath the surface of thin-layer structures such as IC chips.[22]

The structure of the acoustic microscope is shown in Figure 6.11. Input signals of up to about 3 GHz are fed into the piezoelectric transducer consisting of a thin coating of zinc oxide on the back of a sapphire (aluminum oxide) acoustic lens. The ultrasonic waves generated by the transducer are focused by the lens, which has a radius of about 100 μm, on the object under examination. The reflected waves containing information about the object are passed back to the transducer, which converts them into electrical signals for display on a CRT. To obtain a higher resolution very high frequencies must be used, but these high frequencies are attenuated by the material.

Focusing on layers beneath the top layer and reflections from these sublayers can provide useful information about the state of the chips, such as whether any corrosion has taken place. The points where the electrodes make contact with the semiconductor layers can also be examined and the optical and acoustic images compared. These operations can be performed without damaging the device, so that a comparison can be made of the performance of different devices that have various defects in the subsurface structure. A correlation between these defects and performances will help to identify the origin and nature of defects in semiconductors.

6.10. Device Characterization by Microscopy

The characterization of devices of small dimensions requires dimensional, chemical, structural, and functional measurements.[30] The general applicability of the microscopy techniques discussed in this chapter are summarized in Table 2.

TABLE 2. *Summary of Microscopy Techniques*

Measurement	Techniques Available	Range of Usage	Remarks
1. Dimensions	Optical	Dimension $>0.3\ \mu m$	Poor depth of field
	TEM	>10 Å	Requires thin samples
	SEM	>30 Å	Good for surface topography
	EEM	>100 Å	
	FEM	>10 Å	Useful only with
	EMM	>25 Å	special preparation
	FIM	>1 Å	
2. Chemical Composition	ESCA	Measures constituents to a depth of 100 μm	1 ppm
	X-ray microprobe	Measures constituents to a depth of 1 μm	100 ppm
	SIMS		1 ppm
	ISS	Measures constituents of surface layer only	100 ppm
	Auger spectroscopy		
	ELS	Measures constituents through sample thickness	1000 ppm
3. Device Operation	SEM (EBIC)		
	SEM (voltage contrast)		See Table 1
	Laser scanning		
	Acoustic scanning	Subsurface layers	

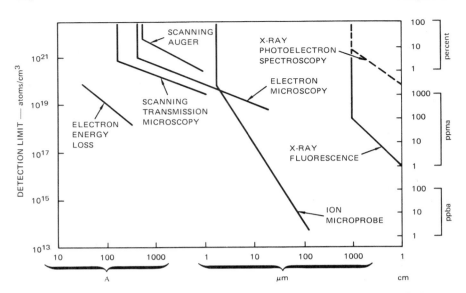

FIGURE 6.12. Effect of diameter of analyzed area on detection limit. Source: ref. 30.

It is apparent that to determine microscale morphology, elemental constituents, molecular species, and lateral and depth distributions, a variety of microanalytical techniques must be employed.[31] It is frequently of great interest to simultaneously determine both the surface and bulk compositions of a sample with submicron resolution. Such requirements are met by combined analytical systems. However, for nanodevice characteriziation it will be necessary to develop new techniques with extremely high signal-to-noise ratios.[32] Better sensitivities at submicron geometries are also needed for chemical analysis. This is apparent from Figure 6.12, which shows that a material must be present in a density greater than 10^{17} atom/cm³ for it to be detected in a pattern with dimensions of under 1000 Å.[30]

Future Directions

7

7.1. Limits and Trends—General

Devices can be made with decreasing linear dimensions until one of two limitations are reached. These are

- Limitations imposed by the physical principles by which the device operates
- Limitations imposed by our ability to fabricate the device to the required dimensions and tolerances

Essentially all the microdevices currently built and explored are based on the bulk properties of materials. That is, the physical property concerned relies on an integrated effect over a large array of atoms: as the dimensions of the array are reduced a critical size is reached less than which the desired physical effect follows different rules or no longer occurs. For example, if we wish to propagate an electromagnetic wave of a given wavelength down a waveguide, there are critical cross-sectional dimensions that cannot be made smaller without cutting off the propagation of waves down the guide.

In Table 1 we list a number of such devices that are of current interest, together with the principal operational characteristic lengths that will limit their further miniaturization.

The main limitations on fabrication using planar technology are in pattern generation and etching (developing) the pattern, since thin-film technology is sufficiently advanced so that the depth dimension can be controlled to any required precision for bulk devices. At the present time electron-beam lithography is capable of attaining linewidths of 3000 Å, with edge sharpnesses of 100 Å. Because of electron scattering in the substrate, further improvements

TABLE 1. *Principal Limiting Factors for Several Microdevices*

Device	Principal Limiting Characteristic Length	Approximate Minimum Dimensions	
		Angstroms	Atoms
MOS Transistor	Channel length determined by breakdown field	2.5×10^3	1000
Bipolar transistor	Base thickness and base doping concentration	2.5×10^3	1000
Magnetic bubble memories	Bubble diameter determined by domain energetics	500	200
Surface acoustic-wave delay lines	Surface wavelength determined by attenuation in surface film	2.5×10^3	1000
Electromagnetic waveguides	Wavelength determined by the photon energy (usually in optical range for integrated optical devices)	4×10^3	1000
Charge storage devices	Number of electrons to be stored on the surface determined by breakdown field and information statistics	1000	300

may not be expected for routine fabrication, although techniques have been evolved to obtain smaller dimensions with electron-beam lithography in special circumstances. Ion-beam lithography, however, can take over where electron beams leave off, and we can reasonably expect that patterning down to 100 Å in linewidth may become routine. Since the ion beam can be used to sputter-etch their own path as the pattern is written, the etching problem is simultaneously brought under control.

Our ability to view devices with electron microscopes has already attained resolutions smaller than the device and fabrication limits, and our ability to analyze the materials of which the devices are made is also approaching similar limits.

It seems clear from the preceding discussions that there is every reason to suppose that the limits for bulk effect devices will be closely approached over the next decade or two. We may also surmise that many new bulk devices will be invented, fabricated, and operated in the decades to come. These devices will not be limited to microelectronics but will include devices for chemical analysis, sensing, and biological functions.

We see, however, a very clear gap between the natural unit of size, namely, the atom, and dimensions up to, say, 1000 atoms, in which bulk descriptions of physical phenomena are no longer valid. This is the molecular scale. Despite the fact that nature, in the form of living cells, has built many diverse and intricate devices based on molecular phenomena, man has not yet been able to carry out such engineering on anything but a trivial level. This is primarily owing to our lack of scientific understanding of the phenomena involved at these dimensional levels. It is our belief that microdevices and instruments operating on known principles will enable us to gain the necessary understanding to penetrate the molecular scale in the not too distant future.

Each bulk effect device has its own set of limits as its size is reduced, depending on its detailed mechanism of operation, and each device must be treated individually. However, in accordance with the emphasis of this book on planar silicon devices, in Section 7.2 we discuss and quantify as a reference example the limits we are approaching in the operation, fabrication, and utilization of the MOS transistor. Limits for other types of devices may be sought using a similar approach.

The limits to planar technology lie in two areas: The first relates to our ability to form layers of a given thickness and with desired chemical, physical and interface properties. Layer thicknesses can be controlled down to atomic dimensions (Chapter 3). The other properties are individual to the device requirements and cannot be discussed generally. The second area relates to our ability to generate patterns in these layers to the smallest tolerances. Such limits are more general and are discussed in Section 7.3.

7.2. Limits for MOS Devices

7.2.1. Scaling Laws

Large integrated systems composed of interconnected MOS field-effect transistors are implemented on a silicon substrate by metal, polysilicon, and diffusion conducting layers separated by intervening layers of insulating material (see Figure 7.1).

In the absence of any charge on the gate the drain-to-source path is like an open switch. If sufficient positive charge is placed on the gate so that the gate-to-source voltage V_{gs} exceeds a threshold voltage V_{th}, electrons flow under the gate between the drain and the source. The basic operation performed by the MOS transistor is to use the charge on its gate to control the flow of electrons between the source and the drain.

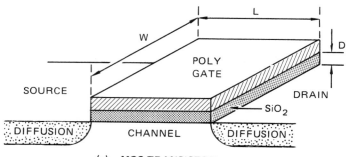

(a)   **MOS TRANSISTOR**

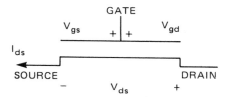

(b)   **SYMBOLIC REPRESENTATION OF MOS TRANSISTOR**

FIGURE 7.1. MOS transistor structure and symbolic representation.

For a small voltage between the source and the drain, V_{ds}, the transit time τ required to move an electron from the source to the drain can be written as

$$\tau = \frac{L}{v} = \frac{L}{\mu E} = \frac{L^2}{\mu V_{ds}} \tag{7.1}$$

where v is the electron velocity and μ its mobility in field $E = V_{ds}/L$. The gate, separated from the substrate by insulating material of thickness D, forms a capacitor. This gate capacitance equals

$$C_g = \frac{\epsilon WL}{D} \tag{7.2}$$

The charge in transit and the current are given by

$$Q = -C_g(V_{gs} - V_{th}) = -\frac{\epsilon WL}{D}(V_{gs} - V_{th}) \tag{7.3}$$

and

$$I_{ds} = \frac{Q}{\tau} = \frac{\mu \epsilon W}{LD} (V_{gs} - V_{th})(V_{ds}) \qquad (7.4)$$

for low V_{ds} the device acts as a resistor ($I_{sd} = \text{const.} \times V_{ds}$) and the gate voltage-controlled resistance is given as

$$R = \frac{V_{ds}}{I_{ds}} = \frac{L^2}{\mu C_g(V_{gs} - V_{th})} \qquad (7.5)$$

However, when the transistor is operated near its threshold, the channel resistance is given by

$$R = R_0 e^{-(V_{gs} - V_{th})q/kT} \qquad (7.6)$$

where T is the absolute temperature, q is the charge on the electron, and k is Boltzmann's constant. This means that below threshold the conductance ($1/R$) has a nonzero value depending on the gate voltage and temperature, as shown in Eq. (7.6).

Now we can examine what happens when the device dimensions are made smaller by a scaling factor n (>1). The new dimensions are then

$$D' = \frac{D}{n}$$

$$L' = \frac{L}{n} \qquad (7.7)$$

$$W' = \frac{W}{n}$$

The electric field cannot be increased when the dimensions are made smaller, since the devices are designed to operate close to the limiting value above which breakdown or quantum-mechanical tunneling would occur. Since the fields are constant, the operating voltages vary as

$$V' = \frac{V}{n} \qquad (7.8)$$

From Eqs. (7.1)–(7.3), (7.7), and (7.8) the physical quantities scale down as follows (using $V = V_{gs} - V_{th}$).

$$\tau' = \frac{1}{\mu}\left[\left(\frac{L}{n}\right)^2 \bigg/ \left(\frac{V}{n}\right)\right] = \frac{1}{\mu}\frac{L^2}{V_n} = \frac{\tau}{n} \tag{7.9}$$

$$C' = \epsilon\left[\left(\frac{L}{n}\right)\left(\frac{W}{n}\right) \bigg/ \left(\frac{D}{n}\right)\right] = \epsilon\frac{LW}{Dn} = \frac{C}{n} \tag{7.10}$$

$$I' = \mu\epsilon\left(\frac{W}{n}\right)\left(\frac{n}{L}\right)\left(\frac{n}{D}\right)\left(\frac{V}{n}\right)\left(\frac{V_{ds}}{n}\right) = \frac{\mu\epsilon V}{nLD}V_{ds} = \frac{I}{n} \tag{7.11}$$

The switching power P_{sw} and dc power scale are as follows:

$$P'_{sw} = \frac{C'V'^2}{2\tau'} = \frac{1}{2}\left(\frac{C}{n}\right)\left(\frac{V}{n}\right)^2\left(\frac{n}{\tau}\right) = \frac{P_{sw}}{n^2} \tag{7.12}$$

and

$$P'_{dc} = I'V' = \left(\frac{I}{n}\right)\left(\frac{V}{n}\right) = \frac{P_{dc}}{n^2} \tag{7.13}$$

and the switching energy

$$E'_{sw} = \tfrac{1}{2}C'V'^2 = \frac{1}{2}\left(\frac{C}{n}\right)\left(\frac{V}{n}\right)^2 = \frac{E_{sw}}{n^3} \tag{7.14}$$

Table 2 summarizes the scaling laws and the present and future operating specifications limited by the physics of the device. Thus, as the dimensions are made smaller, all but one of these variables move in a favorable direction. The operating speed increases, the density increases, and the power density remains constant. However, the current density is increased by the scale-down factor, in this case 10. Therefore the delivery of a dc current presents an important obstacle to scaling. Metal conductors have an upper current density limit imposed by electromigration. Electromigration is a diffusive process in which the atoms of a solid move from one place to another under the influence of electrical forces. This effect limits the maximum current that can be carried by a conductor without its rapid destruction. The current density for aluminum conductors of ICs must be kept lower than 10^6 A/cm^2. Electromigration does not limit the minimum device size but, rather, limits the number of circuit functions that can be carried out by a given number of connected circuit elements per unit time.[1] Essentially, it should be considered as a system limit.

TABLE 2. Scaling Laws for MOS Devices

Physical Quantities	Scale-down Factor (n)	1980	Future (n = 10)
Linear device dimension	$1/n$	5 μm	0.5 μm
Supply and logic voltage	$1/n$	5 V	0.5 V
Current	$1/n$	—	—
Capacitance	$1/n$	—	—
Gate transit (delay)	$1/n$	5×10^{-10}	5×10^{-11}
Power dissipation	$1/n^2$	—	—
Switching energy	$1/n^3$	10^{-13} J	10^{-16} J
Device density	n^2	4×10^6	4×10^8
Impedance	Constant	—	—
Power density	Constant	—	—
Current per unit surface area	n	x	$10x$

7.2.2. Fundamental Limits

Limits that are closely related to basic physical laws are called basic or fundamental limits.[2] Fundamental limits are largely concerned with energetic phenomena related to quantum and statistical physics.

7.2.2.1. Quantum Limits

A basic concept of quantum physics is that a physical measurement performed in a time Δt must involve an energy

$$\Delta E \geq \frac{\hbar}{\Delta t} \tag{7.15}$$

where $\hbar = 1.05 \times 10^{-34}$ J$\cdot$s/radian $= h/2\pi$ and h is Planck's constant. This energy is dissipated as heat. The power dissipated during the measurement (switching) process is

$$P = \frac{\Delta E}{\Delta t} \geq \frac{h}{(\Delta t)^2} \tag{7.16}$$

which can be considered as the lower bound of power dissipation per unit operation.

Using Eq. (7.15), the minimum energy dissipated in a switching device is on the order of 2×10^{-25} J per operation. The actual value for a MOS transistor is 10^{-13} J, which is quite remote from this quantum limit.

7.2.2.2. Tunneling

If very thin (10–100 Å) insulators are placed between two conductors, the decay of the wave function of an electron on one side is not sufficient to give zero amplitude at the opposite side, and there is a finite probability of the electron passing through the dielectric by the process of quantum-mechanical tunneling. In this way a current can pass through a classically forbidden region.

In the case of a MOS transitor operation this "tunneling current" must be much smaller than all of the circuit currents for proper operation. Therefore this quantum effect sets a fundamental size limitation for the thicknesses of the oxide gates and depletion layers at about 10^{-3} μm.[3] The thickness of gate oxides can already be made in the range 10^{-2}–10^{-3} μm and it is not very far from this fundamental size limitation.

7.2.3. Material Limitations

Material properties (critical field for breakdown, doping concentration, dislocation density, and electromigration) also impose limits on device operation.[1,2] Dielectric breakdown limits the electric field in semiconductors and thereby limits the size and speed of the device operation. If the electric field exceeds a critical value ($E_c = 3 \times 10^5$ V/cm in Si) in the semiconductor, a rapid increase of current can occur, owing to avalanche breakdown.

Generally, the breakdown field is not a constant but depends on the doping levels and the doping profile in a junction.[2] For heavy doping, breakdown occurs by tunneling rather than via the avalanche mechanism.

To estimate the limitation set by the material properties of silicon we consider the maximum propagation time in a cube of silicon material of length Δz. Since the potential and the linear dimensions can be written as

$$\Delta V = E \, \Delta z \tag{7.17}$$

and

$$\Delta z = v_{\text{Si}} \, \Delta t \tag{7.18}$$

for $\Delta V = kT/q$, the minimum transit time, using $E_{\text{max}} \cong 3 \times 10^5$ V/cm and $v_{\text{max}} \cong 8 \times 10^6$ cm/s, is

$$\Delta t_{\text{min}} = \frac{kT/q}{E_{\text{max}} v_{\text{max}}} = 10^{-14} \text{ s} \tag{7.19}$$

Thus the breakdown voltage limit imposed by the material properties of silicon limits the minimum transit time to 10^{-14} s.

7.2.4. Device Limits

Device limits are limitations that depend on device operation (turn-on/turn-off voltage, gain, bandwidth, thermal limits, etc.). To illustrate an important example consider the channel conductance $(1/R)$ of a MOS transistor near the threshold voltage. The conductance near, but below, the threshold is not really zero but given from Eq. (7.6) as

$$\frac{1}{R} = \frac{1}{R_0} e^{(V_{gs} - V_{th})/kT/q} \tag{7.20}$$

At room temperature $kT/q = 0.025$ V. If $V_{th} = 1$ and $V_{gs} = 0.5$ V, the device conductance is decreased by a factor of 2×10^8. However, if these dimensions and voltages are scaled down by a factor of 5, then the conductance decreases only by a factor of 1.8×10^2. This is a "leaky" transistor indeed. One way to cope with this problem is to operate the devices at lower than room temperature to reduce kT/q.

Because of thermal fluctuations, the MOS transistor may randomly switch from the conductive to the nonconductive state or vice versa. If the switching energy is E_{sw}, the thermal fluctuation leads to a failure rate of

$$\frac{1}{\tau} \exp\left(\frac{-E_{sw}}{kT}\right) \tag{7.21}$$

per second per device. For a given mean time between errors (T') [Eq. (7.21)] gives the following condition for the switching energy

$$E_{sw} \geq kT \ln\left(\frac{T'}{\tau}\right) \tag{7.22}$$

T' is typically in hours and τ is $\smile 10^{-10}$ s. Then $\ln(T'/\tau) \smile 30$ or $E_{sw}/q \smile V_{sw} \smile 0.75$ V for a single device. It is then possible to reduce the error rate caused by thermal fluctuations to any desired extent by increasing E_{sw}.

It is interesting to note that in human nerve excitation the excitation energy is only a few kT, and for information processing found in biological systems $E_{sw} \smile 20kT$, the same order of magnitude as in a MOS transistor.[1]

This thermal limit for the switching energy is a device and system limit for all the electrically operated microstructures. For this reason the voltage operating the device should be many times kT/q.

Beyond the devices themselves there are circuit and system limits. These limits deal with optimal internal and external partitioning, power dissipation, and chip architecture. These problems, which are outside the present scope, have been reviewed in a classic text on VLSIC (very large scale integrated circuit) design.[4]

Based on the scaling laws and fundamental device limits, we may predict future MOS devices with approximately 0.25–0.5 μm channel lengths and current densities 10 times what they are today. Smaller devices might be built, but at the cost of lowering the voltage or the temperature at which the devices have to be operated. Finally, because of tunneling through the gate, the oxide

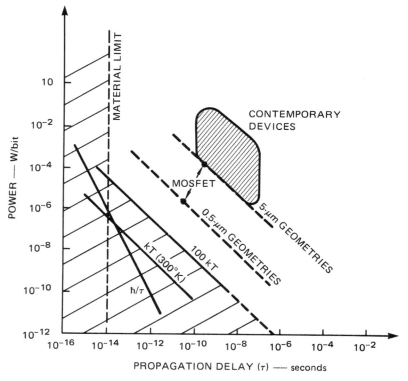

FIGURE 7.2. Power dissipation and transit time of silicon devices. Source: ref. 4.

thickness of the insulator is limited to a value that must be larger than 50 Å.

To give a quantitative comparison of the benefits that can be accrued from improvements in device miniaturization the power required per bit of information is plotted against the transit time (Figure 7.2). The performance improvement that can be achieved by reducing the component dimensions from 5 to 0.5 μm is demonstrated through the diagonal lines. The limits imposed by fundamental ($\hbar/\tau$) thermal and material limitations are also depicted in Figure 7.2.

7.3. Limits in Pattern Generation

The minimum size of a microdevice is determined by the material properties, the device limits, and the fabrication technology available at the given time. Lithography has received a substantial amount of attention as a limit to miniaturization.[5,6] In this section we examine some of these limiting factors.

7.3.1. Limits on Linewidths and Edge Sharpness

To transfer information about a circuit pattern from a computer memory or mask to the wafer a printing tool is needed. Modern printing tools use particle beams, photons, electrons, and ions for lithography. The momentum of these particles in the beam are related to uncertainties in their location, Δl, through the Heisenberg relation

$$\Delta l \, \Delta p \geq \hbar \qquad (7.23)$$

Thus elementary particles, owing to their wave–particle quality, cannot possess both definite coordinates and definite momentum components simultaneously. Because of this, their state cannot be specified by simultaneously giving these observables, as is the case with a classical particle. If in spite of this we want to specify the state of a particle by simultaneously giving these observables, we can do this only with the uncertainties (errors) in the coordinates of the particle and the uncertainties in the momentum components. For photons

$$\Delta l \geq \frac{\hbar c}{E} = \frac{1.23 \times 10^{-6}}{E \text{ (in eV)}} \text{ (in m)} \qquad (7.24)$$

for electrons

$$\Delta l \geq \frac{\hbar}{(2mE)^{1/2}} = \frac{1.22 \times 10^{-9}}{[E \text{ (in eV)}]^{1/2}} \text{ (in m)} \qquad (7.25)$$

and for ions

$$\Delta l \gtrsim \frac{\hbar}{2ME} = \frac{2.738 \times 10^{-11}}{|(M/M_p)E|^{1/2}} \text{ (in m)} \qquad (7.26)$$

where c is the velocity of light, E is the beam energy in electron volts, m is the electron mass, M is the ion mass, and M_p is the mass of the proton.

This position uncertainty (or noise) limits the sharpness of the line edge. For photon beams in the visible range this line edge uncertainty is in the 0.5-μm range; for electrom beams ($E = 10^4$ eV) it is 10^{-1} Å. However, if one includes the low-energy electron contribution to the line edge definition, $\Delta l \backsim 1$–2 nm. If one further includes the proximity effect,[5] the resultant edge uncertainty is on the order of 10 nm for electron beams.

From recent experiments on the electron microscopy of individual atoms[6] it is also clear that the edge of a metal conductor on an insulating substrate cannot be defined sharply because of the movement of the atoms at the edge.

The edge of these thin metallic lines can be considered as a viscous liquid to a depth on the order of 1.0 nm. These edge atoms are in continuous movement in this transition layer, where they are slowly diffusing away from the edge and thereby decreasing the thickness of the conductor. This diffusion process may be characterized by a "liquid-phase" surface diffusion coefficient similar to that involved in high-temperature liquid-phase epitaxy.

7.3.2. Resist Materials

The basic process of resist exposure is the conversion of one kind of molecule into another (lower molecular weight in the case of a positive resist, higher molecular weight for a negative resist) by radiation damage.[6]

Radiation damage is caused by the energy transfer from the energetic particle in the fabricating beam to the resist molecule. In inelastic collisions the average energy loss ΔE is in the range 20–30 eV (see the Bethe formula in Section 2.6.1.5) and it is independent of the energy of the incident particle.

If the resist has an activation energy in the range of $kT \sim 0.025$ eV, an inelastic collision can damage 1000 molecules. In this case the grain-size radius (interaction-volume radius) is 10 times the radius of an individual molecule or

$$R_{\text{damage}} = 10 r_{\text{mol}}$$

From target theory the number of grains that are damaged (N) depends exponentially on the number of incident particles (n) as

$$N = N_0 \left[1 - \exp\left(-\frac{nz}{\Lambda N_0} \right) \right] \qquad (7.27)$$

where N_0 is the number of grains per centimeter before irradiation and z is the thickness of the resist film. Since

$$N_0 = \frac{3z}{4\pi R^3} \qquad (7.28)$$

the "$1/e$" dose is given by

$$n = \frac{\Lambda N_0}{z} = \frac{3\Lambda}{4\pi R^3} \qquad (7.29)$$

where Λ is the mean free path for inelastic collisions ($\Lambda = 400$ Å for PMMA at 20 kV) so that

$$n = \frac{10^{-6}}{R^3} \text{ electron/cm}^2$$

or the dose

$$D = \frac{1.5 \times 10^{-25}}{R^3} \text{ C/cm}^2 \qquad (7.30)$$

From this equation we see that the speed of the resist is proportional to the inverse cube of the grain size. The largest grain size is set by the required edge resolution. To illustrate the role of the maximum grain-size effect for a given exposure we estimate the electron flux for a 1-cm² chip with 0.1-μm² pixel elements (10^8) and a 0.1-μm edge resolution. Here maximum grain size is ≤ 0.1 μm. Thus

$$n = \frac{10^{-6}}{10^{-15}} = 10^9 \text{ electron/cm}^2$$

and the number of electrons per pixel is

$$\frac{n}{\text{number of pixels}} = \frac{10^9}{10^8} = 10$$

We will see in the following section that because of the particle statistics in the beam, to expose 10^8 pixels with a small probability of error (8×10^{-13}) we need at least 200 electrons per pixel, or 2×10^{10} electron/cm^2.

High-resolution resists (PMMA) limit the edge resolution to the range of the grain size, or $R \sim 25$ nm (250 Å).

7.3.3. Exposure Statistics

7.3.3.1 Electron-Beam Exposure

Let D be the dose in C/cm^2 delivered to an electron-beam resist and S the minimum dose (or sensitivity) required to expose the resist for development. Information theory requires that some minimum number of electrons, N_m, must be detected in a pixel[7] in order to be able to assign to it a given probability of being correctly exposed, that is,

$$\frac{D}{e} \geq \left(\frac{N_m}{l_p^2} \right) = \frac{S}{e} \qquad \text{for exposure} \qquad (7.31)$$

where l_p is the pixel linear dimension and e is the electron charge. Equation (7.31) shows that D must increase as l_p decreases for the probability that each pixel will be correctly exposed to remain constant. Stated another way, based on pixel signal-to-noise considerations, the minimum total number of electrons needed to reliably expose a pattern of a given complexity (i.e., with a given number of pixels) is independent of the size of pixels. Since N_m is a constant depending only on the tolerable error rate in assigning a pixel to the exposed state, Eq. (7.31) shows that highly sensitive resists are useful for larger pixels, whereas less-sensitive resists are necessary for smaller pixels.

To ensure that each pixel is correctly exposed a minimum number of electrons must strike each pixel. Since electron emission is an uncorrelated process, the actual number of electrons striking each pixel, n, will vary in a random manner about a mean value $\bar{n}$. Adapting the signal-to-noise analysis to the case of binary exposure of a resist, one can show[7] that the probability of error for

TABLE 3. *Probability of Error of*
Exposure

	Exposure	
$\bar{n}$	Probability of Error	Average Error Rate
50	2.2×10^{-4}	1 in 4.5×10^3
100	3×10^7	1 in 3.3×10^6
150	4.7×10^{-10}	1 in 2.1×10^9
200	7.8×10^{-13}	1 in 1.3×10^{12}

large values of the mean number of electrons per pixel, n, is

$$\frac{e^{-\bar{n}/8}}{[(\pi/2)\,\bar{n}]^{1/2}} \tag{7.32}$$

This leads to Table 3 for the probability of error of exposure.

To be conservative, for the correct exposure of 10^{10} pixels the average number of electrons exposing a pixel should be at least 200, that is, $\bar{n} = 200$. Thus a pixel of dimension l_p with the minimum number of electrons striking it ($N_m = 200$) gives the following charge density:

$$D = \frac{N_m e}{l_p^2} \tag{7.33}$$

Using Eq. (2.49) (with $d_{\text{eff}} = l_p$), the minimum charge on a pixel τi can be expressed as

$$\tau i = N_m e = \frac{3\pi^2}{16} \frac{\beta}{C_s^{2/3}} \tau \, l_p^{-8/3} \tag{7.34}$$

where β is the source brightness and τ is the exposure time:

$$\tau = \frac{16 N_m e}{3\pi^2} \frac{C_s^{2/3}}{\beta} \, l_p^{-8/3} \tag{7.35}$$

For a real resist $Q = \tau i = S l_p^2$. Then from Eq. (7.27) we get

$$\tau_R = \frac{16 S C_s^{2/3}}{3\pi^2 \beta} \, l_p^{-2/3} \tag{7.36}$$

Equation (7.35) expresses the fact that the time τ required to expose a pixel increases as the pixel linear dimension l_p decreases ($N_m = 200$). This equation is plotted in Figure 7.3. For a real resist with sensitivity S (C/cm²) the exposure time is given by Eq. (7.36). A family of curves corresponding to resists with different sensitivities $[10^{-5}-10^{-8}\ (\text{C/cm}^2)]$ is shown in Figure 7.3. These

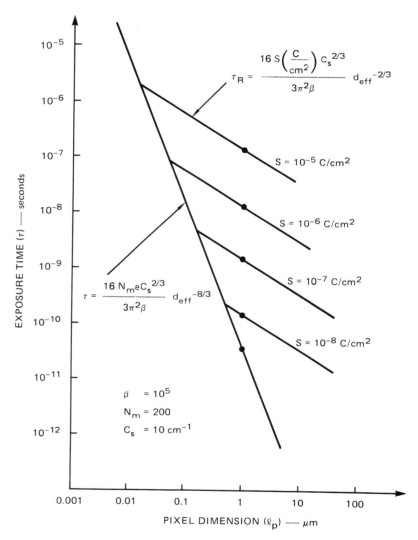

FIGURE 7.3. Exposure time as a function of pixel dimension.

TABLE 4. Comparison of Exposure Times at 0.5-μm Resolution for Different Sources and Different Electron-Optical Parameters[a]

Source and Conditions	Exposure Time τ			
	For 1 pixel	For 10^8 pixels	For 1 cm^2	Typical Wafer
Electron beam				
Vector scan[b]				
Tungsten (hairpin)	9.7 μs	970 s	107 h	—
LaB$_6$ (standard)	0.26 μs	26 s	2.9 h	—
Field effect (gun and lens)	0.018 μs	1.8 s	0.2 h	—
Raster scan[c]				
Tungsten (hairpin)	0.16 μs	16 s	1.7 h	—
LaB$_6$ (standard)	0.021 μs	2.1 s	0.23 h	—
Conventional X ray: 100 mA, 10 kV, 30 cm working distance	—	—	—	1000 s
Synchrotron radiation: 100-mA, 500-MeV beam	—	—	—	1 s

[a]Source: ref. 7.
[b]C_s = 12 cm, C_c = 1 cm, beam voltage = +25 kV.
[c]C_s = 1.8 cm, C_c = 1 cm, beam voltage = 25 kV.

curves illustrate that for electron-beam exposure slow resists are necessary for high resolution.

From Figure 7.3 the exposure times for a given resolution can be calculated using a different source (β) and electron-optical parameters (C_s) (see Table 4).

7.3.3.2. X-Ray Resist Exposure

In X-ray resist exposure the statistical nature of photon absorption in the resist limits the exposure parameters.[8] Because photons arrive at random, the actual number of photons absorbed in different pixels will deviate from the average value. These deviations from the average number of photons (n) are responsible for the limitations in exposure. Since photon emission is uncorrelated, the probability of absorbing n photons in a pixel is given by the Poisson distribution for large n as

$$p(n, \bar{n}) \approx \frac{1}{\sqrt{2\pi n}} \exp\left[\frac{-(n - \bar{n})^2}{2\bar{n}}\right] \qquad (7.37)$$

The X-ray mask consists of areas with two different transmission values. In the clear part (where only the supporting thin film absorbs) the value of the transmission coefficient levels to 1, and in the "opaque" parts we assume that the transmission coefficient is T_a. If n photons are incident on a pixel, n would be counted in a clear part of the mask and $n_a = T_a n$ in an opaque part of the mask.

The probability $p(n)$ of observing n photons in a pixel is shown in Figure 7.4 for $n = 15$ and $n_a = 7.5$, that is, $T_a = 0.5$. Since the two Poisson distributions overlap, one cannot tell for certain from the exposed resist whether the photons came through a transparent or an opaque region. However, one can define n_0 in such a way that if $n \leq n_0$, there is no exposure, and if $n \geq n_0$, the ideal resist will be exposed.

In this case we observe an error in assignment for a clear area when $n \leq n_0$ and in an opaque area when $n \geq n_0$. The probability for such errors (or

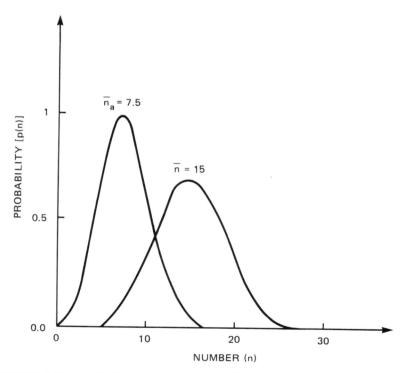

FIGURE 7.4. Poisson distributions for an exposure $\bar{n} = 15$ in the clear and $\bar{n} = 7.5$ in the opaque areas of the mask. Source: ref. 8.

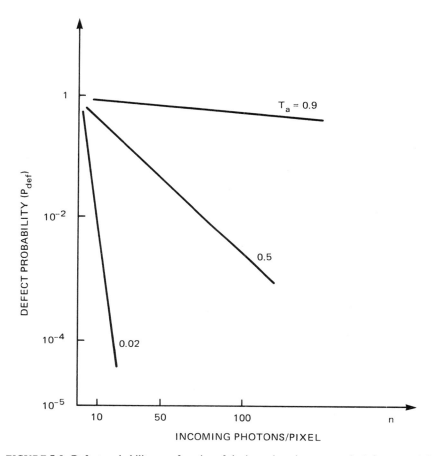

FIGURE 7.5. Defect probability as a function of the incoming photons per pixel. Source: ref. 8.

defects) is given by

$$P_{\text{def}} = \sum_{0}^{n_0} p(n, \bar{n}) + \sum_{n_0+1}^{\infty} p(n, \bar{n}_a) \qquad (7.38)$$

Figure 7.5 shows the probability for a defect (P_{def}) as a function of the incoming photons per pixel with different T_a values.

It is important to note that the probability for obtaining a defect becomes smaller for a larger separation of the two distributions (Figure 7.4). This can be achieved by increasing the number of photons impinging on a pixel and/or by increasing the contrast (C) of the mask ($C = 1/T_a$). The limitation imposes strict design and selection rules for the mask materials and thicknesses when a specific exposure time and resolutions are required.

7.4. *Summary*

Figure 7.6[9-11] summarizes the improvements attained and anticipated in the most important device and fabrication sizes as a function of time and when we may expect to reach the limiting values. Based on physical limitations, the minimum channel length for MOS transistors is about 0.25 μm. Similar limits can be placed on the minimum feature size in bipolar transistors. Since the fabrication tools are available, it is expected that in the next two decades miniaturization will reach these physical limits.

Beam fabrication techniques (electron-beam, ion-beam, X-ray, dry processes) are on the horizon, which would allow devices with characteristic dimensions of 100–250 Å.[12] In this range we will encounter fundamental questions both in the form of new device physics and in the form of limitations to traditional planar approaches.

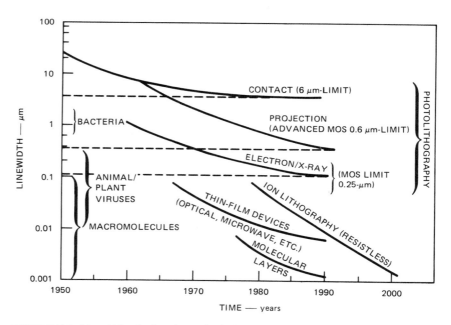

FIGURE 7.6. Linewidth of microelectronic devices and lithographies as a function of time. Source: refs. 9–11.

Because of the high electric fields in devices that small, the transport mechanisms of electrons and holes can change drastically. The velocity of the electrons becomes very large and the time between collisions very short. In these small devices the electrons can pass through the device without collisions. At such very small dimensions the device properties will depend strongly on the environment, that is, on other devices nearby, insulating structures, surfaces, and interfaces. The possibility exists that new cooperative phenomena will arise among the no-longer-isolated devices.

Of course, all of these new phenomena could lead to a new kind of device and structure in this microworld. Clearly, the evolution of fabrication technology will have widespread scientific consequences in device development and opportunities for new physics.[13]

Several materials-related phenomena will limit the advance of microelectronics. The most important effects are the following:

- Electromigration
- Breakdown
- Resistivity
- Power dissipation and heat conduction

In the 100–250 Å range a broad range of new physical and chemical problems will be encountered. New processing (deposition, etching, doping) has to be developed to fabricate structures and new materials, such as intercalated compounds and relief structures.

Surfaces with specific site properties will be an important area for research. Using artificially created nanometer patterns, surface sites for attachment to specific parts of organic molecules could be fabricated. These site-specific surfaces could play an important role in chemical analysis, superconductive studies, and biological research.[14,15]

The Error Function and Some of Its Properties

The error function is expressed as

$$\text{erf } z \equiv \frac{2}{\sqrt{\pi}} \int_0^z e^{-a^2} \, da$$

The definition of the complementary error function is

$$\text{erfc } z \equiv 1 - \text{erf } z$$

Some of their most common properties are the following:

$$\text{erf } 0 = 0$$

$$\text{erf } \infty = 1$$

$$\text{erf } z \approx \frac{2}{\sqrt{\pi}} z \qquad \text{for } z \ll 1$$

$$\text{erfc } z \approx \frac{e^{-z^2}}{\sqrt{\pi} \, z} \qquad \text{for } z \gg 1$$

$$\frac{d \text{ erf } z}{dz} = \frac{2}{\sqrt{\pi}} e^{-z^2}$$

$$\int_0^z \text{erfc } z' \, dz = z \text{ erfc } z + \frac{1}{\sqrt{\pi}} (1 - e^{-z^2})$$

$$\int_0^\infty \text{erfc } z \, dz = \frac{1}{\sqrt{\pi}}$$

TABLE 1. Values of erfc z

z	erfc z	z	erfc z	z	erfc z	z	erfc z
0	1.000 00	1.00	0.157 30	2.00	0.004 68	3.00	0.000 022 09
0.10	0.887 54	1.10	0.119 80	2.10	0.002 98	3.10	0.000 011 65
0.20	0.777 30	1.20	0.089 69	2.20	0.002 86	3.20	0.000 006 03
0.30	0.671 37	1.30	0.065 99	2.30	0.001 14	3.30	0.000 003 06
0.40	0.571 61	1.40	0.047 72	2.40	0.000 689	3.40	0.000 001 52
0.50	0.479 50	1.50	0.033 90	2.50	0.000 407	3.50	0.000 000 743
0.60	0.396 14	1.60	0.023 65	2.60	0.000 236	3.60	0.000 000 356
0.70	0.322 20	1.70	0.016 21	2.70	0.000 134	3.70	0.000 000 167
0.80	0.257 90	1.80	0.010 91	2.80	0.000 075	3.80	0.000 000 77
0.90	0.203 09	1.90	0.007 21	2.90	0.000 041	3.90	0.000 000 35

Appendix B

Properties of Silicon

Atomic number	14
Atomic weight	28.06
Atom density	$5.02 \times 10^{22}/cm^3$
Density	$2.33 \ gm/cm^3$
Crystal structure	Diamond
Atomic shell configuration	$1s^2 \ 2s^2 \ 2p^6 \ 3s^2 \ 3p^2$
Lattice constant (cube edge) a	5.43 Å
Tetrahedral radius r_0	1.18 Å
Spacing between {001} planes	5.42 Å
Spacing between {011} planes	3.83 Å
Spacing between {111} planes	3.13 Å
Melting point	$1412°C$
Energy gap E_g	1.11 eV
Conductivity effective mass (electrons)	$0.26 \ m_0$
Conductivity effective mass (holes)	$0.38 \ m_0$
Diffusion constant	
Electron (D_n)	$34.6 \ cm^2 \cdot s^{-1}$
Hole (D_p)	$12.3 \ cm^2 \cdot s^{-1}$
Young's modulus	
$\langle 111 \rangle$ direction (Y)	$1.9 \times 10^{12} \ dyn \cdot cm^{-2}$
Intrinsic carrier concentration n_i	$1.38 \times 10^{10}/cm^3$
Lattice mobility for electrons μ_n	$1350 \ cm^2/V \cdot s^{-1}$
Lattice mobility for holes μ_p	$480 \ cm^2/V \cdot s^{-1}$
Coefficient of thermal expansion	
Si	$\simeq 2.5 \times 10^{-6}/°C$
SiO_2	$\simeq 0.5 \times 10^{-6}/°C$

Dielectric constant ϵ/ϵ_0

Si	$\simeq 11.7$
SiO$_2$	$\simeq 3.9$

Index of refraction n 3.44

Index of refraction for SiO$_2$ 1.46

Thermal conductivity K $1.412 \text{ W}\cdot\text{cm}^{-1}\cdot{}^\circ\text{K}^{-1}$

Thermal diffusivity k $0.87 \text{ cm}^2\cdot\text{s}^{-1}$

Electric field E at breakdown $\sim 3 \times 10^5 \text{ V}\cdot\text{cm}^{-1}$

Sources

Mead, C., and L. Conway, 1980: *Introduction to VLSI Systems* (Addison-Wesley Publishing Company, Reading, Massachusetts).

Grove, A. S., 1967: *Physics and Technology of Semiconductor Devices* (Wiley & Sons, New York).

Sze, S. M., 1969: *Physics of Semiconductor Devices* (Wiley & Sons, New York).

Hill, D. E., 1971: *Some Properties of Semiconductors* (Monsanto Co., St. Peters, Missouri).

Wolf, H., 1971: *Semiconductors* (Wiley & Sons, New York).

Appendix C
Useful Physical Constants in Microscience

Charge on the electron	$e \simeq 1.6 \times 10^{-19}$ coulomb
Electron volt	$eV \simeq 1.6 \times 10^{-19}$ joule
Planck's constant	$h \simeq 6.6 \times 10^{-34}$ joule·second/cycle
	$\simeq 4.1 \times 10^{-15}$ eV·second/cycle
h-bar	$\hbar = h/2\pi \simeq 1.05 \times 10^{-34}$
	joule·second/radian
	$\simeq 6.6 \times 10^{-16}$
	eV·second/radian
1 cycle/second	1 cps $= 2\pi$ radians/second $= 1$ hertz
Flux quantum	$\Phi_0 = h/2q \simeq 2.1 \times 10^{-15}$ volt·second [or weber]
Boltzmann's constant	$k \simeq 1.4 \times 10^{-23}$ joule/°K
	$\simeq 8.6 \times 10^{-5}$ eV/°K
Permeability of vacuum	$\mu_0 = 4\pi \times 10^{-7}$ henry/meter
Permittivity of vacuum	$\epsilon_0 \simeq 8.85 \times 10^{-12} \simeq 10^{-9}/36\pi$ farad/meter
Velocity of light in vacuum	$c = (\epsilon_0\mu_0)^{-1/2} \simeq 3.0 \times 10^8$ meter/second
Wavelength of visible light in vacuum	0.4–0.7 μm (4000–7000 Å)
Avogadro's number	$A_0, N_A = 6.022 \times 10^{23}$ molecule/gram·mole
Thermal voltage	$V_t = kT/q$
at 80.6°F (300°K)	0.025860 volt
at 68°F (293°K)	0.025256 volt
Free electron mass	$m_0 = 9.11 \times 10^{-31}$ kilogram

Conversion Factors

1 angstrom	$= 10^{-1}$ nm $= 10^{-4}$ μm $= 10^{-8}$ cm $= 10^{-10}$ m
1 mil	$= 10^{-3}$ in. $= 25.4$ μm
1 electron volt	$= 1.602 \times 10^{-19}$ J
1 joule	$= 10^{7}$ erg $= 6.242 \times 10^{18}$ eV $= 2.389 \times 10^{-1}$ cal
1 degree	$= 60$ min $= 0.01745$ radian
λ (nm)	$= 1240/E$ (eV)

References

References for Chapter 1

1. A. R. Von Hippel (Ed.), *The Molecular Designing of Materials and Devices*, MIT Press, Cambridge, Massachusetts (1965).
2. K. B. Blodgett, Films built by depositing successive monomolecular layers on a solid surface, *J. Am. Chem. Soc.* **57**, 1007–1022 (1935).
3. I. Langmuir, Pilgrim Trust lecture—Molecular layers, *Proc. R. Soc. London,* **A170**, 1–39 (1939).
4. A. Y. Cho and J. R. Arthur, Molecular beam epitaxy, *Prog. Solid State Chem.* **10**, No. 3, 157–191 (1975).
5. M. Von Ardenne, *Tabellen der Electronenphysik, Ionenphysik, und Ubermikroskopie Band I und II, Deutscher Verlag der Wissenschaften* Berlin (1956).
6. V. E. Coslett, *Practical Electron Microscopy*, Butterworths, London (1951).
7. C. W. Oatley, W. C. Nixon, and R. F. W. Pease, Scanning electron microscopy, *Adv. Electron. Electron Phys.* **21**, 181 (1965).
8. P. R. Thornton, *Scanning Electron Microscopy: Applications to Materials and Device Science*, Chapman and Hall, London (1968).
9. J. C. Slater, *Predictions by a priori Theory*, in ref. 1, pp.7–8.
10. B. E. Deal and J. M. Early, The evolution of silicon semiconductor technology: 1952–1977, *J. Electrochem. Soc.* **126**, No. 1, 20C–30C (1979).
11. A. E. Anderson, Transistor technology evaluation, *West. Electr. Eng.* **3**, No. 3, 3 (1959); *ibid.* **3**, No. 4, 30 (1959); *ibid.* **4**, No. 1, 14 (1960).
12. J. Bardeen and W. H. Brattain, Physical principles involved in transistor action, *Phys. Rev.* **75**, No. 8, 1208 (1949).
13. W. Shockley, The path to the conception of the junction Transistor, *IEEE Trans. Electron Devices* **ED-23**, No. 7, 597–620 (July 1976).
14. W. Shockley, The theory of $p-n$ junctions in semiconductors and $p-n$ junction transistors, *Bell Syst. Tech. J.* **28**, 435 (1949).
15. G. K. Teal and J. B. Little, Growth of germanium single crystals (paper presented at the Oak Ridge meeting), *Phys. Rev.* **78**, 647 (1950).
16. G. K. Teal, M. Sparks, and E. Buehler, Single-crystal germanium, *Proc. IRE* **40**, No. 8, 906 (August 1952).
17. W. G. Pfann, Principles of zone melting, *Trans. AIME*, **4**, No. 7, 747–753 (July 1952).

18. R. N. Hall and W. C. Dunlap, $p-n$ junctions prepared by impurity diffusion, *Phys. Rev.* **80**, No. 3, 467–468 (November 1950).
19. S. M. Sze, *Physics of Semiconductor Devices,* Wiley-Interscience, New York (1969).
20. C. Mead and L. Conway, *Introduction to VLSI Systems,* Addison-Wesley, Reading, Massachusetts (1980).
21. H. C. Theurer, J. J. Kleimack, H. H. Loar, and H. Christensen, Epitaxial diffused transistors, *Proc. IRE* **48**, 1642–1643 (September 1960).
22. D. Kahng and M. M. Atalla, Paper presented at the IRE International Solid-State Device Research Conference, Pittsburgh, Pennsylvania (June 1960).
23. J. E. Lilienfeld, Method and Apparatus for Controlling Electric Currents, U.S. Patent 1,745,175 (1930).
24. R. N. Noyce, Semiconductor Device-and-Lead Structure, U.S. Patent 2,981,877 (1961).
25. R. M. Wanlass and C. T. Sah, Nanowatt logic using field-effect metal–oxide semiconductor triodes, Paper No. WPM 3.5, presented at the University of Pennsylvania IEEE International Solid-State Circuits Conference (February 1963).
26. W. S. Boyle and G. E. Smith, Charge-coupled semiconductor devices, *Bell Syst. Tech. J.* **49**, No. 4, 587–593 (April 1970).
27. B. E. Deal and A. S. Grove, General relationship for the thermal oxidation of silicon, *J. Appl. Phys.* **36**, No. 12, 3770–3778 (December 1965).
28. A. S. Grove, in: *Physics and Technology of Semiconductor Devices,* Wiley & Sons, New York, Chapters 1–3 (1967).
29. R. S. Ronen and P. H. Robinson, Hydrogen chloride and chlorine gathering. Effective technique for improving performance of silicon devices, *J. Electrochem. Soc.* **119**, No. 6, 747–752 (1972).
30. C. M. Osburn, Dielectric breakdown properties of silicon dioxide films grown in halogen and hydrogen-containing environments, *J. Electrochem. Soc.* **121**, No. 6, 809–815 (1974).
31. E. D. Wolf, The national submicron facility, *Phys. Today* **32**, No. 34(3), 34–36 (November 1972).
32. D. A. Markle, A new projection printer, *Solid State Technol.* **17**, No. 6, 50–53 (1974).
33. M. C. King, Future developments for 1 : 1 projection photolithography, *IEEE Trans. Electron Devices* **ED-26**, No. 4, 711 (April 1979).
34. J. Roussel, Step-and-repeat wafer imaging, *Solid State Technol.* **21**, No. 5, 67–71 (May 1978).
35. G. L. Resor and A. C. Tobey, The role of direct step-on-the-wafer in microlithography strategy for the '80s, *Solid State Technol.* **22** (8) 101 (August 1979).
36. H. E. Mayer and E. W. Loebach, A new step-by-step aligner for very large-scale integration (VLSI) production, *Proc. Soc. Photo-Optical Instrum. Eng. (SPIE)* **221**, Semiconductor Microlithography V, 9 (1980).
37. M. Lacombat, Photorepetition on silicon for large scale integration, *Proceedings of the International Conference on Microlithography,* 21–24 June 1977, Paris, France, Comité du Colloque International de Microlithographie, Paris, France, pp. 83–83 (1977).
38. R. E. Tibbetts and J. S. Wilczvnski, High-performance reduction lenses for microelectronic circuit fabrication, *IBM J. Res. Dev.* **13**, No. 2, 192–196 (1969).
39. J. P. Scott, Recent progress on the electron image projector, *J. Vac. Sci. Technol.* **15**, 3, 1016–1021 (1978).
40. P. R. Malmberg, T. W. O'Keefe, M. M. Sopira, and M. W. Levi, LSI pattern generation and replication by electron beams, *J. Vac. Sci. Technol.* **10**, No. 6, 1025–1027 (November–December 1973).
41. H. I. Smith, Fabrication techniques for surface-acoustic-wave and thin-film optical devices, *Proc. IEEE* **62**, No. 10, 1361–1387 (1974).
42. S. Somekh, Introduction to Ion and Plasma Etching, *J. Vac. Sci. Technol.* **13**, No. 5, 1003–1007 (September–October 1976).

43. E. Bassous, Fabrication of novel three-dimensional microstructures by the anisotropic etching of $\langle 100 \rangle$ and $\langle 110 \rangle$ silicon, *IEEE Trans. Electron Devices* **ED-25**, No. 10, 1178–1185 (1978).

44. R. M. Finne, and D. L. Klein, A water–amine-complexing agent system for etching silicon, *J. Electrochem. Soc.* **114**, No. 9, 965–970 (1967).

45. M. Cantagrel and M. Marchal, Argon ion etching in a reactive gas, *J. Mater. Sci.* **8**, No. 12, 1711–1716 (December 1973).

46. N. Hosokawa, R. Matsuzaki, and T. Asamaki, RF sputter-etching by flouro-chloro-hydrocarbon gases, *Jpn. J. Appl. Phys. Suppl.* **2**, 435–438 (1974).

47. H. R. Kaufman, Technology of electron-bombardment ion thrusters, in *Advances in Electronics and Electron Physics* Vol. 36 (L. Marton, Ed.) 265–373, Academic Press, New York (1974).

48. M. Cantagrel, Comparison of the properties of different materials used as masks for ion-beam etching, *J. Vac. Sci. Technol.* **12**, No. 6, 1340–1343 (1975).

49. N. Hosokawa, R. Matsuzaki, and T. Asamaki, RF sputter-etching by flouro-chloro-hydrocarbon gases, *Jpn. J. Appl. Phys. Suppl.* **2**, 435–438 (1974).

50. A. S. Grove, *Physics and Technology of Semiconductor Devices*, Wiley & Sons, New York (1967).

51. S. K. Ghandhi, in: *The Theory and Practice of Microelectronics*, Wiley & Sons, New York, Chapter 4 (1968).

52. R. S. Muller and T. I. Kamis, *Device Electronics for Integrated Circuits*, Wiley & Sons, New York, Chapter 2 (1979).

53. E. S. Yang, in: *Fundamentals of Semiconductor Devices*, McGraw-Hill, New York, Chapter 3 (1978).

54. J. W. Mayer, L. Ericksson, and J. A. Davies, *Ion Implantation in Semiconductors: Silicon and Germanium*, Academic Press, New York (1970).

55. J. F. Gibbons, Ion implantation in semiconductors—Part I, Range distribution theory and experiments, *Proc. IEEE* **56**, No. 3, 295–319 (March 1968).

56. A. Gat, J. F. Gibbons, T. J. Magee, J. Peng, V. R. Deline, P. Williams, and C. A. Evans, Jr., Physical and electrical properties of laser-annealed ion-implanted silicon, *Appl. Phys. Lett.* **32**, No. 5, 276–278 (March 1978).

57. J. M. Meese (Ed.), *Neutron Transmutation Doping in Semiconductors*, Plenum Press, New York (1978).

58. L. Eckterova, *Physics of Thin Films*, Plenum Press, New York (1976).

59. P. A. Totta and R. P. Sopher, SLT device metallurgy and its monolithic extension, *IBM J. Res. Dev.* **13**, No. 3, 226–238 (May 1969).

60. F. M. d'Heurle, Electromigration and failure in electronics: Introduction, *Proc. IEEE* **59**, No. 10, 1409–1418 (1971).

61. F. Mohammadi, Silicides for interconnection technology, *Solid State Technol.* **24** (1) 65 (January 1981).

62. L. V. Gregor, Thin-film processes for microelectronic application, *Proc. IEEE* **59**, No. 10, 1390–1403 (1971).

63. D. R. Scifres, R. D. Burnham, and W. Streifer, Heterojunctions in integrated optics, *J. Vac. Sci. Technol.* **14**, No. 1, 186–194 (1977).

64. V. Evtuhov and A. Yariv, GaAs and GaAlAs devices for integrated optics, *IEEE Trans. Microwave Theory Tech.* **MTT-23**, No. 1, 44–57 (January 1975).

65. D. R. Scifres, W. Streifer, and R. D. Burnham, Leaky wave room-temperature double heterostructure GaAs : GaAlAs diode laser, *Appl. Phys. Lett.* **29**, No. 1, 23–25 (July 1976).

66. H. L. Garvin, E. Garmire, S. Somekh, H. Stoll, and A. Yariv, Ion beam micromachining of integrated optics components, *Appl. Opt.* **12**, No. 3, 445–459 (March 1973).

67. C. V. Shank and R. V. Schmidt, Optical technique for producing $0.1\text{-}\mu$ periodic surface structures, *Appl. Phys. Lett.* **23**, No. 3, 154–155 (August 1973).

68. W. W. Ng, C. -S. Hong, and A. Yariv, Holographic interference lithography for integrated optics, *IEEE Trans. Electron Devices,* **ED-25,** No. 10, 1193–1200 (1978).

69. H. Koops, Verwendung der Kondensor-Objectiv-Einfeldlinse zur Elektronenoptischen Microminiaturisation, *Optik* **29,** No. 1, 119–121 (April 1969).

70. T. Barbee, Synthesis and properties of sputter-deposited layered synthetic microstructures, in: *Proceedings of the NSF Workshop on Opportunities for Microstructure Science, Engineering, and Technology* (Airlie, Virginia), NSF, Washington, D.C., p. 94 (November 1978).

71. J. R. Schrieffer, *Theory of Superconductivity,* W. A. Benjamin Publishers, New York (1964).

72. J. Clarke, Electronics with superconducting junctions, *Phys. Today* **24,** No. 8, 30–37 (August 1971); D. G. MacDonald, Superconductive electronics, *Phys. Today* **34,** No. 2, 36–37 (February 1981).

73. P. W. Anderson and J. M. Rowell, Probable observation of the Josephson superconducting tunneling effect, *Phys. Rev. Lett.* **10,** No. 6, 230–232 (March 1963).

74. J. Gu, W. Cha, K. Gamo, and S. Namba, Properties of niobium superconducting bridges prepared by electron-beam lithography and ion implantation, *J. Appl. Phys.* **50,** No. 10, 6437–6442 (1979).

75. C. A. Spindt, A thin-film field-emission cathode, *J. Appl. Phys.* **39,** No. 7, 3504–3505 (1968)

76. C. A. Spindt, I. Brodie, L. Humphrey, and E. R. Westerberg, Physical properties of thin-film field-emission cathodes with molybdenum cones, *J. Appl. Phys.* **47,** No. 12, 5248–5263 (December 1976).

77. W. Aberth, C. A. Spindt, and K. T. Rogers, Multipoint field ionization beam source, *Record of the 11th Symposium on Electron, Ion, and Laser Beam Technology* (R. F. M. Thornley, Ed.), San Francisco Press, San Franscisco, California, pp. 631–636 (1971).

78. M. W. Geis, D. C. Flanders, and H. I. Smith, Crystallographic orientation of silicon on an amorphous substrate using an artificial surface-relief grating and laser crystallization, *Appl. Phys Lett.* **35,** No. 1, 71–74 (July 1979).

79. L. L. Chang, L. Esaki, and R. Tsu, Resonant tunneling in semiconductor double barriers, *Appl. Phys. Lett.* **24,** No. 12, 593–595 (June 1974).

80. J. E. Fischer and T. E. Thompson, Graphite intercalation compounds, *Phys. Today* **31,** No. 7, 36–39 (July 1978).

81. C. G. Kirkpatrick, J. F. Norton, H. G. Parks, and G. E. Possin, New concepts for electron–ion beam and electron–electron beam memories, *J. Vac. Sci. Technol.* 15, No. 3, 841–844 (May–June 1978).

82. K. Bulthuis, M. G. Carasso, J. P. J. Heemskerk, P. J. Kivits, W. J. Kleuters, and P. Zalm, Ten billion bits on a disk, *IEEE Spectrum* **16,** No. 8, 26–33 (August 1979).

83. J. J. Chang, Nonvolatile semiconductor memory devices, *Proc. IEEE* **64,** No. 7, 1039–1059 (July 1976).

84. K. E. Petersen, Dynamic micromechanics on silicon: Techniques and devices, *IEEE Trans. Electron Devices,* **ED-25,** No. 10, 1241–1250 (1978).

85. H. C. Nathanson and J. Guldberg, Topologically structured thin films in semiconductor device operation, in: *Physics of Thin Films* (G. Haas, M. H. Francomb, and R. W. Hoffman, Eds.) Academic Press, New York (1975).

86. R. M. Finne and D. L. Klein, A water–amine-complexing agent system for etching silicon, *J. Electrochem. Soc.* **114,** *965–970 (1967).*

87. J. B. Mooney, New microporous membrane blood oxygenator, Internal communication, Physical Electronics Department, Engineering Services Laboratory, SRI International, Menlo Park, California (December 1978).

References for Chapter 2

1. W. B. Nottingham, Thermionic emission, in: *Handbuch der Physik* (S. Flugge, Ed.) Vol. 21, Springer-Verlag, Berlin, pp. 1–175 (1956).
2. R. H. Good, Jr., and E. W. Müller, Field emission, in: *Handbuch der Physik* (S. Flugge, Ed.) Vol. 21, Springer-Verlag, Berlin, pp. 176–231 (1956).
3. E. L. Murphy and R. H. Good, Jr., Thermionic emission, field emission, and the transition region, *Phys. Rev.* **102**, 1464–1473 (1956).
4. L. D. Smullin and H. A. Haus (Ed.), *Noise in Electron Devices*, MIT Press and Wiley & Sons, New York (1959).
5. I. Brodie, Studies of field emission and electrical breakdown between extended nickel surfaces in vacuum, *J. Appl. Phys.* **35**, 2324–2322 (pages reverse numbered) (1964).
6. A. Van Oostrom, Field emission cathodes, *J. Appl. Phys.* **33**, 2917–2922 (1962).
7. F. M. Charbonnier, R. W. Straeger, L. W. Swanson, and E. E. Martin, Nottingham effect in field and T-F emission, *Phys. Rev. Lett.* **13**, 397–401 (1964).
8. I. Brodie, The temperature of a strongly field emitting surface, *Int. J. Electron.* **18**, 223–233 (1965).
9. G. A. Haas, *Electron Sources: Thermionic Methods of Experimental Physics* (C. Marton, Ed.) Vol. 4, Part A, Academic Press, New York, pp. 1–38 (1967).
10. R. G. Murray and R. J. Collier, Thoriated tungsten hairpin filament electron source for high brightness applications, *Rev. Sci. Instrum.* **48**, No. 7, 870–873 (1977.
11. A. N. Broers, Thermal cathode illumination systems for round beam electron pulse systems, *Scanning Electron Microsc.* **1971/I**, (1971).
12. G. Hermann and S. Wagener, *The Oxide Coated Cathod, I & II*, Chapman and Hall, London (1951).
13. R. Levi, Improved barium dispenser cathode, *J. Appl. Phys.* **26**, 639 (1955). I Brodie and R. O. Jenkins, "Impregnated Barium Dispenser Cathodes Containing Strontium or Calcium Oxide," *J. Appl. Phys.* **27**, 417 (1956).
14. P. Zalm and A. J. A. van Stratum, Osmisum dispenser cathodes, *Philips Tech. Rev.* **27**, 69 (1966).
15. R. E. Thomas, T. Pankey, and G. A. Haas, Thermionic properties of BaO on iridium, *Appl. Surf. Sci.* **2**, 187–212 (1979).
16. L. W. Swanson and N. A. Martin, Zirconium/tungsten thermal field cathode, *J. Appl. Phys.* **46**, 2029–2050 (1975).
17. C. A. Spindt, I. Brodie, L. Humphrey, and E. R. Westerberg, Physical properties of thin film field emission cathodes, *J. Appl. Phys.* **47**, 5248–5262 (1976).
18. L. W. Swanson and G. A. Schwind, Electron emission from a liquid metal, *J. Appl. Phys.* **49**, No. 11, 5655–5662 (1978).
19. H. Moss, *Narrow Angle Electron Guns and Cathode Ray Tubes*, Academic Press, New York (1968).
20. O. Klemperer and M. E. Bainett, *Electron Optics*, Cambridge University Press, Cambridge (1971).
21. P. Grivet, *Electron Optics*, Pergamon Press, New York (1972).
22. A. Septier, *Focusing of Charged Particles, I and II*, Academic Press, New York (1967).
23. W. Glaser, *Grundlagen der Electronenoptik*, Springer-Verlag, Vienna, (1952).
24. V. K. Zworykin, G. A. Morton, E. G. Ramberg, J. Hillier, and A. W. Vance, *Electron Optics and the Electron Microscope*, Wiley & Sons, New York (1945).
25. H. Boersch, Experimentelle Bestimmung der Energieverteilung in thermisch ausgelosten Electrononstrahlen, *Z. Phys.* **139**, 115–146 (1954).
26. K. H. Loeffler, Energy-spread generation in electron-optical instruments, *Z. Angew. Phys.* **27**, No. 3 (July 1969).

27. K. H. Loeffler and R. M. Hudgin, Energy spread generation and image deterioration by the stochiastic interactions between beam electrons, *7th Proceedings of the International Congress on Electron Microscopy*, Grenoble, France, 1970, p. 67 (1970).

28. H. C. Pfeiffer, Experimental investigation of energy broadening in electron optical instruments, *IEEE 11th Symposium on Electron, Ion, and Laser Beam Technology*, San Francisco Press, San Francisco, p. 239 (1971).

29. R. Lauer, Ein einfaches Modell fur Elektronenkanonen mit gekrummter Kathodenoberflache, *Z. Naturforsch.* **23a**, No. 2, 100–109, (January 1968).

30. J. R. Pierce, *Theory and Design of Electron Beams*, Van Nostrand, Princeton, New Jersey (1954).

31. T. E. Everhart, Simplified analysis of point-cathode electron sources, *J. Appl. Phys.* **38**, 4944 (1967).

32. A. V. Crewe, J. Wall, and L. M. Welter, A high resolution scanning transmission electron microscope, *J. Appl. Phys.* **39**, No. 13, 5861 (1968).

33. R. G. Wilson and G. R. Brewer, *Ion Beams: With Applications to Ion Implantation*, Wiley & Sons, New York (1973).

34. G. Carter and W. A. Grant, *Ion Implantation of Semiconductors*, Halstead Press, New York (1976).

35. L. Valyi, *Atom and Ion Sources*, Wiley & Sons, New York (1978).

36. L. B. Loeb, *Basic Processes of Gaseous Electronics*, University of California Press, Berkeley, California (1960).

37. M. von Ardenne, *Tabellen der Elektronenphysik, Ionenphysik, und Ubermikroskopie Band I und II*, Deutscher Verlag der Wissenschaften, Berlin (1956).

38. N. B. Brooks, P. H. Rose, A. B. Wittkower, and R. P. Bastide, Production of low divergence positive ion beams of high intensity, *Rev. Sci. Instrum.* **35**, No. 7, 894 (July 1964).

39. J. Orloff and L. W. Swanson, A scanning ion microscope with a field ionization source, *Scanning Electron Micros.* **1977/I**, 57–62 (1977).

40. J. H. Orloff and L. W. Swanson, Study of a field-ionization source for microprobe applications, *J. Vac. Sci. Technol.* **12**, No. 6, 1209–1213 (November–December 1976).

41. L. W. Swanson, G. A. Schwind, and A. E. Bell, Emission characteristics of a liquid gallium ion source, *Scanning Electron Microsc.* **1979/I**, 45–51 (1979).

42. R. Clampitt, K. L. Aitken, D. K. Jefferies, Intense field emission ion source of liquid metals, *J. Vac. Sci. Technol.* **12**, No. 6 1208 (November–December 1976).

43. R. Gomer, On the mechanism of liquid metal electron and ion sources, *Appl. Phys.* **19**, 365–375 (1979).

44. L. W. Swanson, G. A. Schwind, A. E. Bell, and J. E. Brady, Emission characteristics of gallium and bismuth liquid metal field-ion sources, *J. Vac. Sci. Technol.* **16**, No. 6, 1864–1867 (1979).

45. G. I. Taylor, Disintegration of water drops in an electric field, *Proc. R. Soc. London* **280A**, 383 (1964).

46. R. L. Seliger, J. W. Ward, V. Wang, and R. L. Kubena, A high-intensity scanning ion probe with submicrometer spot size, *Appl. Phys. Lett.* **34**, No. 5, 310 (1979).

47. J. R. Pierce, *Theory and Design of Electron Beams*, Van Nostrand, New York (1954).

48. B. J. Thompson and L. B. Headrick, Space-charge limitations on the focus of electron beams, *Proc. IRE* **28**, No. 7, 318 (July 1940).

49. J. W. Schwartz, Space-charge limitation on the focus of electron beams, *RCA Rev.* **18**, No. 1, 3 (1957).

50. E. L. Ginzton and B. H. Wadia, Positive ion trapping in electron beams, *Proc. IRE* **42**, No. 10, 1548 (October 1954).

51. L. A. Harris, Physics of electron beam fundamentals, in: *Electron Beam Technology* (R. Bakish, Ed.), Wiley & Sons, New York (1962).

52. W. Glaser, *Grundlagen der Electronen Optik,* Springer-Verlag, Vienna (1952).
53. *Focusing of Charged Particles I–II* (A Septier, Ed.), Academic Press, New York, (1967).
54. O. Klemperer, *Electron Optics,* Cambridge University Press, Cambridge (1971).
55. N. D. Wittels, Unipotential lens with electron-transparent electrodes, *J. Vac. Sci. Technol.* **12,** No. 6, 1165–1168 (November–December 1976).
56. A. J. F. Metherell, Energy analyzing and energy selecting electron microscopes, in: *Advances in Optical and Electron Microscopy,* (R. Barer and V. E. Cosslett, Eds.), Vol. 4, Academic Press, London (1971).
57. J. C. Tracy, in: *Electron Emission Spectroscopy* (W. Dekeyser, L. Fiermans, G. Vanderkelen, and J. Vennick, Eds.), Reidel, Dordrecht, Netherlands, p. 331, (1973).
58. R. G. E. Hutter, The deflection of electron beams, in: *Advances in Image Pickup and Display, I,* Academic Press, New York, pp. 163–224 (1974).
59. C. C. T. Wang, Analysis of electrostatic small-angle deflection, *IEEE Trans. Electron Devices* **ED-18,** No. 4, 258 (April 1971).
60. L. N. Heynick, High-information-density storage surfaces, Research and Development Technical Report No. ECOM-01261-F, Stanford Research Institute, Menlo Park, California (January 1970).
61. J. Kelly, *Adv. Electron. Electron Phys.* **43,** 43–135 (1977).
62. C. C. T. Wang, Computer calculations of deflection aberrations in electron beams, *IEEE Trans. Electron Devices* **ED-14,** No. 7, 357 (July 1967).
63. C. C. T. Wang, Two-dimensional small-angle deflection theory, *IEEE Trans. Electron Devices* **ED-15,** No. 8, 603 (August 1968).
64. C. B. Duke and R. L. Park, Surface structure—An emerging spectroscopy, *Phys. Today* **25,** No. 8, 23–28 (August 1972).
65. P. F. Kane and G. B. Larrabee (Eds.), *Characterization of Solid Surfaces,* Plenum Press, New York (1974).
66. E. W. Müller and T. T. Tsong, *Field Ion Microscopy,* American Elsevier, New York (1969).
67. M. Isaacson, All you might want to know about ELS (but are afraid to ask): A tutorial, *Scanning Electron Microsc.* **1978/I,** 763–776 (1978).
68. V. E. Cosslett and R. N. Thomas, Multiple scattering of 5–30 keV electrons in evaporated metal films, I: Total transmission and angular distribution, *Br. J. Appl. Phys.* **15,** 883 (1964).
69. J. P. Langmore, J. Wall, and M. S. Isaacson, Collection of scattered electrons in dark field electron microscopy, I: Elastic scattering, *Optik (Stuttgart)* **38,** No. 4, 335–350 (September 1973).
70. R. W. Nosker, Scattering of highly focused kilovolt electron beams by solids, *J. Appl. Phys.* **40,** 1872–1882 (March (March 1969).
71. F. W. Inman and J. J. Muray, Transition radiation from relativistic electrons crossing dielectric boundaries, *Phys. Rev.* **142,** No. 1, 272 (February 1966).
72. H. A. Bethe, M. E. Rose, and L. P. Smith, Multiple scattering of electrons, *Am. Philos. Soc. Proc.* **78,** No. 4, 573–585 (1938).
73. H. Raether, Electron excitations in solids, *Springer Tracts Mod. Phys.* **38,** 85 (1965).
74. T. E. Everhart and P. H. Hoff, Determination of kilovolt electron energy dissipation vs. penetration distance in solid materials, *J. Appl. Phys.* **42,** No. 13, 5837–5846 (December 1971).
75. L. Reimer, Electron–specimen interactions, *Scanning Electron Microsc.* **1979/II,** 111–124 (1979).
76. D. B. Brown and R. E. Ogilvie, An evaluation of the archard electron diffusion model, *J. Appl. Phys.* 35, No. 10, 2793–2795 (1964).

77. T. E. Everhart, Simple theory concerning the reflection of electrons from solids, *J. Appl. Phys.* **31**, No. 8, 1483 (August 1960).
78. H. Kanter, Zur Ruckstreuung von Elektronen im Energiebereich von 10 bis 100 keV, *Ann. Phys.*, (FRG), **20**, 144–166 (1957).
79. L. Reimer, W. Poepper, and W. Broeker, Experiments with a small solid angle detector for BSE, *Scanning Electron Microsc.* **1978/I**, 705–710 (1978).
80. H. Seiler, Determination of the information depth in the SEM, *Scanning Electron Microsc.* **1976/I**, 9–16 (1976).
81. S. A. Blankenburg, J. K. Cobb, and J. J. Muray, Efficiency of secondary electron emission monitors for 70 MeV electrons, *Nucl. Instrum. Methods* **39**, 303–308 (1966).
82. O. Hachenberg and W. Brauer, Secondary electron emission from solids, *Adv. Electron. Electron Phys.* **11**, 413 (1959).
83. H. Seiler, Einige aktuelle Probleme der Sekundarelektronenemission, *Z. Angew. Phys.* **22**, 249 (1967).
84. W. Heitler, *The Quantum Theory of Radiation,* Oxford Press, Oxford (1954).
85. R. D. Evans, *The Atomic Nucleus,* McGraw-Hill, New York (1955).
86. J. T. Grant, T. W. Haas, and J. E. Houston, Quantitative comparison of Ti and TiO surfaces using Auger electron and soft X-ray appearance potential spectroscopies, *J. Vac. Sci. Technol.* **11**, No. 1, 227–230 (January–February 1974).
87. S. J. B. Reed, *Electron Microprobe Analysis,* Cambridge University Press, London (1975).
88. C. R. Worthington and S. G. Tomlin, The intensity of emission of characteristic X-ray radiation, *Proc. Phys. Soc. London* **69A**, No. 5, 401 (1956).
89. P. F. Kane and G. B. Larrabee (Eds.), *The Characteristics of Solid Surfaces,* Plenum Press, New York (1974).
90. C. K. Crawford, Electron beam machining, in: *Introduction to Electron Beam Technology* (R. Bakish, Ed.), Wiley & Sons, New York (1962).
91. J. Lindhard and M. Scharff, Energy dissipation by ions in the keV region, *Phys. Rev.* **124**, 128 (October 1961).
92. G. M. McCracken, The behaviour of surfaces under ion bombardment, *Rep. Prog. Phys.* **38**, No. 2, 241–327 (February 1975).
93. G. Dearnaley, Ion bombardment and implantation, *Rep. Prog. Phys.* **32**, No. 4, 405–492 (August 1969).
94. R. J. MacDonald, The ejection of atomic particles from ion bombarded solids, *Adv. Phys.* **19**, No. 80, 457–524 (July 1970).
95. J. Lindhard, M. Scharff, and H. Schiott, Range concepts and heavy ion ranges, *K. Dan. Vidensk. Selsk. Mat. Fys. Medd.* **33**, No. 14, 39 (1963).
96. J. F. Gibbons, Ion implantation in semiconductors, Part I, Range distribution theory and experiments, *Proc. IEEE* **56**, No. 3, 295–319 (1968).
97. G. Carter and J. S. Colligon, *Ion Bombardment of Solids,* Elsevier, New York (1968).
98. J. Lindhard and A. Winther, Stopping power of electron gas and equipartition rule, *K. Dan. Vidensk. Selsk. Mat. Fys. Medd.* **34**, No. 4, 1 (1964).
99. O. B. Firsov, A qualitative interpretation of the mean electron excitation energy in atomic collisions, *J. Exp. Theor. Phys.*, (USSR), **36**, No. 5, 1517–1523 (May 1959).
100. S. A. Schwarz and C. R. Helms, A statistical model of sputtering, *J. Appl. Phys.* **50**, No. 8, 5492 (August 1979).
101. J. Lindhard, Influence of crystal lattice on motion of energetic charged particles, *K. Dan. Vidensk. Selsk. Mat. Fys. Medd.* **34**, No. 14, 64 (1965).
102. P. Sigmund, Theory of sputtering, I: Sputtering yield of amorphous and polycrystalline targets, *Phys. Rev.* **184**, No. 2, 184 (August 1969).
103. S. A. Schwarz and C. R. Helms, A statistical model of sputtering, *J. Appl. Phys.* **50**, 5492 (1979).
104. E. Spiller and R. Feder, X-ray lithography, *Top. Appl. Phys.* **22**, 35 (1977).

105. D. J. Nagel, R. R. Whitlock, J. R. Greig, R. E. Pechacek, and M. C. Peckeran, Laser-plasma source for pulsed X-ray lithography, *Proc. Soc. Photo-Optical Instrum. Eng. (SPIE)* Development in Semiconductor Micro-Lithography III, **135**, 46 (1978).
106. R. Z. Bachrach, I. Lindau, V. Rehn, and J. Stohr, Report of the beam line III, Stanford Synchrotron Radiation Laboratory Report, SSRL, 77/14 (1977).
107. J. D. Cuthbert, Optical projection printing, *Solid State Technol.* **20**(8) 59 (August 1977).
108. E. M. Breinan, B. H. Kear, and C. M. Banas, Processing materials with lasers, *Phys. Today* **29**, No. 11, 44 (November 1976).
109. A. E. Bell, Review and analysis of laser annealing, *RCA Rev.* **40**, No. 3, 295–338 (September 1979).

References for Chapter 3

1. M. Faraday, Experimental relations of gold (and other metals) to light, *Philos. Trans. R. Soc. London* **147**, 145–181 (1857).
2. L. Holland, *Vacuum Deposition of Thin Films,* Chapman and Hall, London (1956).
3. L. I. Maissel and R. Glang (Eds.), *Handbook of Thin Film Technology,* McGraw-Hill, New York (1970).
4. K. L. Chopra, *Thin Film Phenomena,* McGraw-Hill, New York (1969).
5. L. Eckertova, *Physics of Thin Films,* Plenum Press, New York (1977).
6. J. C. Vossen and W. Kern (Eds.), *Thin Film Processes,* Academic Press, New York (1978).
7. R. E. Honig, Vapor pressure data for the solid and liquid elements, *RCA Rev.* **23**, No. 4, 567–586 (1962).
8. G. M. McCracken, The behavior of surfaces under ion bombardment, *Rep. Prog. Phys.* **38**, 241–327 (1975).
9. G. K. Wehner, Controlled sputtering of metals by low-energy Hg ions, *Phys. Rev.* **102**, No. 3, 670–704 (1956).
10. G. Carter and J. S. Colligon, *Ion Bombardment of Solids,* Elsevier, New York, Amsterdam (1968).
11. L. B. Loeb, *Basic Processes of Gaseous Electronics,* University of California Press, Berkeley (1955).
12. H. S. Butler and G. S. Kino, Plasma sheath formation by radiofrequency fields, *Phys. Fluids* **6**, No. 9, 1346 (September 1963).
13. I. Brodie, C. T. Lamont, and D. O. Myers, Substrate bombardment during RF sputtering, *J. Vac. Sci. Technol.* **6**, No. 1, 124 (1969).
14. C. F. Powell, J. H. Oxley, and J. M. Blocher, Jr. (Eds.), *Vapor Deposition,* Wiley & Sons, New York (1966).
15. R. R. Chamberlain and J. S. Skarman, Chemical spray deposition process for inorganic films, *J. Electrochem. Soc.* **113**, No. 1 (January 1966).
16. L. V. Gregor, Polymer dielectric film, *IBM Res. Dev.* **12**, No. 2, 140–162 (March 1968).
17. R. M. Handy and L. C. Scala, Electrical and structural properties of Langmuir films, *J. Electrochem. Soc.* **113**, 105–115 (1966).
18. H. E. Ries, Monomolecular films, *Sci. Am.* **204**, 152 (1961).
19. H. E. Farnsworth, in: *The Surface Chemistry of Metals and Semiconductors,* H. Gates (Ed.), Wiley & Sons, New York, p. 21 (1959).
20. D. B. Lee, Anisotropic etching of silicon, *J. Appl. Phys.* **40**, 4569–4574 (1969).
21. E. Bassons, Fabrication of novel three-dimensional microstructures by anistropic etching of $\langle 100 \rangle$ and $\langle 110 \rangle$ silicon, *IEEE Trans. Electron Devices* ED 25, No. 10, 1178–1185 (1978).
22. K. E. Bean, Anisotropic etching of silicon, *IEEE Trans. Electron. Devices* ED-25, No. 10, 1185–1193 (October 1978).

23. R. L. Bersin and R. F. Reichelderfer, The dryox process for etching silicon dioxide, *Solid State Technol.*, **20**(41) 78–80 (April 1977).
24. R. G. Poulsen, Plasma etching in IC manufacture—A review, *J. Vac. Sci. Technol* **14**, No. 1, 266–274 (January–February 1977).
25. R. Kumar, C. Tadas, and G. Hudson, Characterization of plasma etching for semiconductor applications *Solid State Technol.*, **19**(10) 54–59 (October 1976).
26. J. W. Coburn and H. F. Winters, Plasma etching—A discussion of mechanisms, *J. Vac. Sci. Technol.* **16** (March–April 1979).
27. J. P. Hirth and G. M. Pound, *Condensation and Evaporation: Nucleation and Growth Kinetics*, MacMillan, New York (1963).
28. D. Walton, T. N. Rhodin, and R. Rollins, Nucleation of silver on sodium chloride, *J. Chem. Phys.* **38**, No. 11 2698 (June 1963).
29. G. Zinsmeister, in: Basic Problems in Thin-Film Physics (R. Neidermayer and H. Mayer, Eds.), Vandenhoeck and Ruprecht, Gottingen, Federal Republic of Germany p. 33 (1966).
30. H. J. Poppa, Heterogeneous nucleation of Bi and Ag on amorphous substrates, *J. Appl. Phys.* **38**, No. 10, 3883 (September 1967).
31. K. L. Chopra, Growth of thin metal films under applied electric field, *Appl. Phys. Lett.* **7**, No. 5, 140 (September 1965).

References for Chapter 4

1. H. I. Smith, Fabrication techniques for surface-acoustic-wave and thin-film optical devices, *Proc. IEEE* **62**, No. 10, 1361–1387 (October 1974).
2. A. N. Broers and T. H. P. Chang, High resolution lithography for microcircuits, IBM Research Report, No. 7403, Engineering Technology/Physics Surface Science, IBM Thomas J. Watson Research Center, Yorktown Heights, New York (November 1978).
3. R. E. Kinzly, Investigations of the influence of the degree of coherence upon images of edge objects, *J. Opt. Soc. Am.* **55**, No. 8, 1002, (August 1965); A. Offner and J. Meiron, The performance of optical systems with quasimonochromatic partially coherent illumination. *Appl. Opt.* **8**, No. 1, 183 (January 1969).
4. B. J. Lin, Deep UV lithography, *J. Vac. Sci. Technol.* **12**, No. 6, 1317 (November–December 1975).
5. M. C. King, Future development for 1 : 1 projection photolithography, *IEEE Trans. Electron Devices* **ED-26**, No. 4, 711 (April 1979).
6. D. A. Markle, A new projection printer, *Solid State Technol.* **17**, No. 6, 50 (June 1974); H. Moritz, High-resolution lithography with projection printing. *IEEE Trans. Electron Devices* **ED-26**, No. 4, 705 (April 1979); M. C. King and E. S. Muraski, New generation of 1 : 1 optical projection mask aligners, *Proc. Soc. Photo-Optical Instrum. Eng. (SPIE)*, Developments in Semiconductor Microlithography IV **174**, 70 (1979); J. W. Bossung and E. S. Muraski, Optical advances in projection lithography, *Proc. Soc. Photo-Optical Instrum. Eng. (SPIE)*, Devices in Semiconductor Microlithography III, **135**, 16 (1978); J. W. Bossung, Projection printing characterization, *Proc. Soc. Photo-Optical Instrum. Eng. (SPIE)* Semiconductor Microlithography II **100**, 80 (1977).
7. H. Binder and M. Lacombat, Step-and-repeat projection printing for VLSI circuit fabrication, *IEEE Trans. Electron Devices* **ED-26**, No. 4, 698 (April 1979); G. L. Resor and A. C. Tobey, The role of direct step-on-the-wafer in microlithography strategy for the '80's, *Solid State Technol.*, 101 (August 1979); H. E. Mayer and E. W. Leobach, A new step-by-step aligner for very large scale integration (VLSI) production, *Proc. Soc. Photo-Optical Instrum. Eng. (SPIE)* Semiconductor Microlithography V **221** (1980).

8. M. C. King and M. R. Goldrick, Optical MTF evaluation techniques for microelectronic printers, *Solid State Technol.* **19**, No. 2, 37 (February 1977).

9. F. H. Dill, A. R. Neureuther, J. A. Tuttle, and E. J. Walker, Modeling projection printing of positive photoresists, *IEEE Trans. Electron Devices* **ED-22**, No. 7, 456 (July 1975).

10. J. D. Cuthbert, Optical projection printing, *Solid State Technol.* **20**, No. 8, 59 (August 1977).

11. F. H. Dill, Optical lithography, *IEEE Trans. Electron Devices* **ED-22**, No. 7, 440 (July 1975).

12. M. J. Bowden and L. F. Thompson, Resist materials for fine line lithography, IEEE Trans. *Electron Devices* **ED-22**, No. 7, 456 (July 1975).

13. F. H. Dill, W. P. Hornberger, P. S. Hauge, and J. M. Shaw, Characterization of positive photoresist, *IEEE Trans. Electron Devices* **ED-22**, No. 7, 455 (July 1975).

14. D. A. McGillis and D. L. Fehrs, Photolithographic linewidth control, *IEEE Trans. Electron Devices* **ED-22**, No. 7, 471 (July 1975).

15. Mann 4800, manufactured by CICA-Mann Corporation, Burlington, Massachusetts.

16. W. W. Ng, C. -S. Hong, and A. Yariv, Holographic interference lithography for integrated optics, *IEEE Trans. Electron Devices* **ED-25**, No. 10, 1193 (October 1978).

17. C. V. Shank and R. V. Schmidt, Optical technique for producing 0.1 μ periodic surface structures, *Appl. Phys. Lett.* **23**, No. 3, 154 (August 1973).

18. H. W. Schnopper, L. P. Van Speybroeck, J. P. Delvaille, A. Epstein, E. Kallne, R. Z. Bachrach, J. Dijkstra, and L. Lantward, Diffraction grating transmission efficiencies for XUV and soft X-rays, *Appl. Opt.* **16**, No. 4 (April 1977).

19. P. L. Csonka, Holographic X-ray gratings generated with synchronton radiation, *J. Appl. Phys.* **52**(4) 2692 (April 1981).

20. D. L. Spears and H. I. Smith, High-resolution pattern replication using soft X-rays, *Electron. Lett.* **8**, 102 (February 1972).

21. M. Green and V. E. Cosslett, Measurement of K,L and M shell X-ray production efficiencies, *Br. J. Appl. Phys.* **1**, 425 (1968).

22. M. Yoshimatsu and S. Kozaki, High brilliance X-ray sources in: *X-Ray Optics, Application to Solids* (H. J. Queisser, Ed.), Springer-Verlag, Berlin, Chapter 2 (1977).

23. D. Maydan, G. A. Coquin, J. R. Maldonado, S. Somekh, D. Y. Lou, and G. N. Taylor, High speed replication of submicron features on large areas by X-ray lithography, *IEEE Trans. Electron Devices* **ED-22**, No. 7, 429 (1975).

24. M. P. Lepselter, Scaling the micron barrier with X-rays, *IEDM Tech. Dig.*, 42 (December 8–10 1980).

25. D. J. Nagel, in: *Advances in X-Ray Analysis* (W. L. Pickles, C. S. Barrett, J. B. Newkirk, and C. O. Ruud, Eds.), Vol. 18, Plenum Press, New York, p. 1 (1975).

26. J. J. Muray, Photoelectric effect induced by high-intensity laser light beam in quartz and borosilicate glass, *Dielectrics,* **2** 221, (February 1964).

27. R. A. Gutcheck, J. J. Muray, and D. C. Gates, An intense plasma X-ray source for X-ray microscopy, to be published in *The Proceedings of the Brookhaven Conference on the Optics of Short Wavelengths,* Nov. 16–20 (1981).

28. H. I. Smith, D. L. Spears, and S. E. Bernacki, X-ray lithography: A complementary technique to electron beam lithography, *J. Vac. Sci. Technol.* **10**, No. 6, 913 (November–December 1973).

29. J. Kirz, D. Sayre, and J. Dilger, Comparative analysis of X-ray emission microscopies for biological specimens, Paper presented at the workshop sponsored by the New York Academy of Sciences on Short Wavelength Microscopy, February 23–25, 1977, *Ann. N.Y. Acad. Sci.* **306**, 291–305 (1978).

30. G. N. Taylor, X-ray resist materials, *Solid State Technol.,* **23**(5) 73 (May 1980).

31. A. R. Neureuther, Simulation of X-ray resist line edge profiles, *J. Vac. Sci. Technol.* **15**, No. 3, 1004 (May–June 1978).

32. J. H. McCoy and P. A. Sullivan, Precision mask alignment for X-ray lithography, *Proceedings of the Seventh International Conference on Electron and Ion Beam Science and Technology*, Electrochemical Society, Princeton, New Jersey, p. 536 (1976).

33. J. M. Moran and D. Maydan, High resolution, steep profile, resist patterns, *Bell Syst. Tech. J.* **58**(5) 1027 (May–June, 1979).

34. G. N. Taylor, T. M. Wolf, Plasma-developed X-ray resists, *J. Electrochem. Soc.* **127**, No. 12, 2665, (December 1980).

35. E. Spiller, D. E. Eastman, R. Feder, W. D. Grobman, W. Gudat, and J. Topalian, Application of synchrotron radiation to X-ray lithography, *J. Appl. Phys.* **47**, No. 12, 5450 (December 1976).

36. H. Winick and A. Bienenstock, Synchrotron Radiation Research, *Annu. Rev. Nucl. Part. Sci.* **28**, 33–113 (1978).

37. G. R. Brewer (Ed.), *Electron-Beam Technology in Microelectronic Fabrication*, Academic Press, New York (1980).

38. P. R. Thornton, Electron physics in device microfabrication, I: General background and scanning systems, *Adv. Electron. Electron Phys.* **48**, 272 (1979).

39. A. N. Broers and T. H. P. Chang, High resolution lithography for microcircuits, IBM Research Report No. 7403, IBM Thomas J. Watson Research Center, Yorktown Heights, New York (November 1978).

40. H. C. Pfeiffer, Recent advances in electron-beam lithography for high-volume production of VLSI devices, *IEEE Trans. Electron Devices* **ED-26**, No. 4, 663 (April 1979).

41. I. Brodie, D. Cone, J. J. Muray, N. Williams, L. Gasiorek and E. R. Westerberg, Electron beam wafer exposure system for high throughput, direct write, sub-micron, lithography, *IEEE Trans. Electron Devices* **ED-28**, No. 11, 1422 (November 1981).

42. W. Krakow, L. A. Howland, and G. McKinley, Multiplexing electron beam patterns using single-crystal thin films, *J. Phys. E* **12**, 984 (1979).

43. P. R. Malmberg, T. W. O'Keeffe, M. M. Sopira, and M. W. Levi, LSI pattern generation and replication by electron beams, *J. Vac. Sci. Technol.* **10**, No. 6, 1025 (November–December 1973).

44. M. P. Scott, Recent progress on electron image projector, *J. VAc. Sci. Technol.* **15**, No. 3, 1016 (May–June 1978).

45. M. B. Heritage, Electron-projection microfabrication system, *J. Vac. Sci. Technol.* **12**, No. 6, 1135 (November–December 1973).

46. H. Koops, On electron projection systems, *J. Vac. Sci. Technol.* **10**, No. 6, 909 (November –December 1973).

47. L. N. Heynick, E. R. Westerberg, C. C. Hartlelius, Jr., and R. E. Lee, Projection electron lithography using aperture lenses, *IEEE Trans. Electron Devices* **ED-22**, No. 7, 399 (July 1975).

48. C. A. Spindt, A thin-film field-emission cathode, *J. Appl. Phys.* **39**, No. 7, 3504–3505 (June 1968).

49. C. A. Spindt, I. Brodie, L. Humphrey, and E. R. Westerberg, Physical properties of thin-film field emission cathodes with molybdenum cones, *J. Appl. Phys.* **47**, No. 12, 5248–5263 (December 1976).

50. W. Aberth, C. A. Spindt, and K. T. Rogers, Multipoint field ionization beam source, *Record of the 11th Symposium on Electron, Ion, and Laser Beam Technology* (R. F. M. Thornley, Ed.), San Francisco Press, San Francisco, California, pp. 631–636 (1971).

51. W. Aberth and C. A. Spindt, Characteristics of a volcano field ion quadrupole mass spectrometer, *Int. J. Mass Spectrom. Ion Phys.* **25**, 183–198 (1977).

52. K. Amboss, Electron optics for microbeam fabrication, *Scanning Electron Microsc.* **1** 699 (1976).

53. D. R. Herriott, R. J. Collier, D. S. Alles and J. W. Stafford, EBES: A practical electron lithographic system, *IEEE Trans. Electron Devices* **ED-22**, No. 7, 385 (July 1975).

54. R. F. W. Pease, J. P. Ballantyne, R. C. Henderson, A. M. Voshchenkov, and L. D. Yau, Application of the electron beam exposure system, *IEEE Trans. Electron Devices* **ED-22**, No. 7, 393 (July 1975).

55. D. E. Davis, R. D. Moore, M. C. Williams, and O. C. Woodard, Automatic registration in an electron-beam lithographic System, *IBM, J. Res. Dev.* **21**, No. 6, 498 (November 1977).

56. A. D. Wilson, T. H. P. Chang, and A. Kern, Experimental scanning electron-beam automatic registration system, *J. Vac. Sci. Technol.* **12**, No. 6, 1240 (November–December 1975).

57. E.D. Wolf, P. J. Coane, and F. S. Ozdemir, Composition and detection of alignment marks for electron beam lithography, *J. Vac. Sci. Technol.* **12**, No. 6, 1266 (November–December 1975).

58. J. S. Greeneich, Developer characteristics of poly-(methyl methacrylate) electron resist, *J. Electrochem. Soc.* **122**, No. 7, 970 (July 1975).

59. T. E. Everhart, P. H. Hoff, Determination of kilovolt electron energy dissipation vs. penetration distance in solid materials, *J. Appl. Phys.* **42**, No. 13, 5837 (December 1971).

60. G. W. Martel and W. B. Thompson, A comparison of commercially available electron beam resists, *Semicond. Int.* **2**(1) 69 (January–February 1979).

61. L. F. Thompson, J. P. Ballantyne, and E. D. Feit, Molecular parameters and lithographic performance of poly(glycidylmethacrylate-co-ethyl acrylate): A negative electron resist, *J. Vac. Sci. Technol.* **12**, No. 6, 1280 (November–December 1975).

62. R. D. Heidenreich, L. F. Thompson, E. D. Feit and C. M. Melliar-Smith, Fundamental aspects of electron beam lithography, I. Depth-dose response of polymeric electron beam resists, *J. Appl. Phys* **44**, No. 9, 4039 (September 1973).

63. J. S. Greeneich and T. Van Duzer, An exposure model for electron-sensitive resists, *IEEE Trans. Electron Devices* **ED-21**, No. 5, 286 (May 1974).

64. A. Barraud, C. Rausilio, and A. Ruaudel-Teixier, Monomolecular resists—A new approach to high resolution electron beam microlithography, *J. Vac. Sci. Technol.* **16**, No. 6, 2003 (November–December 1979).

65. M. S. Isaacson and A. Muray, Nanolithography using *in situ* electron beam vaporization of very low molecular weight resists, Workshop on Molecular Electronic Devices, Naval Research Laboratory (March 23–24, 1981).

66. M. Isaacson, A. Muray, *In situ* vaporization of very low molecular weight resists using $\frac{1}{2}$ nm diameter electron beams, *J. Vac. Sci. Technol.,* **19**, No. 4 (Nov–Dec 1981).

67. T. H. P. Chang, Proximity effect in electron-beam lithography, *J. Vac. Sci. Technol.* **12**, No. 6, 1271 (November–December 1975).

68. R. L. Seliger, J. W. Ward, V. Wang, and R. L. Kubena, A high-intensity scanning ion probe with submicrometer spot size, *J. Appl. Phys. Lett.* **34**, No. 5, 310 (March 1979).

69. B. A. Free and G. A. Meadows, Projection ion lithography with aperture lenses, *J. Vac. Sci. Technol.* **15**, No. 3, 1028 (May–June 1978).

70. G. Stengl, R. Kaitna, H. Loeschner, P. Wolf, and R. Sacher, Ion projection system for IC production, *J. Vac. Sci. Technol.* **16**, No. 6, 1883 (November–December 1979); G. Stengl, P. Wolf, R. Kaitna, H. Loschner, and R. Sacher, Experimental results and future prospects of demagnifying ion-projection systems. Paper presented at the *Third International Conference on Ion Implantation Equipment and Techniques,* Kingston, Canada (July 8–11 1980).

71. L. Csepregi, F. Iberl, and P. Eichinger, Ion-beam shadow printing through thin silicon foils using channeling, *Appl. Phys. Lett.* **37**, No. 7 630, (October 1 1980).

72. L. W. Swanson, G. A. Schwind, A. E. Bell, and J. E. Brady, Emission characteristics of gallium and bismuth liquid metal field ion sources, *J. Vac. Sci. Technol.* **16**, No. 6, 1864 (November–December 1979).

73. R. J. Culbertson, T. Sakurai, and G. H. Robertson, Ionization of liquid metals, gallium, *J. Vac. Sci. Technol.* **16**, No. 2, 574 (March–April 1979).

74. M. Komuro, N. Atoda, and H. Kawakatsu, Ion beam exposure of resist materials, *J. Electrochem. Soc.* **126**, No. 3, 483 (March 1979).
75. T. M. Hall, Liquid gold ion source, *J. Vac. Sci. Technol.* **16**, No. 6, 1871, (November–December 1979).
76. L. W. Swanson, A. E. Bell, G. A. Schwind, and J. Orloff, A comparison of the emission characteristics of liquid ion sources of gallium, indium, and bismuth, Paper presented at the Ninth International Conference on Electron Science and Ion Beam Technology, St. Louis, Missouri, May 11–16 1980.
77. T. M. Hall, A. Wagner, and L. F. Thompson, Ion beam exposure characteristics of resists, *J. Vac. Sci. Technol.* **16**, No. 6, 1889 (November–December 1979).

References for Chapter 5

1. D. W. Sah, Mechanisms in vapour epitaxy of semiconductors, in: *Crystal Growth Theory and Techniques* (C. H. Goodman, Ed.), Vol. 1 Plenum Press, London (1974).
2. R. M. Burger and R. P. Donovan (Eds.), *Fundamentals of Silicon Integrated Device Technology,* Vol. 1, Prentice-Hall, Englewood Cliffs, New Jersey (1967).
3. S. Nielsen and G. J. Rich, Preparation of epitaxial layers of silicon: I. Direct and indirect processes, *Microelectron. Reliab.,* 165–170 (1964).
4. M. L. Hammond, Silicon epitaxy, *Solid State Technol.* **21**, No. 11, 68 (November 1978).
5. R. W. Dutton, D. A. Antoniadis, J. D. Meindl, T. I. Kamins, K. C. Saraswat, B. E. Deal, and J. D. Plummer, Oxidation and epitaxy, Technical Report No. 5021-1, Integrated Circuit Laboratory, Stanford University, Stanford, California (May 1977).
6. T. I. Kamins, R. Reif, and K. C. Saraswat, *Electrochemical Society Fall Meeting, Las Vegas, October 17–22, 1976,* Abstract 230, Electrochemical Society, Princeton, New Jersey pp. 601–603 (1976).
7. R. Reif, T. I. Kamins, and K. C. Saraswat, Transient and steady-state response of the dopant system of an epitaxial reactor: Growth rate dependence, Electrochemical Society Fall Meeting, Atlanta, October 9–14, 1977, Abstract 350, *Electrochem. Soc. J.* **124**, No. 8, 912–923 (1977).
8. J. D. Meindl, K. C. Saraswat, and J. D. Plummer, *Semiconductor Silicon*, Electrochemical Society, Princeton, New Jersey, pp. 894–909 (1977).
9. J. D. Meindl, K. C. Saraswat, R. W. Dutton, J. F. Gibbons, W. Tiller, J. D. Plummer, B. E. Deal, and T. I. Kamins, Final report on computer-aided semiconductor process modeling, Stanford Electronics Laboratories, Report TR-4969-73-F, Stanford University, Stanford, California (October 1976).
10. R. Reif, T. I. Kamins, and K. C. Saraswat, A model for dopant incorporation into growing silicon epitaxial films, *J. Electrochem. Soc.* **126**, No. 4, 644 (April 1979).
11. B. V. Vanderschmitt, Silicon-on-sapphire: An LSI/VLSI technology, *RCA Eng.* **24**, No. 1 (June–July 1978).
12. L. Jastrzebski, Y. Imamura, and H. C. Gatos, Thickness uniformity of GaAs layers grown by electroepitaxy, *J. Electrochem. Soc.* **125**, No. 7, 1140 (July 1978).
13. L. Jastrzebski, J. Lagowski, H. C. Gatos, and A. F. Witt. Liquid-phase electroepitaxy: Growth kinetics, *J. Appl. Phys.* **49**, No. 12, 5909 (December 1978).
14. F. d'Heurle and P. Ho, Electromigration, in: *Thin Films: Interdiffusion and Reactions* (J. M. Poate, J. Mayer, and K. N. Tu, Eds., Wiley & Sons, New York (1978).
15. F. M. d'Heurle and R. Rosenberg, Electromigration in thin films, in: *Physics of Thin Films,* Vol 7, Academic Press, New York, pp. 257–310 (1973).
16. A. Y. Cho, Recent developments in molecular beam epitaxy (MBE), *J. Vac. Sci. Technol.* **16**, No. 2, 275 (March–April 1979); M. B. Parish, Molecular beam epitaxy, *Science* **208**, 916 (May 1980).

17. B. E. Deal and A. S. Grove, General relationship for the thermal oxidation of silicon, *J. Appl. Phys.* **36**, No. 12, 3770–3778 (December 1965).
18. B. E. Deal, The current understanding of changes in the thermally oxidized silicon structure, *J. Electrochem. Soc.* **121**, No. 6, 198C (June 1974).
19. J. Blanc, A revised model for the oxidation of Si by oxygen, *Appl. Phys. Lett.* **33**, No. 5, 424 (September 1978).
20. J. D. Meindl, K. C. Saraswat, R. W. Dutton, J. F. Gibbons, W. Tiller, J. D. Plummer, B. E. Deal, and T. I. Kamins, Computer aided engineering of semiconductor integrated circuits, Stanford University Integrated Circuit Laboratory, Report TR 4969-3, SEL-78-011, Stanford University, Stanford, California (February 1978).
21. M. Maeda, H. Kamioka, and M. Takagi, High pressure steam oxidation of silicon in a sealed quartz tube, in: *Proceedings of the 13th Symposium on Semiconductors and IC Technology,* Tokyo, November, 1977.
22. D. J. Levinthal, Diffusion system trends, *Semicond. Int.,* **2**(5) 31 (June 1979).
23. B. E. Deal, Standardized terminology for oxide charges associated with thermally oxidized silicon, *IEEE Trans. Electron Devices* **ED-27**, No. 3, 606 (March 1980).
24. G. Lucovsky and D. J. Chadi, Bond coordination defects at the Si/SiO_2 interface, in: *Physics of MOS Insulators,* Pergamon Press, New York, pp. 301–305, 1980.
25. A. S. Grove and E. H. Snow, A model for radiation damage in metal-oxide semiconductor structures, *Proc. IEEE* **54**, 894 (June 1966).
26. P. J. Jorgensen, Effect of an electric field on silicon oxidation, *J. Chem. Phys.* **37**, No. 4, 874 (August 1962).
27. N. Cabrera and N. F. Mott, Theory of the oxidation of metals, *Rep. Prog. Phys.* **12**, 163 (1948).
28. A. S. Grove, *Physics and Technology of Semiconductor Devices,* Wiley & Sons, New York (1967).
29. S. K. Ghandhi, *The Theory and Practice of Microelectronics,* Wiley & Sons, New York (1968).
30. B. I. Boltaks, *Diffusion in Semiconductors,* Academic Press, New York (1963).
31. J. H. Crawford, Jr., and L. M. Slifkin (Eds.), *Point Defects in Solids,* Vol. 2, Plenum Press, New York (1972).
32. L. C. Kimerling and D. V. Lang, in: *Lattice Defects in Semiconductors* (J. E. Whitehouse, Ed.) Institute of Physics, London (1974).
33. G. Carter and W. A. Grant, *Ion Implantation of Semiconductors,* Edward Arnold, London (1976).
34. B. L. Crowder (Ed.), *Ion Implantation in Semiconductors and Other Materials,* Plenum Press, New York (1972).
35. J. W. Mayer, L. Eriksson, and J. A. Davies, *Ion Implantation in Semiconductors, Silicon and Germanium,* Academic Press, New York (1970).
36. J. F. Gibbons, Ion implantation in semiconductors—Part I: Range distribution theory and experiments, *Proc. IEEE* **56**, No. 3, 295–319 (March 1968).
37. J. F. Gibbons, Ion implantation in semiconductors—Part II: Damage production and annealing, *Proc. IEEE* **60**, No. 9, 1062 (September 1972).
38. J. Lindhard, M. Scharff, and H. Schioett, Atomic collisions II. Range concepts and heavy ion ranges, *K. Dan. Vidensk. Selsk. Mat. Fys. Medd.* **33**, No. 14, 1 (1963).
39. J. Sansbury, Applications of ion implantation in semiconductor processing, *Solid State Technol.* **19**, No. 11, 31 (November 1976).
40. J. Lindhard, Influence of crystall lattice on motion of energetic charged particles, *K. Dan. Vidensk. Selsk. Mat. Fys. Medd.* **34**, No. 14 (1965).
41. P. Sigmund and J. B. Saunders, Spatial distribution of energy deposited by ionic bombardment, in: *Proceedings of the International Conference on Applications of Ion Beams to Semiconductor Technology, Grenoble, France* (P. Glotin, Ed.), Editions Ophrys, p. 215 (1967).

42. H. S. Rupprecht, New advances in semiconductor implantation, *J. Vac. Sci. Technol.* **15**, No. 5, 1669 (September–October 1978).

43. J. Gibbons, W. S. Johnson, and S. Mylroie, *Projected Range Statistics in Semiconductors,* 2nd ed., Wiley & Sons, New York (1975); D. K. Brice, *Ion Implantation Range and Energy Deposition Distributions,* Plenum, New York (1975); K. Bruce Winterbon, *Ion Implantation Range and Energy Deposition Distributions,* Vol. 2, Plenum, New York, (1975).

44. L. A. Christel, J. F. Gibbons, S. Mylroie, An application of the Boltzmann transport equation to ion range and damage distributions in multi-layered targets, *J. Appl. Phys.* **51**, No. 12, 6176 (December 1980); D. H. Smith and J. F. Gibbons, *Ion Implantation in Semiconductors, 1976,* Plenum, New York, p. 333, New York, (1977); R. A. Moline, G. W. Reutlinger, and J. C. North, *Proceedings of the Fifth International Conference on Atomic Collisions in Solids,* Plenum Press, New York (1976).

45. Z. L. Liau and J. W. Mayer, Limits of composition achievable by ion implantation, *J. Vac. Sci. Technol.* **15**, No. 5, 1629 (September–October 1978).

46. J. W. Cleland, K. Lark-Horovitz and J. C. Pigg, Transmutation-produced germanium semiconductors, *Phys. Rev.* **78**, 814 (1950).

47. H. M. Janus, *Application of NTD Silicon for Power Devices in Neutron Transmutation Doping in Semiconductors* (J. M. Meese, Ed.), Plenum Press (1978).

48. H. M. Janus and O. Malmros, Application of thermal neutron irradiation for large scale production of homogeneous phosphorus doping of float zone silicon, *IEEE Trans. Electron Devices* **ED-23** No. 8, 797 (August 1976).

49. J. M. Meese, *The NTD Process—A New Reactor Technology in Neutron Transmutation Doping in Semiconductors* (J. M. Meese, Ed.), Plenum Press (1978).

50. S. Prussin and J. W. Cleland, Application of neutron transmutation doping for production of homogeneous epitaxial layers, *J. ElectroChem. Soc.* **125**, No. 2, 350 (February 1978).

51. A. E. Bell, Review and analysis of laser annealing, *RCA Rev.* **40**, 295 (September 1979).

52. R. T. Young, C. W. White, G. J. Clark, J. Narayan, W. H. Christie, M. Murakami, P. W. King, and S. D. Kramer, Laser annealing of boron-implanted silicon, *Appl. Phys. Lett.* **32**, No. 3, 139 (February 1978).

53. G. K. Celler, J. M. Poate, and L. C. Kimerling, Spatially controlled crystal growth regrowth of ion-implanted silicon by laser irradiation, *Appl. Phys. Lett.* **32**, No. 8, 464 (April 1978).

54. P. Baeri, S. U. Campisano, G. Foti, and E. Rimini, Arsenic diffusion in silicon melted by high-power nanosecond laser pulsing, *Appl. Phys. Lett.* **33**, No. 2, 137 (July 1978).

55. J. C. Muller, A. Grob, J. T. Grob, R. Stuck, and P. Stiffert, Laser-beam annealing of heavily damaged implanted layers on silicon, *Appl. Phys. Lett.* **33**, No. 4, 287 (August 1978).

56. A. Gat, J. F. Gibbons, T. J. Magee, J. Peng, P. Williams, V. Deline, and C. A. Evans, Jr., Use of a scanning cw Kr laser to obtain diffusion-free annealing of B-implanted silicon, *Appl. Phys. Lett.* **33**, No. 5 389 (1978).

57. T. N. C. Venkatesan, J. A. Golovchenko, J. M. Poate, P. Cowan, and G. K. Celler, Dose dependence in the laser annealing of arsenic-implanted silicon, *Appl. Phys. Lett.* **33**, No. 5, 429 (September 1978).

58. C. W. White, W. H. Cristie, B. R. Appleton, S. R. Wilson, P. P. Pronko, and C. W. Magee, Redistribution of dopants in ion-implanted silicon by pulsed-laser annealing, *Appl. Phys. Lett.* **33**, No. 7, 662 (October 1978).

59. P. Baeri, S. U. Campisano, G. Foti, and E. Rimini, A melting model for pulsing-laser annealing of implanted semiconductors, *J. Appl. Phys.* **50**, No. 2, 788 (February 1979).

60. D. H. Auston, J. A. Golovchenko, and T. N. C. Venkatesan, Dual-wavelength laser annealing, *Appl. Phys. Lett.* **34**, No. 9, 558 (May 1979).

61. S. S. Lau, J. W. Mayer, and W. F. Tseng, Comparison of laser and thermal annealing of implanted-amorphous silicon in laser–solid interactions and laser processing, *AIP Conf. Proc.,* No. 50 (1979).

62. A. Gat and J. F. Gibbons, A laser-scanning apparatus for annealing of ion-implantation damage in semiconductors, Appl. Phys. Lett. **32**, No. 3, 142 (February 1978).

63. R. B. Fair, Modelling laser-induced diffusion of implanted arsenic in silicon, *J. Appl. Phys.* **50**, No. 10, 6552 (October 1979).
64. M. W. Geis, D. C. Flanders, H. I. Smith, and D. A. Antoniadis, Grapho-epitaxy of silicon on fused silica using surface micropatterns and laser crystallization, *J. Vac. Sci. Technol.* **16**, No. 6, 1640 (November–December 1979).
65. A. R. Kirkpatrick, J. A. Minnucci, and A. C. Greenwald, Silicon solar cells by high-speed low-temperature processing, *IEEE Trans. Electron Devices* **ED-24**, No. 4, 429 (April 1977).
66. R. G. Little and A. C. Greenwald, An advancement in semiconductor processing, *Semicond. Int.*, **2**(1) 81 (January–February 1979).
67. K. N. Ratnakumar, R. F. W. Pease, D. J. Bartelink, and N. M. Johnson, Scanning electron beam annealing with a modified SEM, *J. Vac. Sci. Technol.* **16**, No. 6, 1843 (November–December 1979).
68. A. Neukermans and W. Saperstein, Modeling of beam voltage effects in electron-beam annealing, *J. Vac. Sci. Technol.* **16**, No. 6, 1847 (November–December 1979).
69. D. A. Antoniadis, S. E. Hansen, R. W. Dutton and G. Gonzalez, SUPREM 1-A program for I.C. process modelling and simulation, Technical Report No. 5019-1, Integrated Circuit Laboratory, Stanford University, Stanford, California (May 1977).
70. C. Mead and L. Conway, *Introduction to VLSI Systems,* Addison-Wesley Publishing Company, Reading, Massachusetts (1980).

References for Chapter 6

1. A. Septier, The struggle to overcome spherical aberration in electron optics, *Adv. Opt. Electron Microsc.* **1**, 204–272 (1966).
2. A. V. Crewe, Scanning transmission-electron microscopy, *J. Microsc. (Oxford)* **100**, 247–259 (1974).
3. I. Brodie, Visibility of atomic objects in the field-electron emission microscope, *Surf. Sci.* **70**, 186–196 (1978).
4. V. E. Coslett, Radiation damage and chromatic aberration produced by inelastic scattering of electrons in the electron microscope: Statement of the problem, *Ann. N.Y. Acad. Sci.* **306**, 3 (1978).
5. E. W. Müller and T. T. Tsong, *Field Ion Microscopy,* Elsevier Scientific Publishing, New York (1969).
6. B. M. Siegal (Ed.), *Modern Developments in Electron Microscopy*, Academic Press, New York (1964).
7. P. W. Hawkes, *Electron Optics and Electron Microscopy,* Taylor and Francis, London (1972).
8. L. Reimer and G. Pfefferkorn, *Raster-Electronen-Mikroscopie,* Springer-Verlag, Berlin (1977).
9. M. Isaacson, All you want to know about ELS, *Scanning Electron Microsc.* **1978/1**, 763–776 (1978).
10. L. J. Balk, H. P. Feuerbaum, E. Kubalek, and E. Menzel, Quantitative voltage contrast at high frequencies in the SEM, *Scanning Electron Microsc.* **1976/I**, 615–646 (1976).
11. J. R. Banbury and W. C. Nixon, A high-contrast directional detector for the scanning electron microscope, *Brit. J. Sci. Instrum.* (J. Phys.E) **2**, 1055–1059 (1969).
12. D. L. Crosthwait and T. W. Ivy, Voltage contrast methods for semiconductor device failure analysis, *Scanning Electron Microsc.* **1974/I**, 935–940 (1974).
13. A. Gopinath and W. J. Tee, Theoretical limits on minimum voltage change detectable in the SEM, *Scanning Electron Microsc.* **1976/I**, 603–608 (1976).

14. A. J. Gonzales, On the electron beam induced current analysis of semiconductor devices, *Scanning Electron Microsc.* **1974/I**, 941–948 (1974).

15. J. F. Bresse, Electron beam induced current in silicon planar *p–n* junction: Physical model of carrier generation, determination of some physical parameters in silicon, *Scanning Electron Microsc.*, **1972**, 105–112 (1972).

16. J. F. Bresse and D. Lafeuille, SEM beam induced current in planar *p–n* junctions, diffusion length, and generation factor measurements, in: *Proceedings of the 25th Anniversary Meetings of EMAG*, Institute of Physics, London (1971).

17. R. M. Oman, Electron mirror microscopy, *Adv. Electron. and Electron Phys.* **26**, 217–249 (1969).

18. G. A. Haas and R. E. Thomas, Electron beam scanning technique for measuring surface work function variations, *Surf. Sci.* **4**, 64 (1966).

19. G. Mollenstedt and F. Lenz, Electron emission microscopy, *Adv. Electron. and Electron Phys.* **18**, 251–329 (1963).

20. D. E. Yuhas and T. E. McGraw, Acoustic microscopy, SEM and optical microscopy: Correlative investigation in ceramics, *Scanning Electron Microsc.* **1979/I**, 103 (1979).

21. L. Wegmann, The photoemission electron microscope: Its technique and applications, *J. Microsc.* **96**, 1 (August 1972).

22. R. H. Good and E. W. Muller, Field emission, in: *Handbuch der Physik*, (S. Flugge, Ed.), Vol. 21, Springer-Verlag, Berlin, pp. 176–231 (1956).

23. D. J. Rose, On the magnification and resolution of the field-emission electron microscope, *J. Appl. Phys.* **27**, 215 (1956).

24. J. A. Panitz, The 10-cm atom probe, *Rev. Sci. Instrum.* **44**, 1034–1038 (1973).

25. D. G. Brandon, Field ion microscopy, *Adv. Electron Opt. Microsc.* **2**, 343–402, Academic Press, New York (1968).

26. W. K. Chu, J. W. Mayer, and M. A. Nicolet, *Backscattering Spectrometry*, Academic Press, New York (1978).

27. A. Feurerstein, H. Grahmann, S. Kalbitzer, and H. Oetzmann, in: *Ion Beam Surface Layer Analysis* (O. Meyer, G. Linker, and F. Kappeler, Eds.), Plenum Press, New York (1976).

28. J. A. Davies, in: *Material Characterization Using Ion Beams* (J. P. Thomas and A. Cachard, Eds.), Plenum Press, London (1978).

29. D. E. Sawyer, D. W. Barning, and D. C. Lewis, Laser scanning of active integrated circuits and discrete semiconductor devices, *Solid-State Technol.*, **20**(6) 37 (June 1977).

30. G. B. Larrabee, Characterization of materials at small dimensions, in: *Microstructure Science, Engineering and Technology*, National Academy of Sciences, Washington, D.C. (1979).

31. D. E. Newbury, Microanalysis in the scanning electron microscope; progress and prospects, *Scanning Electron Microsc.* **1979/II**, 1–20 (1979).

32. J. Silcox, Microcharacterization, in: *Microstructure Science, Engineering and Technology*, National Academy of Sciences, Washington, D.C. (1979).

References for Chapter 7

1. B. Hoeneisen, and C. A. Mead, Fundamental limitations in microelectronics. I. MOS technology, *Solid State Electron.* **15**, No. 7, 819–829 (July 1972).

2. R. W. Keyes, Physical limits in digital electronics, *Proc. IEEE* **63**, No. 5, 740 (May 1970).

3. Hill, R. M., Single-carrier transport in thin dielectric films, *Thin Solid Films* **1**, 39 (1967).

4. C. Mead and L. Conway, *Introduction to VLSIC Systems*, Addison-Wesley Publishing, Reading, Massachusetts (1979).

5. J. T. Wallmark, A statistical model for determining the minimum size in integrated circuits, *IEEE Trans. Electron Devices* **ED-26,** No. 2, 135 (February 1979).

6. A. V. Crewe, Some limitations on electron beam lithography, *J. Vac. Sci. Technol.* **16,** No. 2, 255 (March–April 1979).

7. I. E. Sutherland, C. A. Mead, and T. E. Everhart, Basic limitations in microcircuit fabrication technology, Report No. R-1956-ARPA, RAND Corporation, Santa Monica, California (November 1976).

8. E. Spiller and R. Feder, X-ray lithography, in: *X-ray Optics; Applications to Solids* (H. J. Queisser, Ed.), Springer-Verlag, Berlin, New York (1977).

9. R. W. Keyes, The evolution of digital electronics toward VLSI, *IEEE Trans. Electron Devices* **ED-26,** No. 4, 271 (April 1979).

10. G. E. Moore, Progress in digital integrated electronics, in: *Tech. Digest, 1975 Int. Electron Devices Meet.* 11–13, IEEE, New York (1975).

11. R. N. Noyce, Large-scale integration: What is yet to come? *Science* **195,** 1102–1107 (1977).

12. D. K. Ferry, J. R. Barker, and C. Jacoboni (Eds.), *Physics of Nonlinear Transport in Semiconductors,* Series B, Vol. 52, Plenum Press, New York and London (1980).

13. J. A. Krumhansl and Y.-H. Pao, Microscience: An overview, *Phys. Today,* **32**(11) 25 (November 1979).

14. Proceedings of the NSF Workshop on Opportunities for Microstructures Science, Engineering and Technology, in cooperation with the NRC Panel on Thin-Film Microstructure Science and Technology (J. M. Ballantyne, Ed.), Airlie, Virginia, November 19–22, 1978, NTIS, Washington, D.C. (1978).

15. *Microstructure Science, Engineering, and Technology; Final Report,* Vol. 63, National Research Council, Washington, D.C. (1979).

Index